Blended Learning in Engineering Education

Blended Learning in Engineering Education

Recent Developments in Curriculum, Assessment and Practice

Edited by:

Ataur Rahman & Vojislav Ilic

Engineering and Construction Management,
Western Sydney University, Sydney, Australia

CRC Press
Taylor & Francis Group
Boca Raton London New York Leiden

CRC Press is an imprint of the
Taylor & Francis Group, an **informa** business
A BALKEMA BOOK

CRC Press/Balkema is an imprint of the Taylor & Francis Group, an informa business

Typeset by Apex CoVantage, LLC

Library of Congress Cataloging-in-Publication Data
Names: Rahman, Ataur, 1959– editor. | Ilic, Vojislav, editor.
Title: Blended learning in engineering education : recent developments in curriculum, assessment and practice / Ataur Rahman & Vojislav Ilic, editors.
Description: First edition. | London : CRC Press/Balkema, [2019] | Includes bibliographical references and index.
Identifiers: LCCN 2018040027 (print) | LCCN 2018046069 (ebook) | ISBN 9781315165486 (ebook) | ISBN 9781138056220 (hardcover : alk. paper)
Subjects: LCSH: Engineering—Study and teaching. | Blended learning. | Distance education.
Classification: LCC T65.5.B54 (ebook) | LCC T65.5.B54 B54 2019 (print) | DDC 620.0071/1—dc23
LC record available at https://lccn.loc.gov/2018040027

Published by: CRC Press/Balkema
Schipholweg 107c, 2316 XC Leiden, The Netherlands
e-mail: Pub.NL@taylorandfrancis.com
www.crcpress.com – www.taylorandfrancis.com

ISBN: 978-1-138-05622-0 (Hbk)
ISBN: 978-1-315-16548-6 (eBook)
Printed and bound by CPI Group (UK) Ltd, Croydon, CR0 4YY

Contents

Preface

It is arguable that engineering is at the forefront of technological changes that are permeating and shaping every segment of our society, including the education sector in general and the education of engineers in particular. This book provides a window into the current practice as seen through the eyes of practicing engineers and educators, spanning a wide range of engineering disciplines, with a specific focus, among others, on statistical hydrology, fluid flow including Newtonian and non-Newtonian fluids, engineering physics, industrial (product) design, management, as well as engineering and built environment. This book contains 19 chapters providing a range of expert views on traditional and new learning and teaching methods, technologies and best practices.

The book covers a wide range of innovative learning and teaching methods in engineering education, including the application of blended learning, hybrid learning, online learning, experimental learning, "cloud" campus concept, gamification and the concept of "learning spaces." All authors are directly involved in delivery of the subjects' matter and offer useful insights into their classroom didactic experiences.

For example, Chapter 4 describes significantly higher pass rates in the subject among the students who performed well in the weekly online quizzes than those who did not. It also describes another approach, used in first-year physics, which found that gamification resulted in an increase in students' engagement with the course, as well as students' understanding of the subject and enhanced performance. Similar results were found in adopting the concept of collaborative learning spaces, as described in Chapter 8.

The importance of student proficiency with software tools used in their course was found to be a significant factor in finding employment. Chapter 9 describes the case of a product design engineer.

Chapter 10 discusses feedback from students and lecturers on the merits of blended learning *versus* the traditional face-to-face approach, in terms of compatibility of the learning outcomes with the professional competencies set out by Engineers Australia.

Chapter 11 suggests the importance of nurturing and strengthening students' self-motivation, perseverance, independence and attitude from an early age as an essential element in teaching.

Chapter 14 provides a few possible solutions to address the common problem of undergraduate students having poor mathematics and science backgrounds.

Chapter 15 discusses the use of experimental learning together with improvements for developing the professional skills of postgraduate engineering students.

Chapter 16 suggests the need for Engineers Australia to provide a position paper and recommendations for its implementation to help in improving the communication skills of engineering graduates.

Chapter 17 provides an overview of the technical education delivery system in Bangladesh and identifies the need for greater capacity building for an effective and viable technical education system there.

Chapter 18 discusses the merits of virtual/remote laboratories in educating engineering students in developing countries as an affordable option.

Chapter 19 explores the advantages and limitations of technology-enhanced learning using online tools in civil engineering education. It also reviews the potential utility of tools available on the internet for teaching basic data analysis and engineering optimization courses. The limitations of online tools for enhancing students' learning and instruction are also discussed. This chapter further critically examines how learning can be supported by technology rather than driven by it, and how distance learning in engineering can be supported by online content.

The book chapters were peer reviewed and evaluated by the editors. The reviewers are highly appreciated for their time and efforts in reviewing the chapters. We believe that this book provides evidence of good practice and ways to identify emerging areas of learning and teaching in engineering and technological education.

Ataur Rahman and Vojislav Ilic
Western Sydney University, Sydney
Australia

Acknowledgments

We are profoundly thankful to the authors who have sacrificed their valuable time in preparing their manuscripts and addressing the comments by the reviewers. We are also grateful to the reviewers for their critical review and constructive suggestions, which have contributed significantly to the enrichment of the contents of this book. We are deeply grateful to Dr M. Ashiqur Rahman and Mr Abdul Mannan for their efforts in editing and improving English of the book chapters. We would also like to thank Waheda Rahman, Jenis Farzana Islam, Preeti Berwal and Amira Rahman for checking the references of the chapters. We are also thankful to Alistair Bright and other CRC Press staff for their patience, understanding and guidance in making this book a reality.

Ataur Rahman and Vojislav Ilic
Western Sydney University, Sydney
Australia

Editorial advisory board

Reviewers

Dr Amir Ahmed, *Daffodil International University, Bangladesh*
Dr Abdullah Al Mamoon, *Western Sydney University, Australia*
Mr Caleb Amos, *Western Sydney University, Australia*
Dr Wilfredo Caballero, *Western Sydney University, Australia*
Dr Dharma Hagare, *Western Sydney University, Australia*
Dr Evan Hajani, *Western Sydney University, Australia*
Dr Md Mahmudul Haque, *Western Sydney University, Australia*
Professor Chin Leo, *Western Sydney University, Australia*
Dr Ranjith Liyanapathirana, *Western Sydney University, Australia*
Associate Professor Samanthika Liyanapathirana, *Western Sydney University, Australia*
Mr Abdul Mannan, *Macquarie University, Australia*
Dr M. Ashiqur Rahman, *Western Sydney University, Australia*
Dr M. Muhitur Rahman, *King Faisal University, Saudi Arabia*
Ms Sumya Rahman, *EnviroWater Sydney, Australia*
Mr Tarek Rahman, *EnviroWater Sydney, Australia*
Associate Professor Ramesh Teegavarapu, *Florida Atlantic University, USA*
Dr Conrad Wasko, *University of Melbourne, Australia*
Dr Abdullah Yilmaz, *University of Sharjah, United Arab Emirates*

Chapter 1

Statistical hydrology teaching using a blended learning approach

A. Rahman[1]*, S. Kordrostami*[1] *and D. Purdy*[2]

[1]*School of Computing, Engineering and Mathematics, Western Sydney University, Sydney, New South Wales, Australia*

[2]*Blended Learning subject, Western Sydney University, Sydney, New South Wales, Australia*

ABSTRACT: This chapter presents the delivery of an advanced statistical hydrology subject through the blended learning approach in Western Sydney University (WSU) as a part of the Master of Engineering degree. It has been found that blended learning is quite effective in delivering the contents of the subject to Master of Engineering students. The subject currently consists of a 5-hour face-to-face workshop and 11 pre-recorded tutorial sessions. The recorded tutorial sessions have been found to be the most useful element of this subject, as they have allowed a student to listen to the materials as many times as needed and at a time that is convenient to him/her. The preparation of a critical literature review has been found to be the most difficult task for the students. It has been found that a 5-hour workshop is inadequate to deliver the subject contents effectively to students, and hence it is proposed that a 2-hour lecture be delivered every week to answer students' queries and facilitate face-to-face engagement among the students themselves, and between the students and lecturer, to enhance learning of this subject. Furthermore, an intra-session and end-of-semester examination are proposed to evaluate student learning in the subject. This will make the blended learning approach more effective for the students of the advanced statistical hydrology subject in WSU in future years.

1.1 Introduction

Water is the vital source of life on planet Earth, and the ever-increasing water demand and its intrinsic relationship with the environment have made water a subject of study in many disciplines of knowledge (Gleick, 1998; Grimmond, 2010). For civil and environmental engineering degrees, water-related subjects constitute a significant part of the degree. Topics generally include fluid mechanics, surface water hydrology, hydraulics, statistical hydrology, hydrogeology and water resources engineering. Hydrology is one of the most important water engineering subjects in civil and environmental engineering degrees all over the world (Kordrostami *et al.*, 2016). Hydrology, a Greek word that means "science of water" (Raudkivi, 2013), is defined as "a branch of natural science concerned with occurrence, properties, distribution, and movement of water in the natural and man-made environment" (Elshorbagy, 2005).

Hydrology is linked with many different disciplines, and hence it has multiple dimensions of understanding (Vogel *et al.*, 2015; Wagener *et al.*, 2010). The conceptualization of hydrological processes requires higher-order reflective, metacognitive and critical thinking skills (Lenschow, 1998). Hydrologists in future decades must understand climate change impacts, human influences on the hydrologic cycle, the greater competition for water resources, complex water-sharing issues for international rivers, greater demand for environmental needs and water pollution issues. In this regard, uncertainty analysis is a vital task, and hence knowledge on statistical hydrology, which covers how uncertainty in the hydrological process modeling can be accounted for, is important for future hydrologists. Future hydrologists should be able to work across disciplines and geographic areas (Hooper, 2009; Torgersen, 2006). One of the major focuses of the hydrological society is to train and educate hydrologists to solve complex hydrological problems by having an interdisciplinary vision to the major hydrological issues (McGlynn *et al.*, 2010). A greater emphasis has been placed on changing the perspectives of hydrology teaching from a traditional approach to one that caters to the contemporary, societal needs characterized by climate change, land use change, increased water demand and complex water governance issues (Uhlenbrook and Jong, 2012; Wagener *et al.*, 2012). Future hydrologists should have a broad knowledge covering subjects such as fluid mechanics, physics, mathematics, ecology, geography, statistics, sociology and software engineering (Popescu *et al.*, 2012).

Grasping the concept of hydrology appears to be a difficult task for numerous engineering students due to its empirical and conceptual nature. The qualitative and judgmental aspects of hydrological problems make it relatively difficult for engineering students who are used to solving problems using deterministic methods. It is essential for engineering students to comprehend the physics of hydrological processes before jumping into calculations (Raudkivi, 2013), which is not well appreciated by many students enrolled in engineering hydrology courses. The general difficulties in learning hydrology for civil engineers have been discussed by a number of previous researchers (e.g. Aghakouchak and Habib, 2010; Elshorbagy, 2005; Ngambeki *et al.*, 2012).

Thompson *et al.* (2012) noted that hydrology teaching is dominated by a teacher-centered approach, mainly based on lectures, readings and assignments. Lecturers deliver the content and students take notes, read textbooks and lecture notes and apply the concepts through a series of exercises. Research suggests that a "chalk and talk" approach to education delivery is often ineffective in fully delivering the content (Thompson *et al.*, 2012). On the other hand, student-centered approaches allow students to take more responsibility in their learning through the use of problem-based learning, project-based learning, inquiry-based learning, case-based learning and discovery learning (Herrington and Oliver, 2000). One of the limitations of student-centered approaches in engineering education is that it demands notable time to develop teaching materials such as interactive exercises and online assessments (Jiusto and DiBiasio, 2006; Prince and Felder, 2006).

It is important to understand the essential elements in education: the learner, the teacher, the syllabus, the teaching and learning method and the assessment method (Smith *et al.*, 2005). The learners in hydrology should have a broad knowledge of science, mathematics, physics and statistics. The hydrology teachers should have expert knowledge plus the appropriate training to deliver the content of the subject. The syllabus should be updated in a timely manner to meet the emerging needs of the profession and society. The teaching method should be student-centered and assessments should be of a mixed type, i.e. it should not be based on examinations only, as is the case in many engineering universities in developing countries.

Blended learning is one of the most efficient learning strategies, as noted by the American Society for Training and Development, which rated it as one of the top 10 trends to emerge in the knowledge delivery industry (Bonk and Graham, 2012). Blended learning takes the strengths from both face-to-face and online learning experiences to form a unique learning experience, which leads to fulfilling the educational purposes and outcomes intended (Garrison and Vaughan, 2008). Blended learning is a holistic approach that involves rethinking and transforming teaching and learning (Garrison and Vaughan, 2008). Blended learning has evolved over the years since its inception. It has become cheaper and more readily available, and as a result has been adopted by many educators. Early use of technology in teaching can be traced back to mainframe computer-based training in the 1960s, with advances since then including video distance learning in the 1980s to CD-ROM resources in the 1990s and on to the early internet in the late 1990s (Bersin, 2004). Considerable improvements have been made to the infrastructure for the internet- and web-based technologies since their introduction into education, which has seen evolving technologies: audio, video, simulations and more (Bersin, 2004).

Over the last 10 years, substantial efforts have been made in integrating mobile computing into education, including laptops and tablets, coupled with the adoption of emerging web-based technologies, such as the introduction of Web 2.0 and Web 3.0 tools and associated technologies. This has led to interactive and synchronous teaching activities. These technologies have helped teachers and instructors to easily develop online resources, which in the past would have been more difficult to produce. The online resources are often integrated into the learning management systems, computer software and tools available to educators in their institutions. In particular, the use of touch screens and screens with a digitizer that enables the user to draw on the screen with a stylus have enabled teachers to produce digital resources for students. The screen technology can distinguish between the stylus making contact on the screen and the user's hand, thus enabling users to write and draw on the screen as they would on a piece of paper. This advancement allows educators to use these types of screens as a substitute to online hand-written notes or drawings, which pedagogically is a very low level use of the technology, but also to enhance the traditional online teaching resources by giving an oral explanation and demonstrating the process of problem solving.

Western Sydney University (WSU) has "blended learning staff" embedded in each of the schools across the university who specialize in developing and implementing educational strategies in collaboration with the academics. Within the School of Computing, Engineering and Mathematics there is a blended learning team of six staff who provide this support, one of whom is the third author of this paper. Rahman (2017) and Rahman and Al-Amin (2014) adopted a blended learning approach to deliver a fluid mechanics subject at WSU. The method was extended to an advanced statistical hydrology subject at WSU as presented in this chapter.

Advanced statistical hydrology is one of the core subjects in the Masters of Engineering (civil and environmental) course at WSU. This subject covers the statistical methods applied to solve hydrological problems. This chapter discusses the opportunities and challenges faced by students and lecturers in the learning and teaching of advanced statistical hydrology through a blended learning approach.

The remainder of this chapter is organized as below. Section 1.2 presents an overview of the advanced statistical hydrology subject at WSU. Section 1.3 presents the perspectives of a student, who is the second author of this chapter. Section 1.4 presents the perspectives of the lecturer, who is the first author of this chapter. Section 1.5 presents the performance of students from the 2017 cohort. Section 1.6 presents further modifications of the subject based on students' feedback. Section 1.7 presents a summary and conclusion of this study.

1.2 Overview of the advanced statistical hydrology subject at WSU

The advanced statistical hydrology subject (Code: 301014) is a Level 7 postgraduate subject in the Master of Engineering (Civil/Environmental) curriculum at WSU, which is coordinated by the first author of this paper. The subject is worth 10 credit points. The delivery of the subject started in 2015, and a blended learning approach was adopted to teach this subject that consisted of 5-hour-long face-to-face workshops during the semester, 11 tutorial videos and hand-made tutorial solutions on papers accessible on vUWS, which is the university learning management system. The subject consists of at-site flood frequency analysis, regional flood frequency analysis, trend analysis of hydrological data, linear regression analysis and multivariate statistical techniques to solve advanced hydrological problems.

In order to pass this subject, students are required to complete three different assessment tasks and achieve an overall 50% mark (shown in Table 1.1). The first assessment task consists of a 2000-word critical literature review report on a specified topic. It is weighted 25% and must be completed individually by students. The second assessment is an oral presentation with the same weighting of 25% on a specified topic on statistical hydrology. The third assessment is a design project (4000-word length) with a weighting of 50%. These three assessments tasks are designed to test the students' knowledge achieved in the 14-week period of study, which must satisfy WSU graduate attributes.

It is expected that students completing this subject would be able to solve complex hydrological problems using advanced statistical methods. In particular, students should be able to apply recommended statistical techniques in the fourth edition of Australian Rainfall and Runoff (ARR) (Ball *et al*., 2016). ARR emphasizes the application of Monte Carlo simulation techniques in rainfall-based flood modeling (e.g. Caballero and Rahman, 2014a; Charalambous *et al*., 2013; Loveridge and Rahman, 2018; Rahman *et al*., 2002a; Weinmann *et al*., 2002), where the stochastic nature of rainfall duration (Haddad and Rahman, 2011a), rainfall temporal patterns (Rahman *et al*., 2006), storm losses (Loveridge *et al*., 2013; Loveridge and Rahman, 2018; Rahman *et al*., 2002b) and runoff routing model parameters (Caballero and

Table 1.1 Details of current assessments in advanced statistical hydrology subject at WSU

Assessment	*Type*	*Weighting*
1	Critical Literature Review (2000-word length). Individual report by students analyzing about 10 refereed journal papers on a topic allocated randomly to a student. Students should learn how to prepare a critical literature review report and how to use references in a scientific document.	25%
2	Oral Presentation. A 30-minute-long oral presentation before the class by individual students on a selected topic. Students should learn how to make a professional presentation on a scientific topic and how to answer questions professionally.	25%
3	Design Project. A 4000-word design project on a real-world hydrological problem that needs knowledge of statistical methods covered in the workshop and tutorial Video Display Operator (VDOs) of the subject. Student should learn how to design a hydrological project and how to write a professional design report.	50%

Rahman, 2014b; Patel and Rahman, 2015) are explicitly accounted for. The Monte Carlo simulation technique is adequately discussed in this subject. The non-stationarity analysis in hydrological data (e.g. Hajani *et al.*, 2017; Hajani and Rahman, 2018; Ishak *et al.*, 2013; Ishak and Rahman, 2015; Mamoon and Rahman, 2017) is also taught in this subject. Both at-site flood frequency analysis (e.g. Haddad and Rahman, 2011b) and regional flood frequency analysis (e.g. Haddad and Rahman, 2012; Haddad *et al.*, 2013; Micevski *et al.*, 2015; Rahman *et al.*, 2016, 2018; Zaman *et al.*, 2012) are covered extensively in this subject.

1.3 A student's perspective

The second author of the paper completed the advanced statistical hydrology subject at WSU in 2015 and achieved a high distinction grade (> 85% mark). He found the blended learning approach to be an effective method to learn the contents of this subject. He found the recorded videos for each of the tutorial problems to be very useful in understanding the fundamental issues with the problem, i.e. what the relevant assumptions associated with the solution are and how much confidence can be placed on the given solution. According to the student, the lecturer took plenty of time in solving a problem using a step-by-step procedure, which can be followed by a student without much difficulty. The lecturer's teaching style was phenomenal, and he made difficult topics quite easy to understand. The lecturer did not leave any stone unturned to make the solution enjoyable for students, and he provided numerous real-world examples to clarify difficult concepts.

The second important aspect of this subject, observed by the student, was the oral presentation, in which the lecturer and students had plenty of time (over 5 minutes) to ask relevant questions on the topic. The feedback by the lecturer on the oral presentation was found to be quite useful, e.g. "do not use long sentences/paragraphs in slides," "give adequate references to each figure and table (taken from other sources)" and "make an interesting background and clear objectives of the presentation."

The third aspect of learning, reported by students, was how to make in-text references and the chapter-end list of references correctly using a consistent style. It was found that many international students enrolled in this subject did not know how to prepare an acceptable list of references. The fourth aspect of the subject was a 5-hour-long workshop, in which students were given the opportunity to ask questions of the lecturer to clarify misconceptions relevant to the subject matter. This also allowed intellectual discussion among fellow students, which were an interesting component of this subject.

1.4 Lecturer's perspective

The first author of the paper was the lecturer and coordinator of the advanced statistical hydrology subject at WSU in 2015. He developed the learning materials of the subject in Autumn 2015 semester so that the delivery of the subject in the following Spring semester could be implemented smoothly. He found that developing the subject in a blended learning approach was quite challenging, as a significant amount of time was required to develop online materials. The blended learning team of WSU assisted in recording the tutorial class videos. A graphics tablet was used to record the tutorials that included an oral explanation and graphical illustration for a given problem. This allowed students to follow the logic and reasoning behind the lecturer's explanation at a real-time pace rather than simply reading simplified tutorial notes and solutions.

The graphics tablet used for the creation of the tutorial videos was a Wacom Cintiq 13HD. This is a computer input device and acts as a secondary display, which enables users to hand-draw images, animations and graphics, with a special pen-like stylus (as shown in Fig. 1.1). It features an active digitizer built into the display which enables interaction only through the stylus. In this way, the user can simulate writing with pen and paper as the interaction from the pen is displayed live on the display. The Wacom requires a computer to function and does not include any specialized software. The recordings in advanced statistical hydrology were

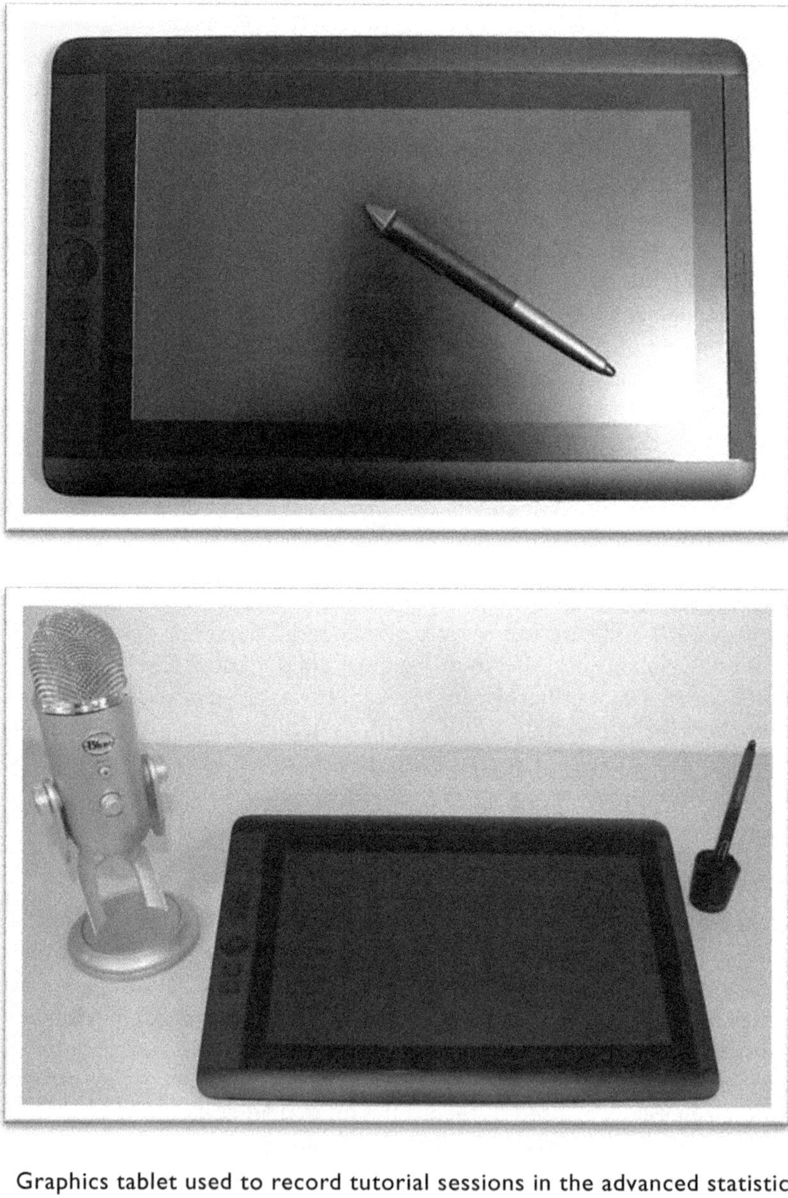

Figure 1.1 Graphics tablet used to record tutorial sessions in the advanced statistical hydrology subject at WSU

a mix of screen captures and recordings using a digital whiteboard application called Explain Everything. Explain Everything is an application available on almost all mainstream devices, including Windows, Android and iOS. It functions as a digital whiteboard, with accompanying oral explanations. Audio was recorded using a Blue Yeti microphone (as shown in Fig. 1.1), which is a high-quality USB studio microphone, connected to the computer.

Eleven tutorials were recorded in this subject, which were then posted to the Web through the vUWS. There were no face-to-face tutorial classes in the subject delivery. A total of 22 hours were spent in recording the tutorials by the development team, consisting of the lecturer and the blended learning team of WSU.

There were six students in the class in 2015. The total watch time of the online tutorials was 2,684 minutes by the six students (average 7.45 hours per student). The total number of views by the six students was 458, i.e. 76 views by a student on average. The tutorial access rate by the students is illustrated in Figure 1.2, which shows that the access rates peaked before the project submission dates.

Students struggled in writing the critical literature review report. The references (both in-text and chapter-end) were not consistent for the majority of the students. In oral presentations, some students included only the basic and intermediate information on the topic without providing advanced/latest materials from high-quality journal papers. Students had a tendency not to quote the references for the figures and tables taken from other sources. The lecturer (first author) felt that the 5-hour workshop was inadequate to deliver the subject content to the students effectively. Also, without any formal examination; the students were reluctant in mastering the topics contained in this subject. In the design project, a good number of students could not apply the right statistical techniques

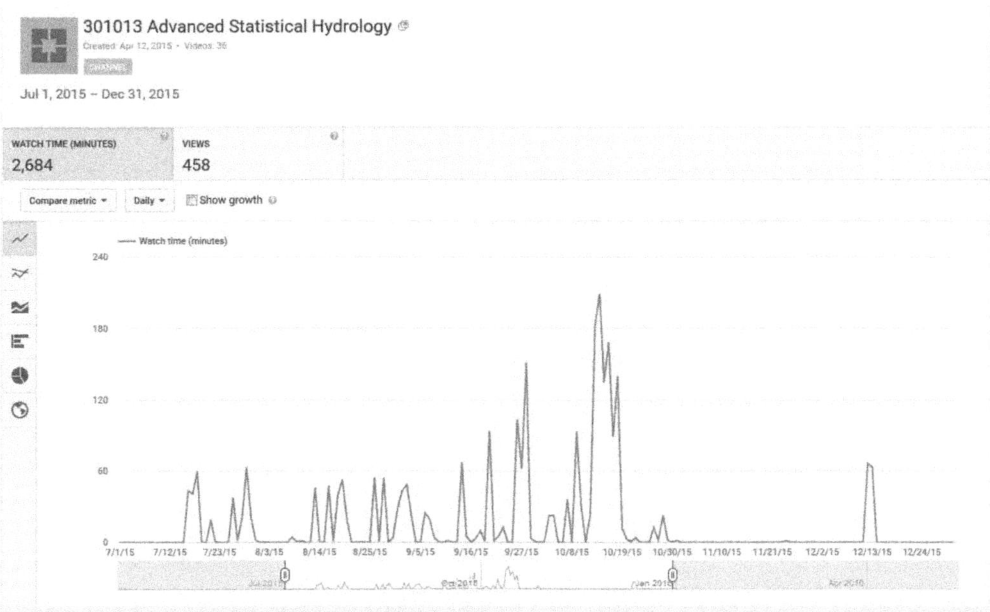

Figure 1.2 Tutorial access rate by students for advanced statistical hydrology subject at WSU (2015)

to solve the given hydrological problem, which indicated a lack of competence in the subject matter.

1.5 Students' performance analysis

The results of 47 students in 2017 are summarized in Table 1.2. The average mark for oral presentations (Assessment 2) is the highest among the three assessment types, and the average mark for the design project (Assessment 3) is the lowest; the difference between these two assessment tasks is 12%. During the oral presentation, students were found to be more actively engaged with the lecturer and fellow students. In the design project, students were found to be relatively weak in discussing the results, in particular the implications of the associated assumptions on the findings and comparison of the findings with similar studies available in scientific literature. The standard deviation for the design project is the highest, which indicates a higher degree of variation among the individual students' performances. Since there is no formal examination in the subject, the overall mark of 66% looks reasonable for the subject. It should be noted that the examination-based subjects generally have smaller overall average marks than the project-based subjects, as experienced by the first author of this chapter.

The correlations of marks achieved by the students in 2017 among the three assessment tasks are presented in Figures 1.3, 1.4 and 1.5. It is interesting to note that the correlations among the marks achieved by individual students in the three assessments tasks are relatively weak, with a coefficient of determination ranging from 24% to 46% only. Figure 1.6 presents

Table 1.2 Summary of student results in 2017

Statistics	*Assessment 1 (Literature review) (%)*	*Assessment 2 (Oral presentation) (%)*	*Assessment 3 (Design project) (%)*	*Overall average mark (%)*
Average	70	73	61	66
Standard deviation	13.9	14.2	21.3	15.3

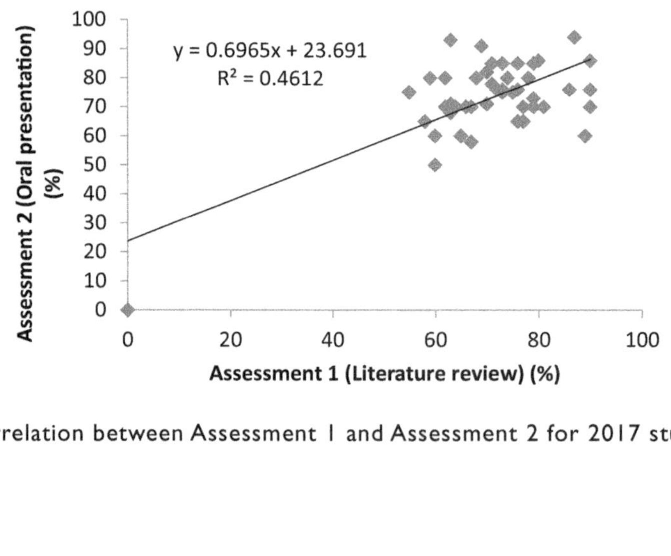

Figure 1.3 Correlation between Assessment 1 and Assessment 2 for 2017 students

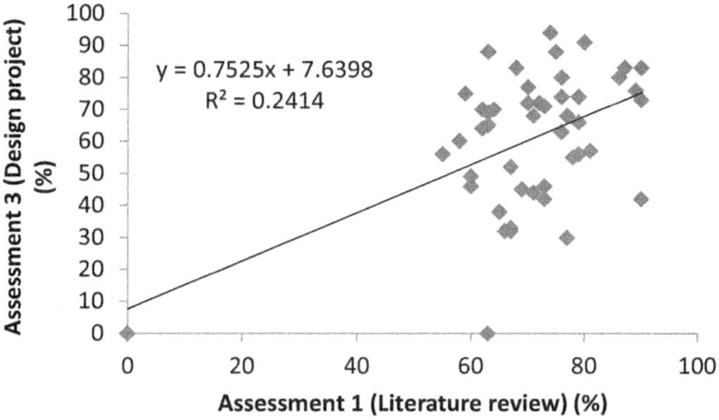

Figure 1.4 Correlation between Assessment 1 and Assessment 3 for 2017 students

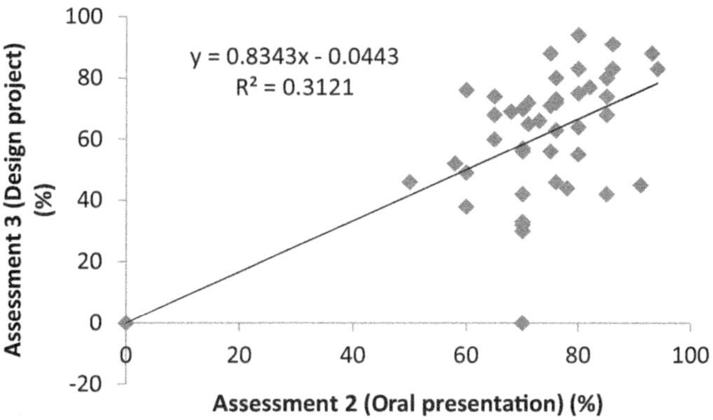

Figure 1.5 Correlation between Assessment 2 and Assessment 3 for 2017 students

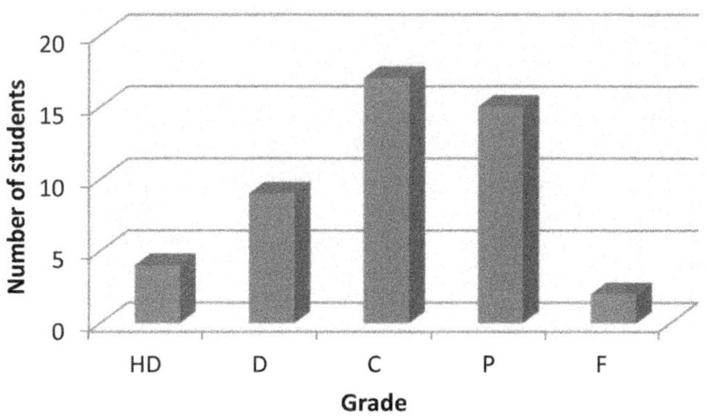

Figure 1.6 Grades for students in 2017 (total number of students: 47)

student grades of the 2017 students, of which only two students failed (i.e. 4.2% students) and four and nine students received high distinction (HD) and distinction (D), which represents 8.5% and 19.1% of the students respectively. Overall, this seems to be an impressive outcome confirming that a blended learning approach assists students in achieving the learning outcomes.

1.6 Upgrading the subject

The current 5-hour-long face-to-face workshop seems to be inadequate for students to grasp the concepts of the subject. This has been reflected in the student feedback on the subject (SFU) and also during the informal discussion among the lecturer and some students. To enhance student learning, a number of changes are proposed. Two hours of face-to-face lecture every week would allow students to engage better with the lecturer and other students through being able to ask more questions and interacting more with other groups of students. An intra-session examination will ensure that the students are working towards achieving the learning outcomes at the earlier part of the semester rather than leaving it to the last moment. By completing the engineering design project as a group, students will be able to develop their teamwork skills. The literature review (Assessment 1 of the current course) will be integrated with the design project, i.e. it will contain both critical literature review, and data analysis and design. The final examination will test the students' competency in the subject in an effective manner at an individual student level. Currently, there is no final examination and intra-session examination, which makes it difficult to assess students' competency adequately in the subject matter. The proposed changes in the assessments are summarized in Table 1.3. If these changes are not made, it will be difficult for students to understand the challenging concepts of the subject due to limited face-to-face contact hours with the lecturer. The current 30-minute oral presentation by each student took about 25 hours oral presentation time in 2017 (for 47 students), i.e. three full days, which appears to be too long. Without a final examination, it is difficult to test students' competency in the subject matter. It is expected that the proposed changes will enhance students' learning in this subject in future years.

Table 1.3 Details of proposed changes in the assessments of advanced statistical hydrology subject at WSU

Assessment	*Type*	*Weighting*
1	Intra-session exam: A closed book 1-hour examination which assesses a student's progressive learning after five face-to-face lectures.	25%
2	Group Project: A 5000-word design project on a real-world hydrological problem that needs knowledge of statistical methods covered in this subject. Students should learn how to design a hydrological project and how to write a professional design report.	30%
3	Final examination: A closed book 2-hour examination which assesses a student's overall learning in the subject matter.	45%

1.7 Conclusion

This chapter outlines the delivery of the advanced statistical hydrology subject in Western Sydney University (WSU) as part of the Masters of Engineering degree. It has been found that blended learning is quite effective in delivering this subject to Master of Engineering students. The recorded tutorial videos were found to be the most useful element of this subject, as these allowed students to use the materials as many times as needed, and more importantly, at the time when it is convenient to him/her. The preparation of critical literature review was found to be the most difficult task for the students. As further development in the delivery of this subject, the assessment should include a 2-hour long formal, closed book examination to ensure that students have learned the subject matter with required competence. Also, an intra-session examination is proposed to ensure progressive learning for the students. It has been found that a 5-hour-long workshop is not sufficient to deliver the subject contents effectively to the students, and hence it is proposed that 2-hour lectures be delivered every week to answer students' queries and to facilitate face-to-face engagement among the students and between the students and lecturer. This will make the blended learning approach more effective for the students in the advanced statistical hydrology subject at WSU in future years.

References

Aghakouchak, A. & Habib, E. (2010) Application of a conceptual hydrologic model in teaching hydrologic processes. *International Journal of Engineering Education*, 26, 4, 963–973.

Ball, J., Babister, M., Nathan, R., Weeks, W., Weinmann, E., Retallick, M. & Testoni, I. (2016) *Australian Rainfall and Runoff: A Guide to Flood Estimation.* Commonwealth of Australia, Canberra.

Bersin, J. (2004) *The Blended Learning Book: Best Practices, Proven Methodologies, and Lessons Learned.* John Wiley & Sons, New York.

Bonk, C.J. & Graham, C.R. (2012) *The Handbook of Blended Learning: Global Perspectives, Local Designs.* John Wiley & Sons, New York.

Caballero, W.L. & Rahman, A. (2014a) Development of regionalized joint probability approach to flood estimation: A case study for New South Wales, Australia. *Hydrological Processes*, 28, 13, 4001–4010.

Caballero, W.L. & Rahman, A. (2014b) Application of Monte Carlo simulation technique for flood estimation for two catchments in New South Wales, Australia. *Natural Hazards*, 74, 1475–1488.

Charalambous, J., Rahman, A. & Carroll, D. (2013) Application of Monte Carlo simulation technique to design flood estimation: A case study for North Johnstone River in Queensland, Australia. *Water Resources Management*, 27, 4099–4111.

Elshorbagy, A. (2005) Learner-centered approach to teaching watershed hydrology using system dynamics. *International Journal of Engineering Education*, 21(6), 1203.

Garrison, D.R. & Vaughan, N.D. (2008) *Blended Learning in Higher Education: Framework, Principles, and Guidelines.* John Wiley & Sons, New York.

Gleick, P.H. (1998) The human right to water. *Water Policy*, 1(5), 487–503.

Grimmond, J. (2010) Special report on water. *The Economist*, 22 May 2010.

Haddad, K. & Rahman, A. (2011a) Regionalization of storm duration for determining derived flood frequency curve: A case study for Victoria in Australia. *Asian Journal of Water, Environment and Pollution*, 8, 3, 37–46.

Haddad, K. & Rahman, A. (2011b) Selection of the best fit flood frequency distribution and parameter estimation procedure – A case study for Tasmania in Australia. *Stochastic Environmental Research & Risk Assessment*, 25, 415–428.

Haddad, K. & Rahman, A. (2012) Regional flood frequency analysis in eastern Australia: Bayesian GLS regression-based methods within fixed region and ROI framework – Quantile regression vs. Parameter regression technique. *Journal of Hydrology*, 430–431(2012), 142–161.

Hajani, E. & Rahman, A. (2018) Characterising changes in rainfall: A case study for New South Wales, Australia. *International Journal of Climatology*, 38, 1452–1462.

Hajani, E., Rahman, A. & Ishak, E. (2017) Trends in extreme rainfall in the State of New South Wales, Australia. *Hydrological Sciences Journal*, 62, 13, 2160–2174.

Haddad, K., Rahman, A., Zaman, M. & Shrestha, S. (2013) Applicability of Monte Carlo cross validation technique for model development and validation using generalised least squares regression. *Journal of Hydrology*, 482, 119–128.

Herrington, J. & Oliver, R. (2000) An instructional design framework for authentic learning environments. *Educational Technology Research and Development*, 48(3), 23–48.

Hooper, R.P. (2009) Towards an intellectual structure for hydrologic science. *Hydrological Processes*, 23(2), 353–355.

Ishak, E. & Rahman, A. (2015) Detection of changes in flood data in Victoria, Australia over 1975–2011. *Hydrology Research*, 46, 5, 763–776.

Ishak, E., Rahman, A., Westra, S., Sharma, A. & Kuczera, G. (2013) Evaluating the Non-stationarity of Australian Annual Maximum Floods. *Journal of Hydrology*, 494, 134–145.

Jiusto, S. & DiBiasio, D. (2006) Experiential learning environments: Do they prepare our students to be self-directed, life-long learners? *Journal of Engineering Education*, 95(3), 195–204.

Kordrostami, S., Purdy, D. & Rahman, A. (2016) Teaching and learning of advanced statistical hydrology using a blended learning approach: A case study in Western Sydney University. *Proceedings of International Conference on Engineering Education and Research*, 21–24 Nov 2016, Sydney, Australia.

Lenschow, R.J. (1998) From teaching to learning: A paradigm shift in engineering education and lifelong learning. *European Journal of Engineering Education, 23*(2), 155–161.

Loveridge, M., Rahman, A., Hill, P. & Babister, M. (2013) Investigation into probabilistic losses for design flood estimation: A case study for the Orara River catchment, NSW. *Australian Journal of Water Resources*, 17(1), 13–24.

Loveridge, M. & Rahman, A. (2018) Monte Carlo simulation for design flood estimation: A review of Australian practice. *Australasian Journal of Water Resources*. doi:10.1080/13241583.2018.1453979.

Mamoon, A.A. & Rahman, A. (2017) Selection of the best fit probability distribution in rainfall frequency analysis for Qatar. *Natural Hazards*, 86(1), 281–296.

McGlynn, B., Wagener, T., Marshall, L., McGuire, K., Meixner, T. & Weiler, M. (2010) Challenges and opportunities for hydrology education in a changing world-The modular curriculum for hydrologic advancement. *EGU General Assembly Conference Abstracts*, 12, 7544.

Micevski, T., Hackelbusch, A., Haddad, K., Kuczera, G. & Rahman, A. (2015) Regionalisation of the parameters of the log-Pearson 3 distribution: A case study for New South Wales, Australia. *Hydrological Processes*, 29, 2, 250–260.

Ngambeki, I., Thompson, S.E., Troch, P.A., Sivapalan, M. & Evangelou, D. (2012) Engaging the students of today and preparing the catchment hydrologists of tomorrow: Student-centered approaches in hydrology education. *Hydrology and Earth System Sciences Discussion*, 9(1), 707–740.

Patel, H. & Rahman, A. (2015) Probabilistic nature of storage delay parameter of the hydrologic model RORB: A case study for the Cooper's Creek catchment in Australia. *Hydrology Research*, 46, 3, 400–410.

Prince, M.J. & Felder, R.M. (2006) Inductive teaching and learning methods: Definitions, comparisons, and research bases. *Journal of Engineering Education*, 95(2), 123–138.

Popescu, I., Jonoski, A. & Bhattacharya, B. (2012) Experiences from online and classroom education in hydroinformatics. *Hydrology and Earth System Sciences*, 16(11), 3935.

Rahman, A. (2017) A blended learning approach to teach fluid mechanics in engineering. *European Journal of Engineering Education*, 42, 3, 252–259.

Rahman, A. & Al-Amin, M. (2014) Teaching of fluid mechanics in engineering course: A student-centered blended learning approach. In: Alam, F. (ed) *Using Technology Tools to Innovate Assessment, Reporting, and Teaching Practices in Engineering Education*. IGI Global Publisher, Hershey, PA, USA. pp. 12–20.

Rahman, A., Charron, C., Ouarda, T.B.M.J. & Chebana, F. (2018) Development of regional flood frequency analysis techniques using generalized additive models for Australia. *Stochastic Environment Research & Risk Assessment*, 32, 123–139.

Rahman, A., Haddad, K., Kuczera, G. & Weinmann, P.E. (2016) Regional flood methods. In: Ball et al., (ed) *Australian Rainfall & Runoff*, Chapter 3, Book 3. Commonwealth of Australia, Canberra.

Rahman, A., Islam, M., Rahman, K., Khan, S. & Shrestha, S. (2006) Investigation of design rainfall temporal patterns in the gold coast region Queensland. *Australian Journal of Water Resources*, 10, 1, 49–61.

Rahman, A., Weinmann, P.E., Hoang, T.M.T. & Laurenson, E.M. (2002a) Monte Carlo simulation of flood frequency curves from rainfall. *Journal of Hydrology*, 256(3–4), 196–210.

Rahman, A., Weinmann, P.E. & Mein, R.G. (2002b) The use of probability-distributed initial losses in design flood estimation. *Australian Journal of Water Resources*, 6(1), 17–30.

Raudkivi, A.J. (2013) *Hydrology: An Advanced Introduction to Hydrological Processes and Modelling*. Elsevier, Amsterdam.

Smith, K.A., Sheppard, S.D., Johnson, D.W. & Johnson, R.T. (2005) Pedagogies of engagement: Classroom-based practices. *Journal of Engineering Education*, 94(1), 87–101.

Thompson, S.E., Ngambeki, I., Troch, P.A., Sivapalan, M. & Evangelou, D. (2012) Incorporating student-centered approaches into catchment hydrology teaching: A review and synthesis. *Hydrology and Earth System Sciences*, 16(9), 3263–3278.

Torgersen, T. (2006) Observatories, think tanks, and community models in the hydrologic and environmental sciences: How does it affect me? *Water Resources Research*, 42(6).

Uhlenbrook, S. & Jong, E.D. (2012) T-shaped competency profile for water professionals of the future. *Hydrology and Earth System Sciences*, 16(10), 3475–3483.

Vogel, R.M., Lall, U., Cai, X., Rajagopalan, B., Weiskel, P.K., Hooper, R.P. & Matalas, N.C. (2015) Hydrology: The interdisciplinary science of water. *Water Resources Research*, 51(6), 4409–4430.

Wagener, T., Kelleher, C., Weiler, M., McGlynn, B., Gooseff, M., Marshall, L., Meixner, T., McGuire, K., Gregg, S., Sharma, P. & Zappe, S. (2012) It takes a community to raise a hydrologist: The Modular Curriculum for Hydrologic Advancement (MOCHA). *Hydrology and Earth System Sciences*, 9, 2321–2356.

Wagener, T., Sivapalan, M., Troch, P.A., McGlynn, B.L., Harman, C.J., Gupta, H.V., Kumar, P., Rao, P.S.C., Basu, N.B. & Wilson, J.S. (2010) The future of hydrology: An evolving science for a changing world. *Water Resources Research*, 46(5).

Weinmann, P.E., Rahman, A., Hoang, T., Laurenson, E.M. & Nathan, R.J. (2002) Monte Carlo simulation of flood frequency curves from rainfall – The way ahead. *Australian Journal of Water Resources*, 6(1), 71–80.

Zaman, M., Rahman, A. & Haddad, K. (2012) Regional flood frequency analysis in arid regions: A case study for Australia. *Journal of Hydrology*, 475, 74–83.

Chapter 2

Blended learning

Online and hands-on activities in an applied fluid flow course

P.R. Piergiovanni

Department of Chemical and Biomolecular Engineering, Lafayette College, Easton, PA, USA

ABSTRACT: Online and hands-on activities were blended in an applied fluid flow course to help students understand and apply the mechanical energy balance to Newtonian and non-Newtonian fluids. Few formal lectures were given during class. Students remained engaged and interested in the classroom and participated fully in the online assignments. The hands-on activities included a ranking exercise, two jigsaw activities, a gallery walk and in-class experiments. Students learned how to characterize non-Newtonian fluids and modify the Hagen Poiseuille equation, Reynolds number and friction factors for their flow. They also learned how centrifugal and positive displacement pumps work and the conditions for choosing the best type. The instructor coached the students and encouraged them to reflect and write after each activity. Assessment of student learning shows they met the student learning objectives.

2.1 Introduction

Blended learning has many definitions and can be accomplished in many forms, thus the Online Learning Consortium defined a spectrum of course-level definitions as shown in the list below (Mayadas *et al.*, 2015).

- The classroom course where all activity takes place in face-to-face meetings;
- The synchronous distributed course where technology is used to extend the classroom course to students at remote sites in real time;
- The web-enhanced course where online course activity complements the classroom sessions without replacing them;
- The hybrid course where online activity replaces a significant percentage of classroom meetings, but not all of them, and
- The fully online course where all course activities take place online.

The Christensen Institute further defines each category, including four models of web-enhanced courses (Horn and Staker, 2014). For example, in the "flex" model, online learning provides the "backbone" while teachers are in the classroom guiding students to different offline projects on an individualized schedule. Alternatively, students in a traditional school can take online courses to supplement their learning in the "a la carte" model. The course described in this chapter follows the "rotational" model, in which students rotate between online, classroom and laboratory activities. The online activities are provided via the classroom management system (CMS) and are designed for students to recall previous information, learn basic definitions and begin to understand figures and diagrams. All course notes are available via the CMS after class meetings. Additionally, the instructor provides support information for topics in folders on the CMS. In the classroom, students practice applying the equations, drawing conclusions and gathering data to analyze. Both classroom time and laboratory time provide opportunities for active learning. This chapter describes the online components and hands-on activities that were used to fully engage the students in their own learning.

2.2 Active learning in the classroom

Socrates asked his students leading questions, forcing them to be intellectually active – this exemplifies active learning. According to Prince (2004), any method of teaching that engages students in the learning process constitutes active learning and develops students who become responsible for their own learning. Active learning:

- Increases the percentage of students who understand concepts (Prince, 2004);
- Improves long-term retention of the material (Felder *et al.*, 2000);
- Has been linked to more enjoyment of the course and higher graduation rates (Rich *et al.*, 2015) and
- Leads to increased confidence (Kirkham and Seymour, 2005).

As noted by Chickering and Gamson (1989), students do not learn much by listening to a professor talking, memorizing pre-packaged assignments and spitting out answers. They learn far more when they discuss and write about what they are learning and relate it to past experiences or apply it to their daily lives.

Many techniques for active learning have been developed, and a pictorial spectrum has been proposed by O'Neal & Pinder-Grover (n.d.). Simple techniques, such as a pause for reflection and asking students to write and perhaps share their thoughts with a neighbor, can be used several times in one class setting. Other techniques, such as the jigsaw exercise or gallery walk, require some advance preparation and more class time, but all students are engaged and thus retain more knowledge. More involved techniques such as the use of case studies, games or simulations and field trips are also effective but not available for all subjects or courses.

One hundred years ago, Dewey (1916) described experiential learning, one type of active learning, as a process where a student participates in an activity, then reflects on what took place. He stressed that the reflection process is key to retaining the concepts from the activity. The literature shows a consensus that experiential learning is effective (Smith *et al.*, 2005; Smith, 2011). The instructors, however, must be active listeners and coaches in order to maximize student learning (Lalley and Miller, 2007).

Problem-based learning (PBL), another type of active learning, provides students with a real problem to solve. A relevant problem is introduced at the beginning of an instruction cycle and is used to provide context and motivation for the students to learn (Prince, 2004). The problem is open ended and ambiguous. PBL has been shown to:

- Increase students' problem solving and critical thinking skills (Krajcik and Blumenfeld, 2006; Reynolds and Hancock, 2010; Savery, 2006);

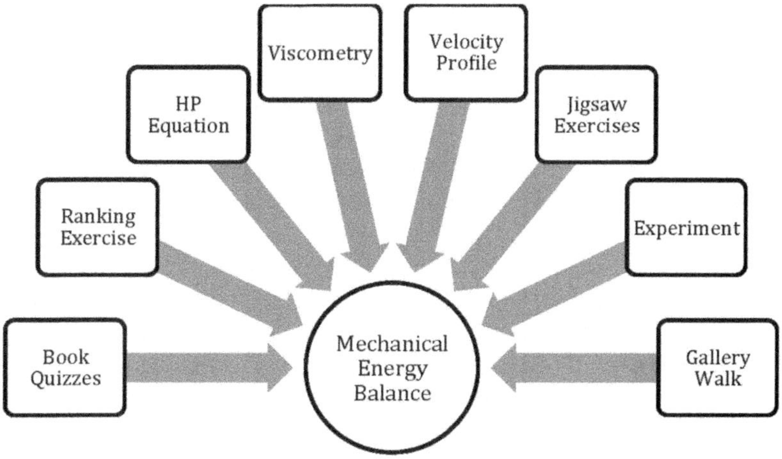

Figure 2.1 Diagram of online, active and experiential exercises designed to support students as they applied the mechanical energy balance to Newtonian and non-Newtonian fluids

- Increase students' conceptual understanding (Yadav *et al.*, 2011) and knowledge retention (Prince, 2004) and
- Develop more positive attitudes and a deeper approach to learning (Prince, 2004).

A small negative effect of PBL occurs if students have not had much experience with problem solving (Prince, 2004).

Providing learning support through scaffolding can increase student learning from projects (Greening, 1998). Scaffolding makes it easier for students to remember concepts as they apply them to the larger project and can negate some of their inexperience. Scaffolding can be provided through online instruction or active learning activities.

This chapter describes a module from an applied fluid dynamics course covering the mechanical energy balance and pump selection for Newtonian and non-Newtonian fluids. The module was configured as a rotational model blended learning class. Learning objectives were first determined, and then online and offline activities were developed to support them. The module was presented to the students in terms of a project: could the students turn a former soda-bottling plant into a ketchup-bottling factory?

Before each class period, students completed an online quiz on the required reading. During class, students participated in a variety of active learning exercises, including time in the laboratory collecting data. The components, shown in Figure 2.1, are described in Section 2.3. An understanding of each component was necessary to apply the mechanical energy balance to the project.

2.3 The course and the project

"Applied Fluid Flow and Heat Transfer" is a third-year course taught to chemical engineering students. All students have completed a theoretical fluid dynamics course in the previous semester. The students meet twice a week for a 1-hour period and once a week for a 2-hour problem-solving period.

As in most fluid flow courses, the mechanical energy balance is a central theme. One of its main applications is pump choice and sizing. Most courses have an accompanying laboratory where students run water through a pump loop and measure pump power, head and efficiency. However, many fluids are non-Newtonian, and pumping these fluids can be significantly different from pumping water. A PBL scenario with multiple supporting activities was designed to illustrate the differences and help students learn how to design systems for non-Newtonian fluids and to determine when to use the various types of pumps.

An effective project for PBL should be realistic, and scenarios familiar to the students are especially effective (Thomas, 2000). In this project, the students determined if a recently abandoned local soda-bottling factory could be used to bottle ketchup. The problem statement and a detailed description on this task can be seen in Piergiovanni (2017a). The student learning objectives (SLOs) for this module are described below.

By the end of the module, the students should be able to:

1. Calculate velocity profiles and wall shear rates for flow in a tube;
2. Understand how to characterize a non-Newtonian fluid's viscosity;
3. Explain how to modify fluid flow equations for non-Newtonian fluids;
4. Evaluate the use of different pumps for different applications;
5. Explain why different pump types are used for non-Newtonian fluids;
6. Create a pump characteristic curve and calculate power and efficiency;
7. Explain the difference in the curves for Newtonian and non-Newtonian fluids;
8. Compare friction losses for the two types of fluids and
9. Apply the mechanical energy balance to various pumping situations for both Newtonian and non-Newtonian fluids.

Several active learning supports or scaffolding were developed to surround the project, as depicted in Figure 2.1 and described in the next section. The online components and experiment covered all nine student learning objectives. Each support activity provided support for one or two of the SLOs.

The PBL scenario was introduced during class. The students were provided with the problem statement and specific guidelines for a final report. After reading the statement, small groups spent about 10 minutes brainstorming questions they had about the process and the project, and what information they would need to complete it. Links to relevant information were provided, and the instructor circulated around the classroom, answering some questions and guiding and coaching students as needed. The active learning activities were scheduled over the following weeks. Descriptions, example student results and a summary of student learning are supplied in the following pages for each support activity.

2.4 Support description, example results and student learning

The support activities were developed to aid student learning. At the end of each activity, the students were required to reflect (as recommended by Dewey, 1916) and discuss (as recommended by Chickering and Gamson, 1989).

2.4.1 Online book quizzes (SLO 1–9)

For the active learning exercises to be effective, the students needed to come to class with the necessary background information (Prince, 2004). Some of the information had been taught in the previous course, and all of it was provided in the course textbook. In order to encourage the students to read the textbook before class, short online quizzes were provided via the CMS, and completion before class began was worth 5% of the course grade. Many questions were multiple choice, which required close reading of short passages of text (page numbers were provided). Some questions required students to drag and drop labels on diagrams, to ensure they understood them (see Fig. 2.2 for some examples). The

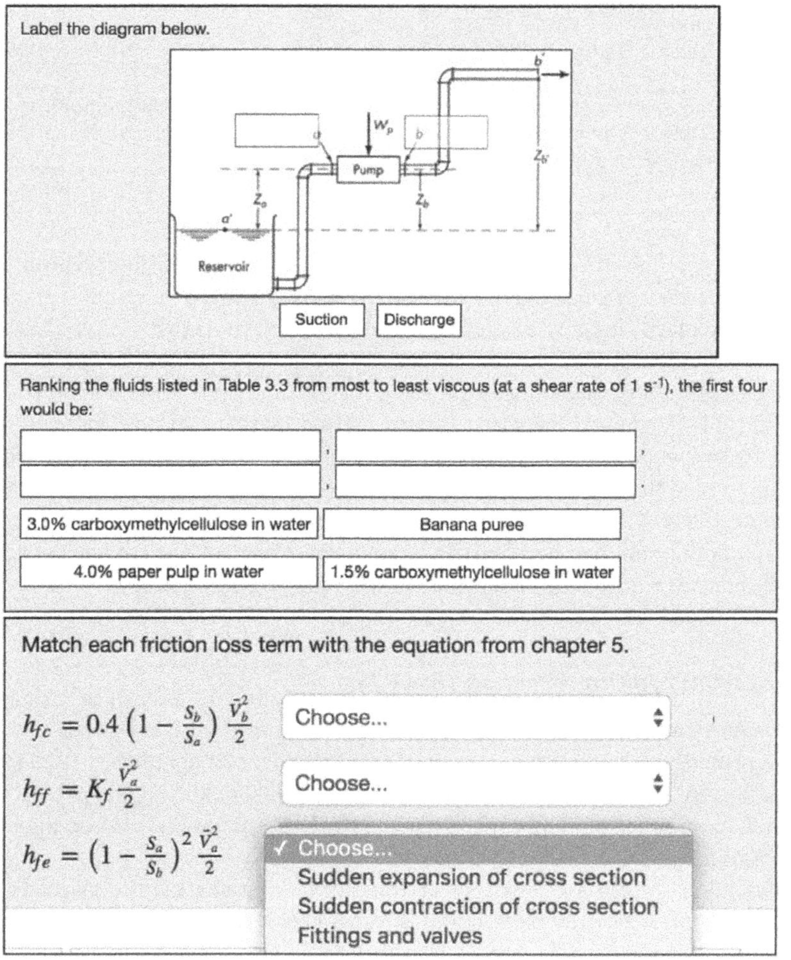

Figure 2.2 Sample questions from the online book quizzes. Some questions had the students drag and drop text boxes to label figures or put items in order, other questions were multiple choice or true/false. The quizzes listed the relevant text pages or other literature for the students to read.

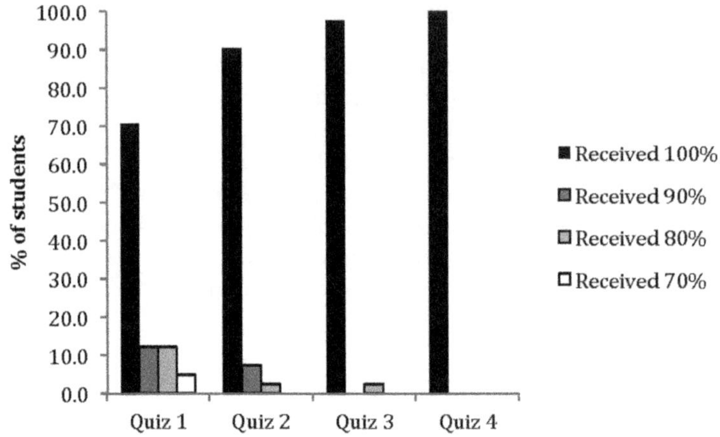

Figure 2.3 Summary of results from the first four online book quizzes. More than 70% of the students answered all questions correctly, and more than 90% of the students completed every quiz.

book quizzes were automatically graded, and required no instructor intervention once they were created. The instructor began each class by asking if there were questions about the book quizzes and clarifying any concerns. Occasionally, a short discussion of the concepts followed.

With the exception of one student for one quiz, all students completed the book quizzes that supported this portion of the class. Figure 2.3 summarizes the student scores from the first four quizzes. Over 70% of the students always earned perfect scores, showing that they took the time to read the material and answer the questions carefully. By the fourth quiz, all students scored 100%. The effect of the book quizzes was noted in class – the instructor could sketch a graph or diagram that had been covered on the quiz, and the students were comfortable using it to answer questions. The students were prepared for class.

2.4.2 Hands-on: ranking exercise (SLO 2)

The students participated in a ranking exercise shortly after the PBL was assigned. Student groups were provided with toothpicks, popsicle sticks and small containers of 10 fluids, such as mayonnaise, olive oil, corn syrup and ketchup. They poked, stirred and poured them, and then ranked them in order of increasing viscosity. Each group put their list on the blackboard to compare rankings.

The students had used a Brookfield viscometer to characterize the viscosity of motor oil the previous semester. They determined it was a non-Newtonian fluid, but most had not thought deeply about what that meant. During the ranking exercise, as they poked and stirred various fluids, they recognized that some fluids became easier to move the faster they were stirred. Ranking the fluids was not an easy task – and there was no unique solution – but they were able to categorize the fluids into groups. This led to a discussion of apparent viscosity and shear rates present in pipes and pumps, which are important when pumping non-Newtonian fluids.

2.4.3 Hands-on: Hagen Poiseuille equation (SLO 3)

In their previous course, students had derived the Hagen Poiseuille (HP) equation (Equation 2.1).

$$\Delta P = \frac{8\mu L Q}{\pi R^4} \tag{2.1}$$

where ΔP is the pressure developed, μ is the solution viscosity, L is the length of the tube, Q is the volumetric flow rate and R is the radius of the tube. The derivation assumes the fluids are Newtonian and moving in laminar flow. In this course, they applied the HP equation to estimate the viscosity of xanthan gum solutions. The flow rates of water and three xanthan gum solutions (2000, 4000 and 8000 ppm) were simultaneously measured while pumped by a peristaltic pump with four heads. Using the viscosity of water and assuming the four heads operated identically, the HP equation was used to estimate the apparent viscosity of the xanthan gum solutions. The students also calculated the wall shear rate, γ_w, according to Equation 2.2:

$$\dot{\gamma}_w = \frac{4Q}{\pi R^3} \tag{2.2}$$

The results are summarized in Table 2.1. Students had little trouble with the calculations. They were surprised that the pump could move the solid-appearing 8000 ppm xanthan gum solution but accepted the large increase in viscosity. They calculated the Reynolds number to confirm laminar flow, but few students questioned if the xanthan gum solutions were Newtonian. As the starch solution exited the tube after the pump, they could see that the moving starch solution was less viscous than the stationary one – an indication that the starch was shear thinning, but not recognized by students.

2.4.4 Hands-on: Brookfield viscometer (SLO 1, SLO 2)

In their previous lab course, the students had used the Brookfield viscometer with Newtonian liquids and motor oils. They measured the viscosity of the oils at two spindle speeds and concluded they were non-Newtonian. In this course, they measured the apparent viscosity of the three xanthan gum solutions at many spindle speeds and discovered that the solutions were not Newtonian fluids, but were shear thinning (see Fig. 2.4 for sample data).

A shear-thinning fluid can be modeled by Equation 2.3:

$$\eta_{\text{app}} = K\dot{\gamma}^{n-1} \tag{2.3}$$

Table 2.1 Calculations from the Hagen Poiseuille equation

Fluid	*Q [mL/s]*	*ΔP [Pa]*	*μ [Pa s]*	*γ_w [s^{-1}]*
Water	3.02	3000	0.00089	1900
2000 ppm	0.44	3000	0.0061	286
4000 ppm	0.078	3000	0.034	50
8000 ppm	0.018	3000	0.15	11

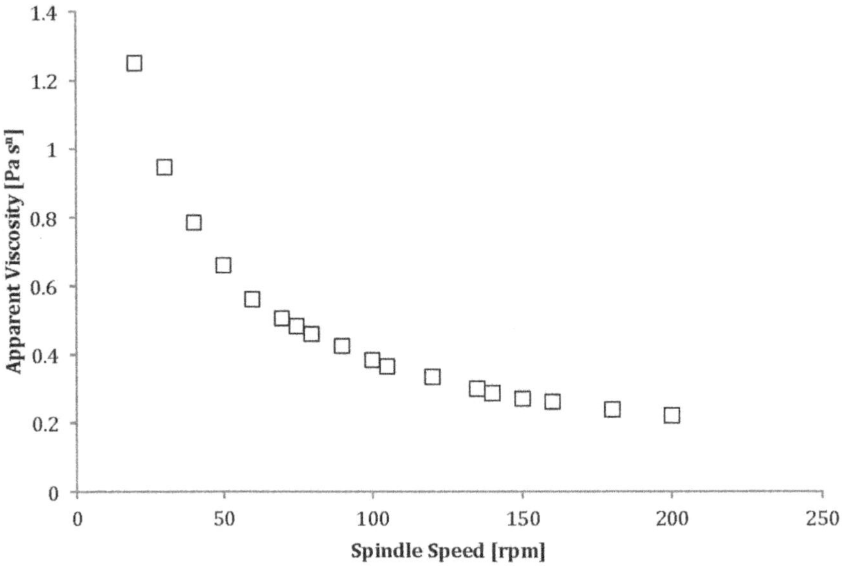

Figure 2.4 Viscosity data from the Brookfield viscometer for an 8000 ppm xanthan gum solution. As the spindle speed increases, the viscosity decreases, indicating a shear-thinning fluid.

where η_{app} is the apparent viscosity at the shear rate γ. K and n are the flow consistency and flow behavior indices, respectively. The students were asked to use their data to find K and n for the xanthan gum solutions. In order to do this, the spindle speeds must be converted to shear rates, γ. The standard Brookfield disc spindles do not have a directly definable shear rate (Brookfield Engineering, 2016). A journal article that provides a procedure to convert from spindle speeds to shear rates was provided online via the CMS (Mitschka, 1982). The procedure requires the flow behavior index, n. Each group developed a method to estimate n, calculate the shear rates and finally plot Equation 2.3 to calculate K and n. Through this exercise, the students carefully read a journal article to understand the procedure and develop their own method. Figure 2.5 shows an example plot for 8000 ppm xanthan gum. This solution has a flow behavior index of 0.23 and clearly is non-Newtonian. The results for all three concentrations are summarized in Table 2.2.

Now, however, the students reflected and questioned their previous activity: the HP equation assumes the fluids are Newtonian. The xanthan gum solutions had flow behavior indices below 0.4, all clearly non-Newtonian. Acting as a coach (Lalley and Miller, 2007), the instructor asked the students to carefully examine their original plots (Fig. 2.4). They eventually recognized that at higher shear rates, the non-Newtonian xanthan gum solutions behave in a Newtonian fashion (i.e. constant viscosity as the shear rate increases, which occurred around 150 rpm). Perhaps, if shear rates were high enough in the pump tubing, the HP equation could be applied. The shear rate calculations from Equation 2.2 supported this for the 2000 ppm solution, but not the higher concentrations.

The instructor provided a few pages from a food engineering text on the CMS for students to read (Singh and Heldman, 2014). This book provides an equation for the wall shear

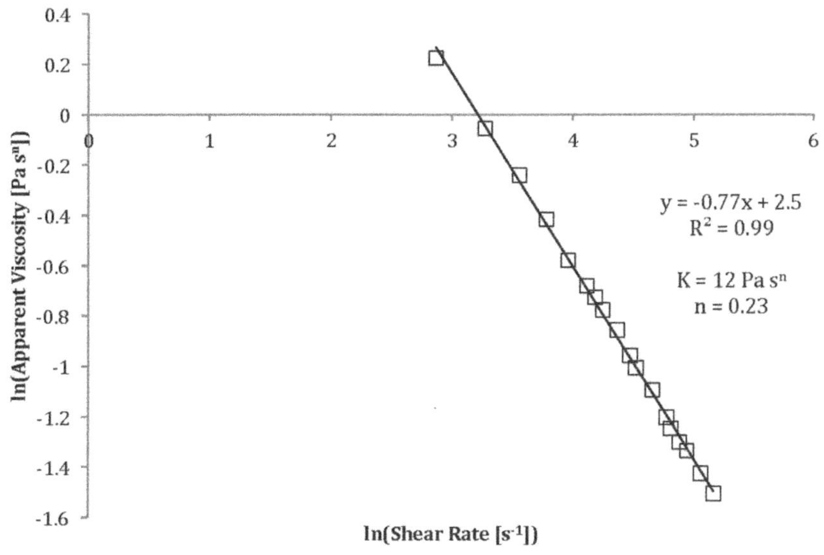

Figure 2.5 Linearized plot of Equation 2.3 for 8000 ppm xanthan gum solution used to calculate the flow behavior and flow characterization indices

Table 2.2 Non-Newtonian characterization of the xanthan gum solutions

Fluid	*n*	*K [Pa s^n]*	*γ [s^-1] (Non-Newtonian)*	*γ [s^-1] (Newtonian)*	*Difference in γ*
2000 ppm	0.36	1.4	410	286	31%
4000 ppm	0.32	7.4	78	50	35%
8000 ppm	0.23	12	21	11	46%

rate for non-Newtonian fluids (Equation 2.4). The text also provides the HP equation modified for non-Newtonian fluids.

$$\dot{\gamma}_w = \frac{3n+1}{4n}\frac{4Q}{\pi R^3} \tag{2.4}$$

The shear rates calculated using the non-Newtonian equation were slightly higher than those calculated assuming Newtonian flow as shown in Table 2.2. The differences were less at the higher shear rates.

Through this activity, the students practiced reading and applying material from a journal article and increased their confidence in working with non-Newtonian fluids.

2.4.5 Calculation: velocity profile (SLO 1)

The final assignment, designed to increase students understanding of non-Newtonian flow, was a homework assignment. The students knew that Newtonian fluids in laminar flow exhibit a parabolic velocity profile. In this assignment, they created the velocity profiles of non-Newtonian fluids according to Equation 2.5:

$$v(r) = \left(\frac{\Delta P}{2KL}\right)^{1/n} \left(\frac{n}{n+1}\right)\left(R^{\frac{n+1}{n}} - r^{\frac{n+1}{n}}\right) \tag{2.5}$$

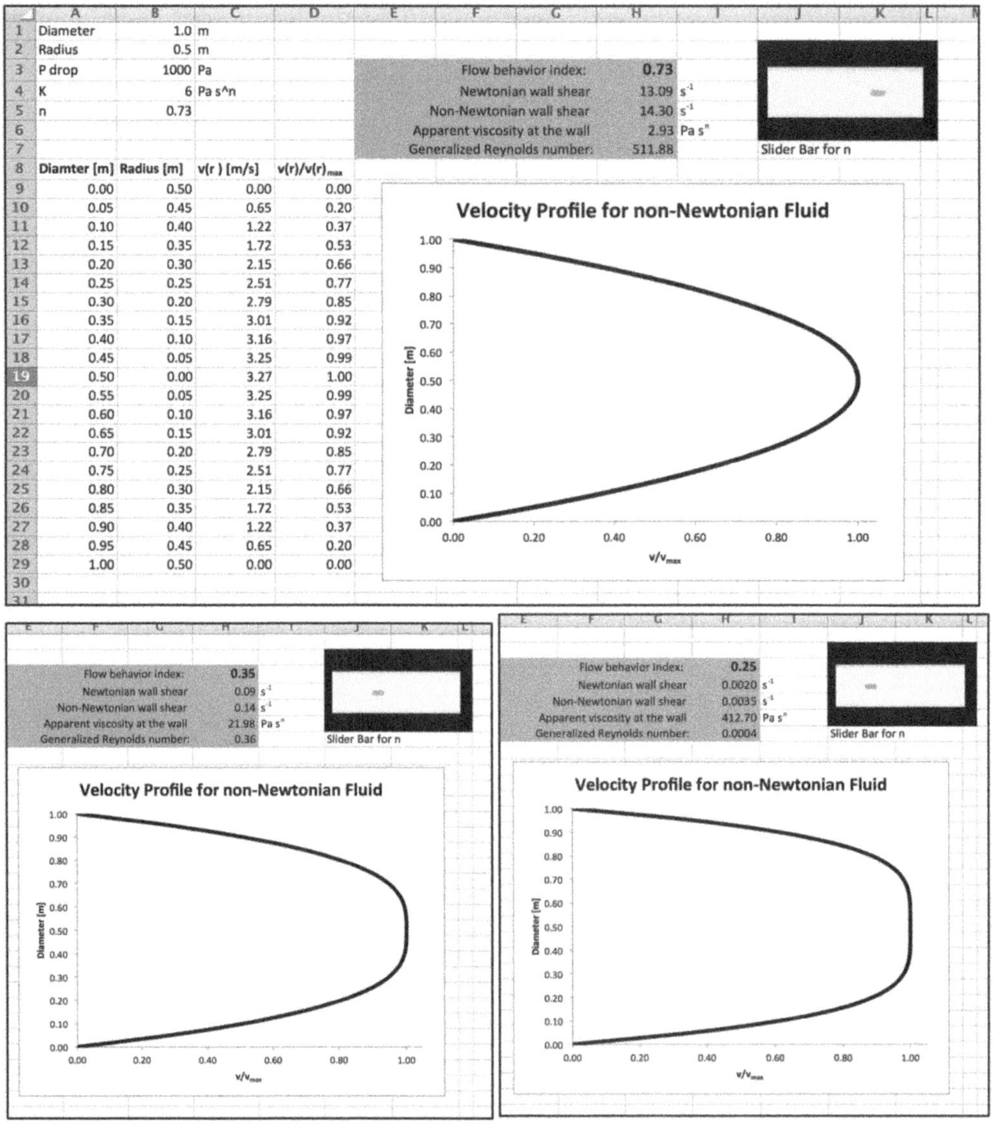

Figure 2.6 Non-Newtonian velocity profiles created by students. By moving the slider bar for the flow characterization index, the students could see how the profile changed as the fluid became more non-Newtonian. The top graph is for $n = 0.73$, and the bottom graphs are for $n = 0.35$ (left) and $n = 0.25$ (right).

The velocity profile is a function of the flow consistency index, K, and the flow behavior index, n. Because the value of n can vary between 0 and 1 for a shear-thinning fluid, the students incorporated a slider bar to change its value. They could then observe how the profile changed as the fluid became more non-Newtonian. Inserting the slider bar and setting up the columns for the chart to show the profile across the entire pipe were not trivial, and the students gained confidence in their Excel skills as the instructor coached them on various ways to accomplish this. Finally, by using the slider bar, they could easily observe how the profile flattens as n decreases. Figure 2.6 shows the spreadsheet and profiles for three values of n.

2.4.6 Group jigsaw: mechanical energy balance (SLO 9)

During an early class period, the students used a jigsaw activity to review and expand what they had learned in their previous class about the mechanical energy balance components. One person from each "home" group joined to form an "expert" group on a topic. The expert groups completed worksheets on three topics: (i) applications of the Bernoulli equation; (ii) calculation of friction losses and (iii) classification and calculations for power law fluids. After 45 minutes in the expert groups, the students returned to their home groups and taught what they had learned to the other members.

During the jigsaw activity, the students were fully engaged in learning the material assigned to them. Their strong theoretical background made it possible – they were not beginning learners (Prince, 2004). Using their textbook and the handouts provided, they figured out how to set up and solve applications of the Bernoulli equation and calculate friction losses in a typical system. The third group learned how to characterize a power law fluid and how to adjust the Reynolds number and friction factor equations for these fluids. As the students returned to their home group, the professor circulated and eavesdropped as each student taught the other students what he or she had learned, and carefully corrected the few misconceptions, acting as a coach (Lalley and Miller, 2007). The method of instruction appeared effective as the students completed a final problem encompassing all concepts before they left the classroom.

2.4.7 Group jigsaw: pump types (SLO 4)

At a later class date, the students used a jigsaw activity to learn about various pump types and their applications. Brochures from pump manufacturers were distributed to the student groups. The brochures covered pump types (positive displacement and centrifugal pumps) and pumping non-Newtonian fluids. Each group listed the important points from their reading, and then the individual students reported to their home groups the characteristics and limitations of the different pumps and fluid combinations.

2.4.8 Hands-on: experiment (SLOs 1–9)

The students had characterized the viscosity of xanthan gum solutions and read about pumping viscous and non-Newtonian liquids with different pump types. In the experiment, they used class time to measure capacity as a function of head for the solutions in a centrifugal pump, and capacity as a function of power for a positive displacement (diaphragm) pump. Collecting the data for pumping the four fluids in the two pumps took about 3 hours. Supplies and equipment were purchased online for about US$100. Details of the experiment are provided in Piergiovanni (2017b). Typical plots are provided in Figure 2.7.

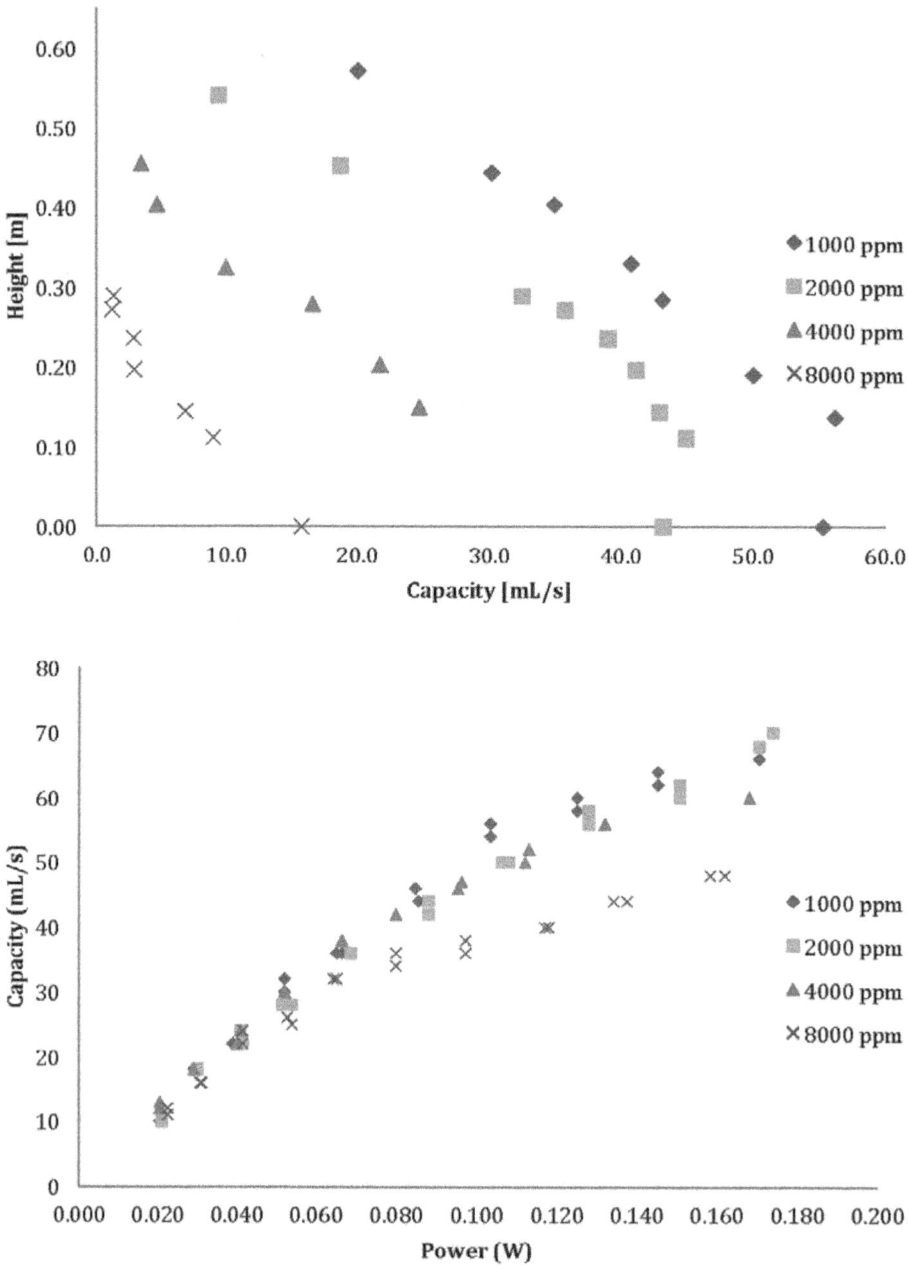

Figure 2.7 Pump curves for the centrifugal pump (top) and diaphragm pump (bottom). The students could easily observe the impact of the more viscous and non-Newtonian fluids.

The students clearly recognized that the centrifugal pump had more trouble pumping the viscous solutions than the diaphragm pump did and could explain why, based on what they learned during the jigsaw exercise. However, when they observed the plot for the diaphragm pump, initially they contacted the instructor complaining that the pump was not working, or that the method for collecting data was not precise enough. They expected that the positive displacement pump would also show effects of viscosity. Instead of answering their questions, the instructor reminded them of the literature that had been provided. Positive displacement pumps have little trouble moving viscous liquids until the liquids become highly non-Newtonian.

Students were asked to use their data from pumping one of the xanthan gum solutions to solve for the pressure developed by the small pumps. This required calculating the non-Newtonian generalized Reynolds number and friction losses. They then used this pressure drop and applied the mechanical energy balance to the system to predict the flow rate of water by the system. The most common difficulty students had with the exercise was with unit conversions. They found that for the same amount of pump work, the volumetric flow rate of water would be significantly higher.

To increase learning, students were asked to draw conclusions and reflect on what they learned during the experiment as recommended by Dewey (1916) and Chickering and Gamson (1989). A few of their comments are included in Tables 2.3 and 2.4. Due to the open-ended nature of the assignment, the comments varied widely. Several students created new plots to support their conclusions, as shown in the bottom line of Table 2.4.

Table 2.3 Student comments from the reports on the diaphragm and centrifugal pumps

Diaphragm Pump Comments	
"As voltage to the diaphragm pump increases, fluid flow becomes turbulent."	"As power is increased, the capacity increases proportionally. The xanthan gum concentration didn't seem to have an effect on the capacity until the 8000 ppm solution when both capacity and efficiency dropped significantly."
"Frictional losses dominate the mechanical energy balance since the change in height was small. In this calculation, form friction was ignored so most of the energy provided by the pump was needed to overcome the skin friction losses between the xanthan gum solution and the tube walls. The frictional losses were normally at least an order of magnitude greater than the kinetic energy term and at least double the losses caused by the potential energy term."	
Centrifugal Pump Comments	
"It is clear that an increase in height results in a decrease in capacity."	"Concentration plays a bigger role in determining capacity than pump size (with the three small impellers). The trend with viscosity coincides with the diagram in 'When to use a PD Pump.' "
"At low head values, friction losses are similar to potential energy effects. However, as the head increases, the potential energy effects dominate the mechanical energy balance. The kinetic energy terms remain small."	
"Friction losses due to elbows were 2.1% of the pump work, thus neglecting form friction losses is verified."	

Table 2.4 Student comments comparing results between the centrifugal and diaphragm pumps.

Comparison Between Pump Types	
"For the same power, the diaphragm pump delivers higher capacity, especially with more viscous solutions."	"A fluid's viscosity impacts the output and performance of the pump. Overall, diaphragm pumps seem less affected by the high viscosity fluids than centrifugal pumps."
"For the 2000 ppm xanthan gum and a head of 0.3 m, the centrifugal pump required 0.16 W and delivered 32 mL/s, while the diaphragm pump required 0.13 W and delivered 29 mL/s. This suggests that the diaphragm pump is more efficient as indicated by the literature we read."	
"We created a plot of Power * Head as a function of Flow rate for the 2000 ppm solution. At low Power * Head values, the efficiencies for the two pumps are similar, but then the centrifugal pump seems to drop off significantly."	

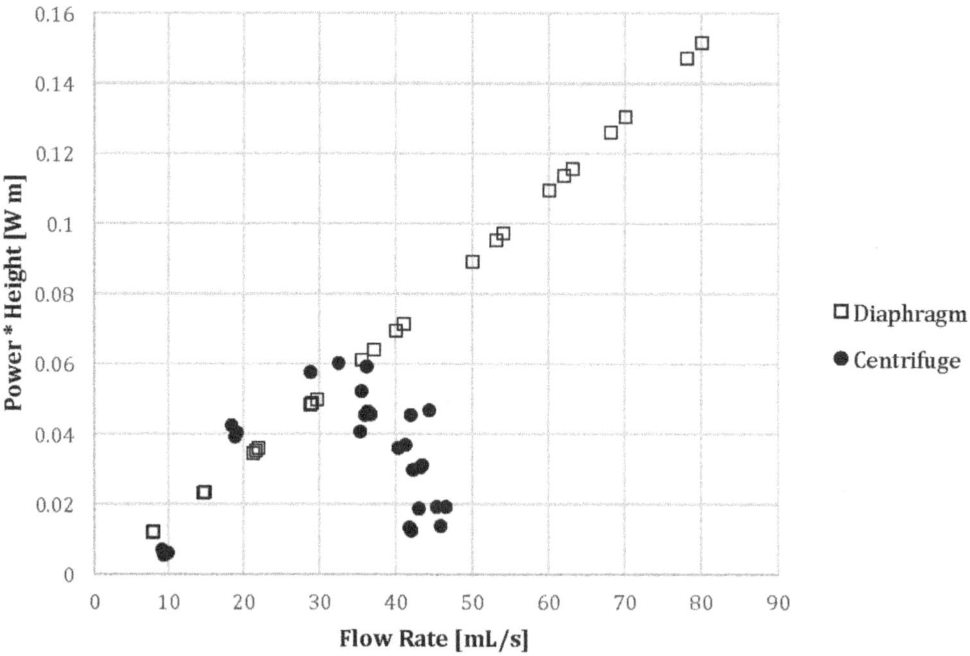

2.4.9 Group: gallery walk (SLOs 5–7)

On the day the final project was due, the students participated in a gallery walk. During the first 30 minutes of class, the students prepared posters (70 × 90 cm) describing the main elements of what they had learned during the module. Each group explained their poster during the final 20 minutes of class.

Creating a poster requires students to prioritize their results. Some groups compared how different pumps handle viscous solutions, while other groups focused on how one pump handled solutions of different viscosities. Much discussion occurred between group members as they designed their posters. During a quiet gallery walk, students read posters and wrote down questions they had while comparing others' work with their own. Each group then presented a 3-minute pitch for their poster. Overall, through the gallery walk, it was clear that the groups had met the learning objectives stated in Table 2.1.

2.5 Assessment of student learning

Using multiple types of delivery modes in the classroom generally leads to more learning (Moor and Piergiovanni, 2010). In this module students learned from five different types of presentation activities (online, hands-on, calculation, jigsaw and gallery walk). After each exercise, the students were given time to reflect and note what they had observed and learned. These reflections were an important part of the final report. A word cloud created from the student reflections after using the centrifugal pump is shown in Figure 2.8. Clearly the students remembered that the increase in pressure comes from kinetic energy.

In previous offerings of the course, online journal questions were posted periodically to help students remain engaged in their learning. These questions asked students to relate the concept discussed in class to a past experience or apply it to their daily lives. An analysis

Figure 2.8 Word cloud created from student reflections on learning after using the centrifugal pump

of the reflective writing showed that when students understood the purpose of the journal, responses improved their critical thinking skills (Piergiovanni, 2014). Currently, these reflective writings were incorporated into the activities.

Table 2.5 lists the assessment questions and scores from a mid-term exam following the module to measure individual student learning. The questions are listed in order of increasing difficulty according to common definitions of Bloom's taxonomy levels. The first question, pumping honey, is simply remembering and understanding – making sense of what the student has learned. The second question, barbecue sauce viscosity, applies the information about viscosity in a slightly new way, and the third question requires the students to break concepts into parts and understand their relation: analyzing. The students must evaluate information for the fourth question, and the fifth question requires the students to put information together in an innovative way. Thus the questions in the table are arranged in order of increasing levels of learning.

Table 2.5 Assessment of student learning on mid-term exam

Assessment Question	*Average Score*	*% of students with score above 75%*
If you needed to pump honey, what type of pump would you use? Why? Why are centrifugal pumps not used for highly viscous liquids? Sometimes, however, a centrifugal pump is perfect for a shear-thinning liquid – why might this be?	95.0%	92%
Barbecue sauce is a non-Newtonian liquid. At 20 rpm, its measured viscosity is 75.2 Pa s, and at 90 rpm, its viscosity is 20.5 Pa s. Estimate its flow behavior index and characteristic viscosity in SI units.	85.4%	85%
A chemical plant needs to pump 100 L/min of ethanol from the bottom of a storage tank, open to the atmosphere, to a packaging plant according to the diagram below. The ethanol has a density of 700 kg/m^3 and a viscosity of 0.8 cP. What power is needed for the pump if it is 70% efficient? The same chemical plant is being used to pump a shear-thinning liquid with K = 2 Pa s and n = 0.7 and the same density. What power is needed now?	91.5%	89%
Sketch a plot of pump capacity as a function of viscosity for both centrifugal and positive displacement pumps.	84.6%	85%
The efflux tank is filled with a shear-thinning liquid. At time = 0, the plug is removed and the fluid starts to exit the tank. What happens to the shear rate during the first few seconds? What happens to the viscosity?	87.2%	77%

As might be expected, the average scores generally decrease with level of difficulty, as does the percentage of students who scored above 75%. However, even as the numbers decrease, the average score is nearly 85% or higher for all questions, and a large majority of the students scored above 75%. The students learned the concepts, even though few lectures were provided on the material.

The PBL project scores were also compared with scores on the final exam mechanical energy balance problem where the students calculated the amount of pump work required to move a viscous liquid. While the distributions were different (see Fig. 2.9), the average scores of the project grades and exam problem grades were within 1 point of each other (88% and 89%, respectively). About 46% of the students received the same or a higher grade on the final exam, showing that they had learned and retained the information. However, 54% of the students received a lower score on the exam than they had received on the project. Figure 2.10 looks at this information classified by final exam score. The top students did the same or better on the exam. The students who scored 75% or less on the final exam had all earned higher scores on the project. This was likely because the projects were completed in groups, and "social loafing" – a negative aspect of group work – occurred (Lee and Lim, 2012). As with all group projects, the instructor must be aware of this potential problem.

This course generally enrolls approximately 50 students. With larger class sizes, the hands-on activities might be difficult to manage. However, Vigeant *et al.* (2017) recently showed that offering the activities as demonstrations (live or online) allows nearly the same level of learning. Thus, this rotational model of blended learning could be used for larger classes.

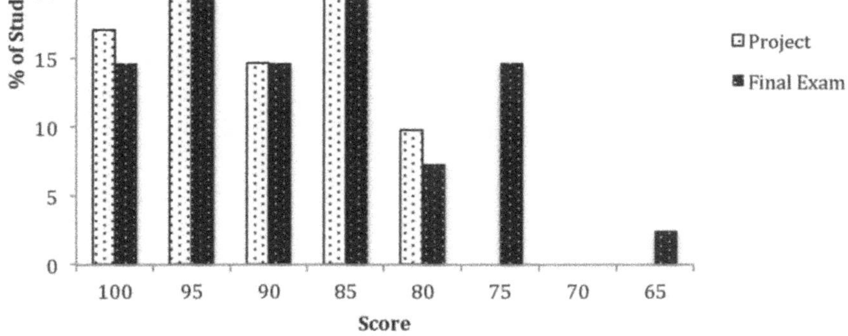

Figure 2.9 Comparison of students' grades on the project and on a mechanical energy balance problem on the final exam

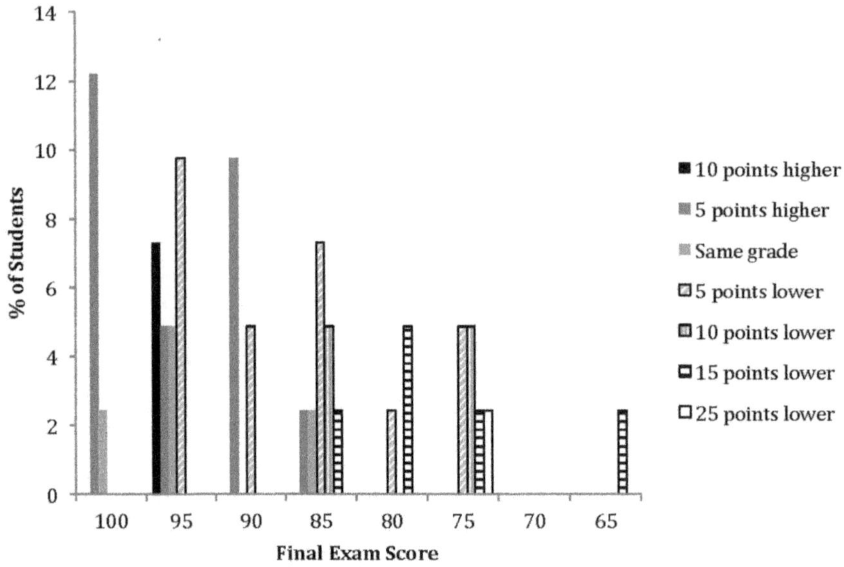

Figure 2.10 Comparison between students' project and exam grades, classified by their final exam score. Solid bars show students who performed the same or better on the exam. In general, higher performing students improved their score on the exam, indicating they learned from the project.

2.6 Conclusions

The rotational model of a web-enhanced blended learning course was applied successfully in an applied fluid flow course. The online and hands-on support structures aided student learning and understanding of the mechanical energy balance with Newtonian and non-Newtonian fluids, while few formal lectures were provided. Students remained engaged and interested in the classroom and participated fully in the online assignments. The instructor coached the students, and assessment of their learning shows that they met the student learning objectives.

2.7 Acknowledgements

The author acknowledges the contributions made by the laboratory assistant Ms Xinyi (Violet) Guo and the students in the subject ChE 321. Summer support for the development of the activities was supported through a grant from the KEEN Network.

References

Brookfield Engineering (2016) *More Solutions to Sticky Problems*. [Online]. Middleborough, MA, AMETEK Brookfield. Available from: www.brookfieldengineering.com/-/media/ametekbrookfield/tech%20sheets/more%20solutions%20to%20sticky%20problems%202016.pdf?la=en [Accessed 5 September 2017].

Chickering, A.W. & Gamson, Z.F. (1989) Seven principles for good practice in undergraduate education. *Biochemical Education*, 17(3), 140–141.

Dewey, J. (1916) *Democracy and Education; An Introduction to the Philosophy of Education*. The Macmillan Company, New York, USA.

Felder, R.M., Woods, D.R., Stice, J.E. & Rugarcia, A. (2000) The future of engineering education: II. Teaching methods that work. *Chemical Engineering Education*, 34(1), 26–39.

Greening, T. (1998) Scaffolding for success in PBL. *Medical Education Online* [serial online] 3(4). Available from: www.med-ed-online.org/f0000012.htm [Accessed 6 September 2017].

Horn, M.B. & Staker, H. (2014) *Blended: Using Disruptive Innovation to Improve Schools*. Jossey-Bass, San Francisco, CA, USA.

Kirkham, K. & Seymour, L. (2005) The value of teaching using a live ERP system with resource constraints. In: *World Conference on Computers in Education (WCCE)*, Cape Town, South Africa.

Krajcik, J. & Blumenfeld, P. (2006) Project based learning. In: Sawyer, R.K. (ed) *The Cambridge Handbook of the Learning Sciences*. Cambridge University Press, Cambridge. pp. 317–333.

Lalley, J.P. & Miller, R.H. (2007) The learning pyramid: Does it point teachers in the right direction? *Education*, 128(1), 64–79.

Lee, H-J. & Lim, C. (2012) Peer evaluation in blended team project-based learning: What do students find important? *Journal of Educational Technology & Society*, 15(4), 214–224.

Mayadas, F., Miller, G. & Sener, J. (2015) *Definitions of E-Learning Courses and Programs*. version 2.0. [Online]. Available from: Online Learning Consortium, Inc., https://onlinelearningconsortium.org/updated-e-learning-definitions-2/ [Accessed 5 September 2017].

Mitschka, P. (1982) Simple conversion of Brookfield R.V.T. readings into viscosity functions. *Rheologica Acta*, 21, 207–209.

Moor, S.S. & Piergiovanni, P.R. (2010) A multimodal approach to classroom instruction: An example from a process control course. *International Journal for Engineering Education*, 26(6), 1428–1444.

O'Neal, C. & Pinder-Grover, T. (n.d.) *Spectrum for Active Learning*. [Online] Available from: Center for Research on Learning and Teaching, University of Michigan, https://und.edu/academics/center-for-instructional-and-learning-technologies/_files/docs/active-learning-continuum.pdf [Accessed 2 August 2017].

Piergiovanni, P.R. (2014) Reflecting on engineering concepts: Effects on critical thinking. In: *Frontiers in Education Conference, Madrid, Spain. 22–25 Oct 2014*. doi:10.1109/FIE.2014.7044300

Piergiovanni, P.R. (2017a) Students learn without lectures: Scaffolded problem based learning in an applied fluid flow and heat transfer course. *Chemical Engineering Education*, 51(3), 99–108.

Piergiovanni, P.R. (2017b) Laboratory experiment: Pumping power law fluid. *Chemical Engineering Education*. 51(2), 53–60.

Prince, M. (2004) Does active learning work? A review of the research. *Journal of Engineering Education*, 93(3), 223–231.

Reynolds, M.J. & Hancock, D.R. (2010) Problem-based learning in a higher education environmental biotechnology course. *Innovations in Education and Teaching International*, 47(2), 175–186.

Rich, E., Sloan, T. & Kennedy, B. (2015) *The Effect of Problem-based, Experiential Learning on Undergraduate Business Students*. [Online] Available from: Athens State University College of Business Journal, www.athens.edu/business-journal/spring-2015/the-effect-of-problem-based-experiential-learning-on-undergraduate-business-students/ [Accessed 23 May 2016].

Savery, J. (2006) Overview of problem-based learning: Definitions and distinctions. *The Interdisciplinary Journal of Problem-based Learning*, 1(1), 9–20.

Singh, R.P. & Heldman, D.R. (2014) *Introduction to Food Engineering*, 5th ed. Academic Press, London.

Smith, K. (2011) Cooperative learning: Lessons and insights from thirty years of championing a research-based innovative practice. In: *Frontiers in Education Conference*, Rapid City, SD, USA, 12–15 Oct 2011.

Smith, K., Sheppard, S.D., Johnson, D.W. & Johnson, R.T. (2005) Pedagogies of engagement: Classroom-based practices. *Journal of Engineering Education*, 94, 87–101.

Thomas, J.W. (2000) *A Review of Research on Project-Based Learning*. The Autodesk Foundation. Available from: htttp://www.bie.org/index.php/site/RE/pbl_research/29 [Accessed 7 September 2017].

Vigeant, M.A., Prince, M.J., Nottis, K.E.K., Koretsky, M., Bent, E.C., Cincotta, R. & MacDougall, K.A. (2017) Why not just run this as a demo? Differences in students' conceptual understanding after experiments or demonstrations. In: *American Society for Engineering Education Conference*, Columbus, Ohio, June 2017. Available from: https://peer.asee.org/why-not-just-run-this-as-a-demo-differences-in-students-conceptual-understanding-after-experiments-or-demonstrations.pdf [Accessed 7 September 2017].

Yadav, A., Subedi, D., Lundeberg, M.A. & Bunting, C.F. (2011) Problem-based learning: Influence on students' learning in an electrical engineering course. *Journal of Engineering Education*, 100(2), 253–280.

Chapter 3

Transitioning staff, students and course materials to blended and online learning environments

A.B. Wedding[1,], A.M. Cousins[2] and D. Quinn[3]*

[1]School of Engineering, University of South Australia, Australia
[2]Future Industries Institute, University of South Australia, Australia
[3]Teaching Innovation Unit, University of South Australia, Australia
**Corresponding author: E-mail: bruce.wedding@unisa.edu.au*

ABSTRACT: The University of South Australia has been delivering fully online and blended engineering programs to undergraduates since 2011. This endeavor has brought much change – at the organizational level in the refocusing of courses and programs, at the staff level in their approaches to teaching, and at the students' level in their approaches and engagement with online and blended learning environments, and with the materials themselves. In this chapter we consider the process of transition within this changing system. We provide working definitions of online and blended engineering and explain how it is distinctly different from what happens in face-to-face teaching. As such, materials, staff and students need to transition to new ways of teaching and learning. Examples of how to systematically support this transition are explicated using two case studies that show the centrality of the time budget for making necessary changes for staff, students and materials to new modes of delivery. Throughout any such transition process that impacts on others, the goal must always be '*primum non nocere*,' to do no harm. Engineers have always used technology to support their endeavors – why should teaching be any different?

3.1 Introduction

Change is the new norm in higher education (Ernst and Young, 2012). There have been changes in student engagement levels in our engineering classrooms, and this can provide an opportunity to revitalize teaching and learning. However, our courses, programs, materials, teachers and students make a complex and interrelated system, where changes in one area can have significant ramifications in others. This has sometimes been referred to as a wicked problem (Borko *et al.*, 2009), in which there are a large number of complex variables – all of which are dynamic, contextually bound and interdependent. Recognition of this has meant that academics may be reluctant to change the way that things are done in their courses in fear of ending up in a worse position due to unforeseen consequences flowing onto other parts of the system. We argue that to successfully make positive changes in student experience and outcomes, it can be helpful to take a more holistic view of what we are doing in course development, ensuring that we transition the whole system – staff, students and materials – to new ways of sustainable operation (Ellis and Goodyear, 2013).

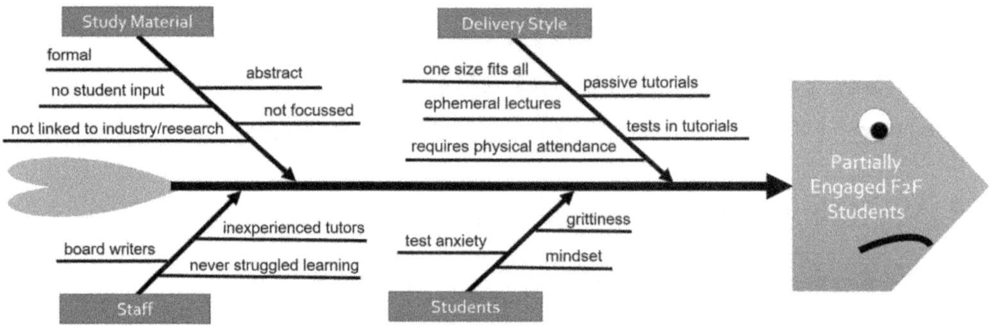

Figure 3.1 Fishbone diagram of the various factors contributing to declining engagement in face-to-face (F2F) students in first-year engineering

3.1.1 The problem

Face-to-face (F2F) engineering students are often partially engaged in their studies (Fig. 3.1). They find the study material to be rather formal and abstract, as often in the first year there are few ties to industry or professional practice. Students often find the material to be not focused – that is, broad rather than deep. This can happen as teachers find it easier to add new content each year as they come across it, but are less compelled to take anything away from the materials presented to students.

Students can also find the F2F delivery style itself to be less engaging. F2F classes require physical attendance, and students will express their frustration if they feel that they have been unnecessarily committed to be present at a set time or place. Although lecture recordings are the norm, teacher gesticulations and spontaneous drawings on whiteboards are often not captured, making attendance at some sessions essential. Materials are often presented as "one size fits all" rather than considering the various learning preferences of students. F2F tutorials in engineering can be quite passive or relegated to a time when testing is carried out (to ensure attendance). All this puts the academic in control of how, when and where learning happens, with little input from the student.

First-year students may have the mindset that their academic potential is limited (Dweck, 2012), and when they make their first mistakes at university they are confronted and some may not have the necessary tenacity or grit to unpack their error (Duckworth *et al.*, 2007). A significant proportion of learners can be classified as "Roberts," choosing only to engage deeply with their studies when it is not possible for them to get away with superficial engagement (Biggs, 1999). Together, these issues and others have meant that our classrooms have partially engaged students trying to balance their anxieties, work, family and social lives with their studies and professional ambitions (Fig. 3.1). The challenge exists for higher education educators and professionals to change their F2F curriculum to focus more on what the student does (Biggs, 1999), and online learning can play a role in this process.

3.1.2 The global push for online learning

Today's universities compete in a global marketplace which has been enabled by innovation and digital technologies (Ernst and Young, 2012). Increased flexibility in STEM within tertiary learning environments is included as a part of a national strategy to secure Australia's prosperity (Chubb, 2013). One driver for this change is students. Contemporary university students have greater access to information, are more mobile and are juggling more competing commitments (Coates and McCormick, 2014). Despite hurdles, such as the practical component and aging academia reluctant to engage with the online modality (Mayadas *et al.*, 2009), it has been predicted that online engineering programs, that allow learning anywhere and anytime, will continue to develop and improve (Bourne *et al.*, 2005). As this has been a slow and steady transition, the terminology associated with the many shades of online learning and how it differs from the F2F experience need to be clarified.

3.1.3 Definitions

As terminology around university offerings such as F2F, blended, flipped and online can vary, here are our definitions. We refer to bachelor engineering degrees and associate degrees as programs. Programs are made up of courses, where full-time study is four courses per study period (semester). At the University of South Australia (UniSA), a course contributes 4.5 units towards the completion of a 144 unit engineering program. According to our Assessment Policy, the time students need to spend on all activities should not exceed 35 hours per unit within a course. Thus, a 4.5 unit course × 35 hours = 157.5 hours of student workload to complete all the learning activities and assessment for the course (UniSA, 2017).

Face-to-face offerings of engineering courses at university usually involve lectures and tutorials, with or without a practical component. F2F students are classified as internal students. Online learning is when all course curriculum and learning activities are completed online, and students are not required to attend a physical campus. Online students are classified as external. Blended learning is defined as the convergence of online technologies with F2F instruction to optimize student engagement in the learning process and provide greater flexibility in study (Dawson, 2015), allowing cooperation rather than competition to bring together the best of both F2F and online modalities (Ellis and Goodyear, 2013). These students are still classified as internal.

Flipped learning is a subset of blended learning and has been defined as consisting of two parts: interactive group learning activities inside the classroom and direct computer-based individualized instruction outside the classroom (Bishop and Verleger, 2013). Flipping classrooms at university can make students more active and in control (Hamdan *et al.*, 2013). Common themes for development of flipped classrooms include the creation of pre-learning materials (e.g. lecture recordings), active in-class learning and revised assessment (McLaughlin *et al.*, 2016). However, introductory courses may not be well suited to flipped classrooms, as students have varying incoming foundational knowledge (Strayer, 2012) and it is feared that without "pedagogical integrity," flipped classroom approaches may "wither on the vine" (O'Flaherty and Phillips, 2015).

3.2 Preparing people for a change in teaching and learning

One of the key pedagogical shifts needed as we change course modalities is the shift from teacher-centered to student-centered teaching and learning. This change involves people, as well as materials. In this section we look at what consideration needs to be given to support staff and students through this transition.

3.2.1 Staff preparation

Universities often engage institution-wide strategies to increase the number and quality of online courses available, sometimes with teams of people descending on selected courses or an entire program and completing "extreme makeovers" in a relatively short period (James *et al.*, 2011). Course design is seen as an "enabling agent" (Goodyear, 2013) to allow pedagogical changes such as constructive alignment (Biggs and Tang, 2011), online resource creation, activity and assessment refocus and design, and more sophisticated website creation to support a switch from teacher-led to student-centered pedagogies (McQuiggan, 2007).

In a review of the staff development approaches used to support the implementation of blended learning at 11 institutions (Porter *et al.*, 2014), staff workshops that support technological skills were found to be the most common strategy. In some cases, pedagogical workshops were also used to investigate the wide variety of instructional methods afforded by blended approaches. In addition, incentives were often used to allow time release for staff, to hire assistants or to provide stipends or specialized equipment. Staff workload calculations were also adjusted in some cases to incentivize staff to adopt blended learning (Porter *et al.*, 2014). Changes in blended courses evolve over time, so ongoing staff development opportunities need to be available (Quinn *et al.*, 2015; Wingard, 2004).

In addition, we argue that it is imperative to develop a holistic view of courses during the development process using tools such as the time budget (Quinn and Wedding, 2012). Time budgets are a visual map of all that the student needs to do, hour by hour, week by week, across the length of the study period, to be successful in a course (Fig. 3.2). Time budgets are the centerpiece for the negotiation of course redevelopment between course coordinators and development staff. Through this negotiation, a shared vernacular related to the course can be developed between the course coordinator and development staff (Hedberg, 2004). By focusing on what students need to do, rather than what teachers need to teach, the time budget appears to be a more meaningful mechanism for staff to identify if the proposed learning experience of a blended or online course parallels that of its F2F counterpart. Also, the limitations of existing courses, such as heavy workload, assessment timing or an overly didactic focus, become clearly evident to all concerned when mapped as hours of student workload on a time budget.

3.2.2 Staff approaches to student engagement

Teaching staff can be supported in shifting approaches to teaching by developing a better understanding of student learning and engagement. Student learning has been described as three interrelated components: presage, or what happens before the classroom, process, or how the student approaches the learning activity, and product, the learning outcomes achieved (3P model; Biggs, 1989; Trigwell and Prosser, 1997). When the mode of delivery changes the context of learning, this can have flow-on effects to how students perceive the context of their learning and the approach they take to their learning.

To engage and motivate adult learners to learn, it is important to establish inclusion, help them develop a favorable attitude to learning, enhance meaning for their learning experiences and engender competence by creating or affirming something that students value and perceive as authentic to their real world (Wlodkowski, 2011). Traditional learning approaches in lecture halls and tutorial rooms can be distant from how learning is achieved in professional engineering settings. Setting realistic problems with authentic contexts for students to tackle

TIME BUDGET

PHYS	2011	**Engineering Physics N**													
														Contact	Load
SP5	Hrs	1	2	3	4	5	6	7	8	9	10	11	12	hrs/wk	hrs/wk
Week	1	Lectures			Read Course Info		LR	Ex Q						3	8
	2	Lectures			T	WS	Prac OH&S		LR	Ex Q				5	10
	3	Lectures			T	WS	Prac 1		LR	Ex Q				7	10
	4	Lectures			T	WS	Prac 2		LR	Ex Q				7	10
	5	Lectures			T	WS	Prac 3		LR	Ex Q				7	10
	6	Lectures			T	WS	Prac 4		LR	Ex Q				7	10
	7	Lectures			T	WS	Prac 5		LR	Ex Q				7	10
	8	Lectures			T	WS	Prac 6		LR	Ex Q				7	10
		Expect approx. 8 hrs Lecture Review, Problem Solving & General preparation for 2nd half of semester													8
	9	Lectures			T	WS	Prac 7		Report/Assignment					7	12
	10	Lectures			T	WS	Prac 8		LR	Ex Q				7	10
	11	Lectures			T	WS	Prac 9		Report/Assignment					7	12
	12	Lectures			T	WS	Lecture Review		LR	Ex Q				5	10
	13	Lectures			T	WS	Lecture Review		Ex Q		Exam Prep			5	10
	SVac	Exam Preparation													10
	Exam1														
	Exam2												Contact hrs	81	
		University guidelines: Expected load = 150 hrs = 10 to 12 hrs per week per 4.5 point course >>>											Total hrs		**150**
NOTE		T	Tute		WS	Workshops (voluntary)			Report	Approx. 4-5 hrs expected					
		Ex Q	Work on Questions & Problems from Exercise List						LR	Lecture Review					

Figure 3.2 An initial "time budget" for a typical traditionally delivered course. Time budgets help staff re-conceptualize their courses in different modalities.

can help motivate them to learn by helping to see the relevance of what they are learning (Herrington *et al.*, 2014).

Learning is cumulative – that is, learning is built on interest and what the student already knows (Tobias, 1994). If the new material can be related to what students already know, it is easier for learners to build the mental schema to incorporate the new knowledge. This can include pre-learning of key terms prior to engaging with the main topic being studied (Pollock *et al.*, 2002). As the diversity of our student body increases, it is important to ensure that all our students have the foundation knowledge to be successful in first-year engineering courses (Dekkers *et al.*, 2011) – so the open door of university does not become a revolving door.

3.2.3 Students understanding self

Presage factors, such as student preconceptions and pre-developed learning styles coming into tertiary study, can have a significant impact on their ability to embrace and be successful with rapidly changing and various delivery modalities. While first-year students may be exposed to learning style assessments, most are unable to appreciate the impact on their learning, and hence the educators should consider the learning styles of their students especially when considering online learning (Brown *et al.*, 2009).

Increased student diversity means that changes in learning modalities will better suit some students, but not necessarily all. Although GPA can be used as a predictor of success in STEM courses (Hachey *et al.*, 2015), personality factors such as GRIT – perseverance and passion for long-term goals (Duckworth *et al.*, 2007; Duckworth and Quinn, 2009) have been shown to be better predictors of academic success. In a study of students' views of their transition from secondary to tertiary mathematics, engineering students identify this as a time of changing identity (Hernandez-Martinez *et al.*, 2011), and thus the first year is an ideal time to support students to form a clearer understanding of their new selves.

Mathematics is at the core of engineering, and skill levels are correlated with student outcomes (Harris *et al.*, 2014). Math anxiety is a condition that mainly affects adult learners who, through previous negative math learning experiences, have been preconditioned to become stressed when presented with mathematics and will display avoidance behavior, such as putting off studying until the last minute (Marshall *et al.*, 2017). The core of the issue may be the mindset of the student being fixed, that is they operate with a belief in statements such as "you are either good at math, or you are not," rather than a growth mindset, which encourages you to seize on any mistakes and see them as an opportunity to learn and grow (Dweck, 2012). Lessons about the cognitive science of neuroplasticity to students with a fixed mindset can be transformative – rekindling a long-lost love of mathematics (Boaler, 2013). Pre-learning opportunities, which allow students to self-test and learn from their mistakes in the privacy of an online environment, have been shown to ease anxiety (Marshall *et al.*, 2017). Structured active learning processes and sufficient time on task have also been identified as key principles for contributing to student success in university-level mathematics (Thiel *et al.*, 2008).

3.3 Approaches to transitioning delivery

Course materials need to be transitioned for new modalities, but there are several approaches that can be taken.

3.3.1 Taking a staged approach

While universities may employ the services of internal "renovation teams" to help a course transition from traditional F2F to online delivery, some courses require high levels of input from individual coordinators or engineering course teams to adapt their content (e.g. due to size or periodicity of the course, complexity in adapting the content, etc.). In these instances, it is possible to independently change from F2F to blended learning, to flipped classrooms and eventually to high-quality fully online courses, via a progressive roll-out over several years. To do so effectively and efficiently requires understanding the cohort, creating online media, developing relevant and engaging F2F learning activities and then later translating these to online equivalents (Fig. 3.3). By closely monitoring the impact of this staged introduction on students, high-quality robust learning environments can be progressively created. The progressive roll-out also reduces the burden on academic staff, by spreading the development of new or refined course materials over time, as opposed to an "all at once" approach.

While the above transition is useful for managing the progressive change in course structure over time, there is also need for managing the "experiential change" that may be experienced by a cohort within a course. Consider a cohort at the start of their degree program: the transition to university study that includes online environments involves a shift in learning practices for students – a change that needs to be carefully managed (Kift, 2009; Quinn *et al.*, 2012). Compared to high school, students are given much more freedom, including voluntary participation in learning activities; they choose how and when they participate. Similarly, students with experience in the tertiary setting but limited exposure to courses utilizing online delivery of content will require a similar shift in learning practices (Anderson *et al.*, 2011; Blackmore *et al.*, 2010).

Time budgets are particularly useful to make explicit the student workload on various learning activities in and out of the classroom. Time budgets capture not only what the students need to do, and when, but also the expectations of the time to be spent on those activities (Figs. 3.2, 3.5 and 3.8). Time budgets can play an important role in the design and development of tertiary courses and also be a key academic development tool, as they help academics shift from thinking about what the teacher does to thinking about what the student does, a key aspect of the required pedagogical change (Biggs, 1999).

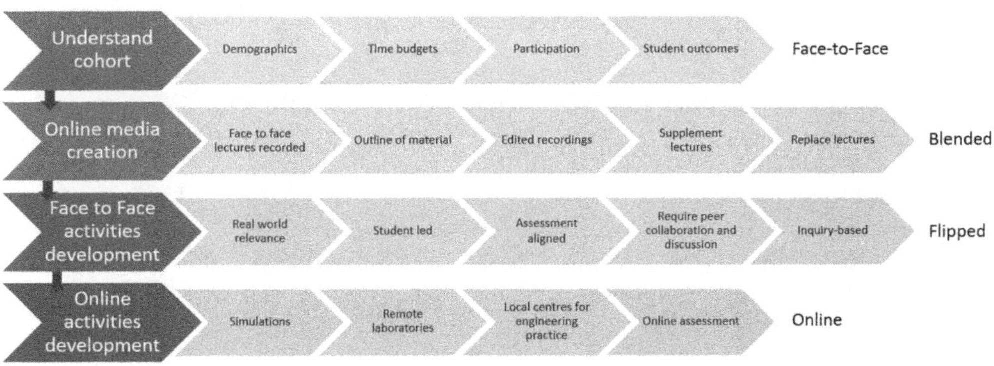

Figure 3.3 Model of progressive course development from face-to-face to online delivery

3.3.2 The importance of study options

Central to any good course design is the objective to ensure students have the necessary tools to achieve the goals of the course (Wiggins and McTighe, 2005). By flipping a course and providing online materials, the structure is opened up to provide a wealth of varying study options for the students (Wingard, 2004). Some approaches are common across many courses, e.g. the use of textbooks to guide the student through content, and some more tailored, e.g. the nature of assessment, practical activities and engagement during tutorials.

In general, providing a range of study options accommodates for a range of learning styles, prior knowledge, GRIT and study habits amongst students, and it can also swamp students with too much choice. Like sheep grazing in a field, access to online content allows students to roam and navigate their way through content in whatever way they see fit; however, issues may arise with students who may not be able to see the "bigger picture" or understand how this free-roaming might impact their success. As with sheep, some students require an external influence to guide them through the study options, to find those that work best for them, so that they can successfully pass the course and make it through to "greener pastures" (Fig. 3.4).

An overarching concept to guiding students through the course is the use of time budgets, described previously. However, when it comes to more open-ended exploration of lecture content, for example, there might be too many options (lecture videos, textbook, external references or further readings, etc.). Certainly, some methods of approach will suit some students better than others. As such, it is important the students are given the tools to help them navigate through these options – mainly an understanding of their own relative strengths and limitations, understanding their learning style, taking into account their level of GRIT and providing feedback throughout the course to help redirect them if they are astray.

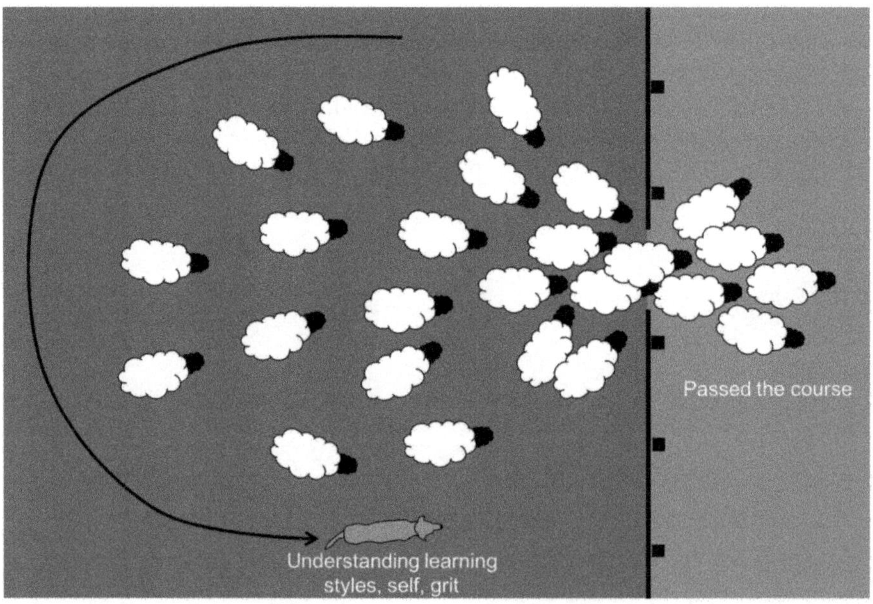

Figure 3.4 "Herding sheep" analogy to describe the importance of ensuring students can effectively navigate their way through the available study options towards passing the course (and to greener pastures)

3.3.3 Supplementing lectures for enhanced engagement

As universities and courses adopt manual or automated recording of lectures, the quantity of materials to aid students studying the course increases. The use of these recorded lectures (which may be 20–40 hours of recorded content, depending on the course!) may be open ended, e.g. to ensure absent students remain up-to-date with lecture content; to allow ESL (English as a second language) students additional opportunities to revise the course content; or, perhaps optimistically from the students' perspective, provide a revision tool for mid-term or end-of-course examination.

Lecture recordings have a justifiable presence in the modern classroom, and for a flipped environment they may be the only alternative to traditional F2F pulpit-style lectures. However, there are clear downsides to taking a traditional lecture and simply recording the content. The sheer volume of recordings at the end of a semester may limit effective use as a revision tool to the student's ability to isolate key content or topics from the rest of the material. If re-recorded each semester, the content may vary in quality between cohorts – e.g. quality may depend on student-teacher interaction during the lecture, interruptions (such as fire drills) and, realistically, on the state of mind of the lecturer on the day. These potential issues open up opportunities for supplementary content that provide students with a focused summary of the lecture content, provide relevance to the content in easy-to-digest sessions, are time efficient and are easy to adapt if the course content changes. Two approaches discussed in this section are the development of pre-recorded "lecturettes" to accompany the full lecture recordings and the integration of "tech talk" sessions into lectures or tutorial sessions (F2F or online delivery).

3.3.3.1 Lecturettes: the need for something more (or less)

Many textbooks include a summary section at the end of chapters, which may cover the core content with a focus on the main equations, definitions and themes discussed throughout the chapter. One would not benefit greatly from only reading the summary section instead of the whole chapter, but as revision or a study aid, these summary sections are often short and refined enough to serve as a more efficient and direct method of study. A lecturette takes this same approach, by contracting down each lecture slide (or groups of lecture slides) into "three key points." Lecturettes were developed as a way of truncating the content of a full lecture (~50 minutes) into a shorter, more succinct video (~10–15 mins). The lecturettes are recorded without students present and typically to a script, and as such can be edited, refined and reused each time a course is run. Table 3.1 presents some practical hints and tips – the product of trial and error over some hundreds of hours creating lecturettes.

The idea behind a lecturette is to provide students with a summary of lecture content, as well as direct them to the main concepts of the particular topic and thus also give the student focus to help study or revise content. Typically, lecturette recordings use the same lecture slides presented in the full-length lecture videos. Aside from the length of the videos, other defining differences in lecturettes may be:

- Use of digital emphasis/graphics (e.g. to underline or highlight equations, figures, etc.);
- Absence of complete derivations/mathematical explanations;
- Sequencing by the addition of an introduction slide and "in the next video" slide; and
- Truncation of slides (e.g. if content is similar across numerous slides, use a slide that best covers the key ideas you wish to discuss).

Table 3.1 A selection of lecturette tips and hints

Use a script, as it can help . . .	• Keep the content focused on the key themes (number one priority) • Gauge pacing of the video (e.g. 200 words per lecture slide) • Reduce variability in style between videos • Serve as transcript for students with a hearing impairment
Use up-to-date recording software . . .	• Ensures files can be easily transferred or edited in following years • Ensures quality between videos remains constant • Low-resolution files "age" quicker and may need to be re-recorded
Warm up for each recording . . .	• Naturally you'll get more comfortable the more you record • Go through the first few slides at least twice, it will help you get into a rhythm for the rest of the slide deck • The more you do in one day, the more consistent the quality will be
Don't spend too much time on a topic . . .	• A lecturette that condenses 50 mins into 30 mins is not much of a summary. Goal is < 20 mins. • Ensure you're covering the core concepts – if more explanation is needed, students can refer to the full lecture recording or textbook • Feel comfortable referring students to other sources, like full-length lecture recordings

Through the use of existing lecture slides, any student notes or copies of lecture slides can still be used to follow the lecturettes, and the amount of content requiring production (hence, staff time) is minimized. Lecturettes can help students form the mental schemata for learning by introducing the terminology and big ideas that will be presented in the lectures.

3.3.3.2 Tech talk sessions (authenticity)

While there is a clear need for the development of students' theoretical knowledge throughout engineering courses, given both the practical application of the profession and the rate at which some applications can change, an opportunity emerges to reinforce student understanding of the material and provide relevance to the course content.

Including an interactive tech talk session in lectures or online tutorials provides real-world examples that supplement the lecture content and offer a chance to explore novel applications of theoretical content. These sessions can also provide a platform for relating lecture content to professional practice; for example, impacts of uncertainty or error in calculations, pushing physical limits to systems and the impact on risk/hazards of a system, development of critical and scientific thinking, etc.

Another benefit to tech talk sessions is the ability to update the examples as frequently as required or even integrate guest speakers. The application of theory could be based on "classical examples" or borrow from popular culture and media to relate to current "hot topics." These might include:

- Recent advances in artificial intelligence and the flow-on ethical considerations;
- Electric or self-driving vehicles as a practical alternative to conventional transportation;

- Exploding phone/laptop batteries and the impacts of design on hazards of a product and
- Plot-holes or impossibilities in movies, YouTube videos, science fiction books, etc.

Some examples may lend themselves to isolated application of knowledge, but as the course progresses, more holistic examples can be implemented to revisit and reinforce prior knowledge as well as move away from the specific content of the lecture towards the integration of systems. An example of this might be the technology that is applied to modern cars (electric or combustion-powered), such as parking sensors, regenerative braking, audio system design, integrated computer systems with feedback loops, power management and more. In doing so, these examples can provide a broader context to not only the content being learned, but the course as a whole (show the "bigger picture").

3.4 Transitioning traditional content, activities and assessment

While flipping lecture content may be a fairly linear transition, some aspects to teaching engineering could benefit from online delivery and others could suffer. One clear example is a practical component to a course, where hands-on experience may outweigh any benefits to having virtual access to laboratory equipment. Similarly, tutorials can benefit greatly from flipping components, such as quizzes and other assessment activities, but still have a place for F2F interaction between staff and students.

3.4.1 Practical program

As both a tool to reinforce theoretical content from tutorials and lectures and a process of acquiring a necessary skill set, practical programs may suffer if the hands-on approach is exchanged with a virtual one (Henry and Knight, 2003). This may not be true of all practical programs, such as computational or programming practicals that have limited hardware integration, or where the constraints of online simulation/virtual interaction (e.g. isolation, limit to creativity, reduced appreciation for real-world risks and hazards) are supplemented by the benefits (e.g. time and resource flexibility, repeatability, focus on theory) (Balamuralithara and Woods, 2009). As such, before transitioning components of a practical program online, a review of the expected outcomes should be performed and the benefits weighed against the barriers presented by this mode of delivery.

As an example, a hands-on practical program that introduces electronic circuitry to students may demand that a portion of the program be completed in a traditional sense – in a classroom with the hardware in front of the student. However, a compulsory work health and safety (WHS) module at the beginning of the course that introduces the students to the hazards and risks associated with working with electricity could be performed online. This may have additional benefits, such as providing on-the-spot marking of completion as well as a digital record of that student's completion of the module. It may also free up time during the semester for practicals or project work, if the module is completed online prior to the first practical class.

Similarly, student competency could be tested in a virtual setting by presenting them with a scenario and getting the student to follow a set of actions to safely deal with potential hazards associated with working in that environment. In addition to WHS or competency modules, supplementary items such as a "practical report writing guide," "how to use

Microsoft Excel," etc. can be provided online without detracting from valuable F2F time. However, all these expected online activities must be recognized in the time budget.

3.4.2 Tutorials

Like practical programs, tutorials are traditionally grounded in F2F interaction between students and staff. While the structure and purpose of tutorials may vary, generally they serve to reinforce and build upon student knowledge, help develop problem solving skills and provide a medium for assessing student knowledge. As with practical programs, the "flipping suitability" of a tutorial will depend on the exact nature of the course, but benefits exist to both online delivered content (where appropriate) and F2F interaction. In general, the balance should be decided based on the outcomes for the student and on ensuring the time interacting with staff is as meaningful as possible.

One aspect that should remain central to tutorials is the development of problem-solving strategies and deeper understanding of course content through questioning, discussion and discovery. One effective use of student and staff time is in group problem-solving activities, such as board tutorials (Liljedahl, 2016; Seaton *et al.*, 2014). Board tutorials require a special room where whiteboards have been installed around the room and furniture (tables and chairs) can be quickly relocated. Students form pairs, are given problems and a marker, and are allocated a whiteboard as they arrive. Students take turns solving the set problems, talking and asking questions of each other about the approaches that are being used. The students are encouraged to look at others' work and discuss any differences in strategies and weigh the pros and cons of such choices. Students are also encouraged to photograph and share their work.

The role of the tutor in board tutorials is to engage students through deep questioning and facilitate the problem-solving process. The tutors do not have a marker and use questioning to help students move through the various stages of the problem-solving process. Answers are available in the session for students to check, and at the end of the week full solutions are shared using the course website's forum. While the number of problems students complete is reduced when using this technique, the learning process is richer and more active (Liljedahl, 2016; Seaton *et al.*, 2014). Such activities may not translate well to the online space or may be technically difficult to facilitate for large class sizes – but given the value of such sessions to learning (Liljedahl *et al.*, 2016), it would be effective use of both staff and student time to continue providing such a valuable F2F interaction.

The use of quizzes and tests in tutorials may serve a dual purpose – on one hand, it is an efficient way of assessing basic levels of student knowledge, while on the other hand it is a simple (not always effective) solution to dwindling attendance as the semester progresses! As such, while it is relatively straightforward to replace pen-and-paper quizzes with an online counterpart (e.g. randomized multiple choice questions taken from a larger pool), apprehension from staff towards removing incentives for showing up to tutorials is understandable. The solution may lie in the addition of other compulsory assessment items to tutorials (e.g. a monthly test that requires more thorough problem solving which may be difficult to effectively transition online) or providing a small grade contribution for attendance; however, providing tangible value to tutorial attendance should be the ultimate goal. Providing clear evidence that tutorial attendance is worth the student's time and effort will be an important overarching objective. It should be noted that understanding the cohort is also of importance here – overall tutorial attendance and engagement of students in a compulsory introductory course in most instances is expected to differ wildly from later-year elective courses.

As with quizzes, assessment items can be entirely directed online. Instead of students being guided through the assessment item by F2F tutors, the students may instead be directed to the course homepage, where they can access the assessment criteria, rubrics and time budget. One must be conscious that removing the F2F interaction may also remove or limit opportunities for students to raise concerns, ask questions for clarification, etc. In this case, feedback or Q&A sessions can be hosted online through virtual tutorials or an online forum. The latter has the benefit that staff can directly address any questions or concerns of a particular student but have it be visible to the whole cohort (unlike direct email correspondence), and it allows instances for students to discuss amongst themselves, provided the discussions are facilitated by a staff member to prevent propagation of misinformation.

3.5 Case studies

Individual academics can progressively develop and prepare their engineering courses for student-centered online delivery without the "extreme makeover team" descending upon them. A key first step is to understand their cohort and how to measure and track student workload both in and out of the classroom using tools such as time budgets (Quinn and Wedding, 2012).

To demonstrate this model in action, we present aspects of two transition projects; an engineering physics and a foundation mathematics course. The case studies examine how students engage with the various components of the courses at different stages of the transition in delivery and consider how students' incoming grade point average (GPA) and personality factors impact on student success. These two descriptive case studies provide "thick" descriptions (Merriam, 1998) of the implementation of blended learning in first-year engineering courses.

3.5.1 Engineering Physics N

This case study describes the transition of a F2F foundation engineering physics course first to blended delivery for local/internal students (flipped lectures), through the creation of online support materials and then finally to an online version of the course for external students (with a practical intensive) while monitoring the impact on different student groups behavior, engagement and learning outcomes.

3.5.1.1 Background

The course, Engineering Physics N, is a second-semester first-year course offered as a part of several electrical and electronic engineering bachelors programs. There are no prerequisites for this course, however stage 2 (year 12) physics is assumed knowledge. While the intake is predominantly local students from the high school system, a number of students are international, from bridging and access pathways, repeat or mid-year entry students. Delivery was traditionally 3 hours of lectures and a 1 hour tutorial per week (over the 13-week study period) in parallel with a 10-week × 2-hour practical program. Assessment for local F2F students is via continuous assessment using in-class quizzes and numerical problem-solving tests (25%), the 10-week practical experience (25%) and a final examination (50%).

3.5.1.2 Preparation of the time budget

The first step for transitioning any course is to generate a holistic view of the course, its components and assessment items and most importantly to appreciate the time expectations involved. The academic needs to appreciate the student experience; allocate time for synthesis, time management and balance, timing of assessment and time allocated *versus* quality of submitted work; have realistic expectations, envisage the roles individuals will play within any group work, and also show they meet any University guidelines. All of this is captured in a time budget, a visual mapping of a course. The initial time budget for Engineering Physics (see Fig. 3.2) was discussed with academic staff, tutors and practical demonstrators leading to a clear consensus enabling an appreciation of any changes in the various components of the course and its delivery. This simple tool is front and center during weekly teaching team meetings, providing a much needed clarity for both staff and students and a strong sense of discipline in the timely delivery of the course.

3.5.1.3 Transitioning materials

Prior to 2014, the course was offered only in F2F mode. As indicated in Table 3.2, from 2014 online media has been sequentially added to allow blended learning and adjustments made to F2F activities to improve their relevance to real-world contexts, e.g. tech talk sessions. Tech talk sessions are 15–20 minute interactive sessions at the end of weekly lectures to relate the lecture topics/theory to real-world examples and professional practices. An example might be critiquing YouTube videos of jump-starting a car with a flat battery; discussing and analyzing the procedure, properties of the cables, batteries and fault finding in circuits; relating aspects back to the lecture topics; discussing ohmic contacts etc. These helped to improve the perceived relevance of the material being studied. In 2015, a "transitional hybrid flipped" classroom model was employed, where only the first 4 or 5 weeks of traditional lectures were presented in parallel with online media available to prepare and support students for the flipped environment, which may be new to some. The lectures were then replaced by a single hour lecture-workshop for the remainder of the semester. In 2016, new F2F activities and online activities were developed along with the first external offering of the course (which included an on-campus "intensive" for the practical component). The changes for internal

Table 3.2 Staged roll-out of Engineering Physics N lecture support

Delivery Component	*2013*	*2014*	*2015*	*2016*
Lectures				
Traditional F2F	✓	✓	Hybrid Flip	Hybrid Flip
Lecture Videos		✓ Raw	✓ Edited	✓ Edited
Tech Talk		✓	✓	✓
Lecturettes			✓	✓
Online version				✓
Tutorials & Workshops	✓	✓	✓	Modified I
Practicals				
F2F – Internal	✓	✓	✓	✓
Intensive – External				✓

students included problem solving using the board tutorial format and (short) fortnightly formative online review quizzes before each summative in class quiz.

While lecture theaters are routinely equipped for automated recording of lectures, we deliberately bypassed making the unedited, automated recordings of computer-projected lecture content immediately available to students, as such recordings did not convey the lecture dynamics, or whiteboard and demonstration activities, which make the lectures more interactive. Instead, all components of the lectures were video recorded and spliced together with the audio before being made available to students in 2014. These "raw" lecture videos still represented minute-for-minute versions of the lectures and were made available on the course website at the end of the week after the scheduled traditional lectures were delivered. These lecture videos were then edited by the academic to remove speech mannerisms, class interruptions, announcements and incidental direction to the class, etc., which typically resulted in the standard 50-minute lecture class being reduced to a typically 40–45 minute edited lecture video. These edited lecture videos were used from 2015 on.

In 2015, lecturettes were also created. Lecturette videos identify and focus only on three key points of each developed traditional lecture slide and are significantly shorter (10.6 ± 2.5 minutes, $n = 37$) than edited lecture videos. The release of each resource (edited lecture and lecturette) video was timed throughout the semester to coincide with the traditionally timetabled F2F lectures (e.g. three resources per week to replace 3 hours of lectures per week).

As mentioned earlier, in 2015 the lecture component was transitioned to a hybrid flipped format. Rather than expose first-year students to a dramatically changed learning environment (i.e. completely flipped) from the outset of the study period, we proposed a transitional approach whereby F2F lectures were delivered in parallel with the provision of online video resources for the first 4 weeks of the semester. This enables awareness and preparation for the change to online lecture resources and a 1-hour lecture-workshop for weeks 5–13. Based on student feedback from a previous project on the development of a collaborative blended delivery model between universities (Quinn *et al.*, 2012), the lecture-workshop was provided to guarantee student exposure with the "expert" (the academic) when flipping the lectures online. The academic is to be open to address student questions and concerns with any aspect of the course, but the lecture-workshop also enables the academic opportunity to reinforce and elaborate on selected topics from the syllabus. The lecturettes and lecture-workshops are representative of media to supplement lectures indicated in Figure 3.3. In the modified flipped classroom model, the lecture-workshops (weeks 5–13) were also used as opportunities to clarify concepts students had been exposed to through online experience. Flipping lectures with video and in particular shortened concise versions of lecture experiences such as lecturettes is not a means to free up time in a time budget.

Since the online video resources were rolled out over the semester, comparing the number of views per resource may not reveal much information about interaction with the course website throughout the semester (i.e. videos released later in the semester are likely to have fewer views), but it can indicate the relative popularity of edited lecture videos *vs.* lecturettes for each topic/resource. In all but two cases, lecturettes received a greater number of views compared to lecture videos. This indicates a student preference for lecturettes, which may result from the method of delivery (shorter and more direct) or because these resources were used preferentially as a revision tool leading up to the exam.

In 2016, the course was also delivered online for external students (see Table 3.2) as per the evolved time budget (Fig. 3.5).

Week	1	2	3	4	5	6	7	8	9	10
	HOURS OF STUDY									
1 24/7	Orientation Activities	Introduction and Electrostatics; Electric Field lines; Electric Flux		Giancoli (2008) Ch21&22	Partial lecture notes	Key words Q&A Forum	Meet eTutor	Exercises **Ch21** (41): (Q) 2, 3, 18, 26 (P) 1, 8, 13, 21, 29, 35, 59, 65		Read Assessment requirements
2 31/7	Gauss' Law; Electric Potential		Giancoli (2008) Ch 22&23	Partial lecture notes	Key words Q&A Forum	Meet eTutor	**Ch22** (Q) 12 (P) 2, 4, 7, 13, 25, (GP) 52; **Ch23** (Q) 1, 4, 8, 9, 10, 16, 18 (P): 1, 4, 5, 18, 25, 29, 34, 42, 51, 58, *(GP) 75*			
3 7/8	Potential from point charges; Capacitance; Dielectrics & capacitors		Giancoli (2008) Ch 23&24	Partial lecture notes	Key words Q&A Forum	Meet eTutor	**Ch24** (Q): 3, 4, 9, 10; (P) 1, 5, 9, 17, 21, 29, 35, 41, 53			Practice Quiz 1
4 14/8	Electrical current; Ohm's law; Simple circuits		Giancoli (2008) Ch 25&26	Partial lecture notes	Key words Q&A Forum	Meet eTutor	**Ch25** (Q): 1, 3, 4, 7, 20; (P) 1, 7, 8, 15, 25, 33, 40, *41, (GP) 94*		**SUBMIT** Quiz 1 (3%)	**SUBMIT** Test 1 (4%)
5 21/8	Kirchhoff's rules; RC circuits; Magnetic fields		Giancoli (2008) Ch 26&27	Partial lecture notes	Key words Q&A Forum	Meet eTutor	**Ch26** (Q): 1, 4, 10, 11, 23, 15, 18, (P): 1, 5, 7, 29, 33, 45, *(GP) 92 20*			Practice Quiz 2
6 28/8	Charged particles in electric fields; The Hall effect & Ampere's law; Bio-Savart Law		Giancoli (2008) Ch27&28	Partial lecture notes	Key words Q&A Forum	Meet eTutor	**Ch27** (Q): 1, 22, 23; (P): 3, 13, 47 **Ch28** (Q): 2, 5 (P): 1, 6, 25, *(GP) 53*			**SUBMIT** Quiz 2 (3%)
7 4/9	Solenoids & Helmholtz Coils; EMI; Inductance		Giancoli (2008) Ch29&30	Partial lecture notes	Key words Q&A Forum	Meet eTutor	**Ch29** (Q): 2, 4, 15, 17, 18; (P): 1, 2, 3, 12, 18, 47, 49; **Ch30** (Q): 1, 4, 7; (P): 1, 5, 11, 15, 23			Practice Quiz 3
8 11/9	RLC circuits; AC circuits; Filters & transformers		Giancoli (2008) Ch 30	Partial lecture notes	Key words Q&A Forum	Meet eTutor	**Ch30** (Q): 9, 12, 14, 17, 19, 20, (P): 29, 39, 40, 47, 49, 62, 66 Read through Practicals		**SUBMIT** Quiz 3 (3%)	**SUBMIT** Test 2 (4%)
	3-Day **INTENSIVE** 27-29 September (Practicals at Mawson Lakes F1-37, F1-41, F2-59) **SUBMIT** Practical worksheets x3 and log book entries x2									
9 2/10	Blackbody radiation and Plank's hypothesis; Photoelectric effect; Bohr's model of atom		Giancoli (2008) Chapter 37	Partial lecture notes	Key words Q&A Forum	Meet eTutor	**Ch37** (Q) 1, 2, 15, 20, 24; (P), 2, 6, 10, 12 13, 19, 28		Work on Practical reports	
10 9/10	Photons & EM waves; Uncertainty principle, Quantum numbers		Giancoli (2008) Ch37,38&39	Partial lecture notes	Key words Q&A Forum	Meet eTutor	**Ch38** (Q): 7, 10, 11; (P) 7, 9; **Ch39** (Q)1, 14; (P):5, 8, 41, 44		Practice Quiz 4	**SUBMIT** Quiz 4 (3%)
11 16/10	Lasers, X-ray spectra; Bonds & conduction; P-N junction; semiconductors		Giancoli (2008) Ch 39&40	Partial lecture notes	Key words Q&A Forum	Meet eTutor	**Ch39** (Q) 26, 27; (P): 54, 55, 61, 64, 73; **Ch40**: (Q): 6,9,10,13; (P): 1,43,46,48,50		Practice Quiz 5	**SUBMIT** Practical Report 1
12 23/10	Magnetism in matter; Mechanical properties of matter; Speed of Sound		Giancoli (2008) Ch 28&16	Partial lecture notes	Key words Q&A Forum	Meet eTutor	**Ch28** (Q) 16, 18, 23; (P) 47		**SUBMIT** Quiz 5 (3%)	**SUBMIT** Test 3 (4%)
13 30/10	Nuclear Physics; Radioactive decay		Giancoli (2008) Ch41&42	Partial lecture notes	Key words Q&A Forum	Meet eTutor	**Ch41** (Q): 1, 21, 24; (P): 2, 6, 41, 42, 57, 60, 65, 69		Practice Quiz 6	**SUBMIT** Practical Report 2 Journal (25%)
	SWOT VAC and **EXAMINATION** (Closed Book; 2.5 hours; 50%)									

Key
Book Work Usually can be done offline
Lectures Suitable for mobile viewing
Online Activities Various learning activities online
Virtual Classroom Requires online participation at a set time
Computer Work Requires access to special software not available on
Online Quiz Requires online participation for assessment purposes
Assessment Work on, or **SUBMIT** larger assessments
Face to Face Examination or on-campus residential

Figure 3.5 A time budget for the external delivery of the Engineering Physics N course

3.5.1.4 Transitioning students

From 2015, students were offered a learning style assessment based on the VAK (visual, audible, kinesthetic) modalities (Frender, 1990) chosen, as students could then be directed to online descriptions and study guides associated with each modality.

To understand when students used which online media, the ratio of lecturette to lecture video views was compared. A trend can be seen (Fig. 3.6(a)), where the students increase their engagement with the lecturettes for topics covered later in the semester when compared

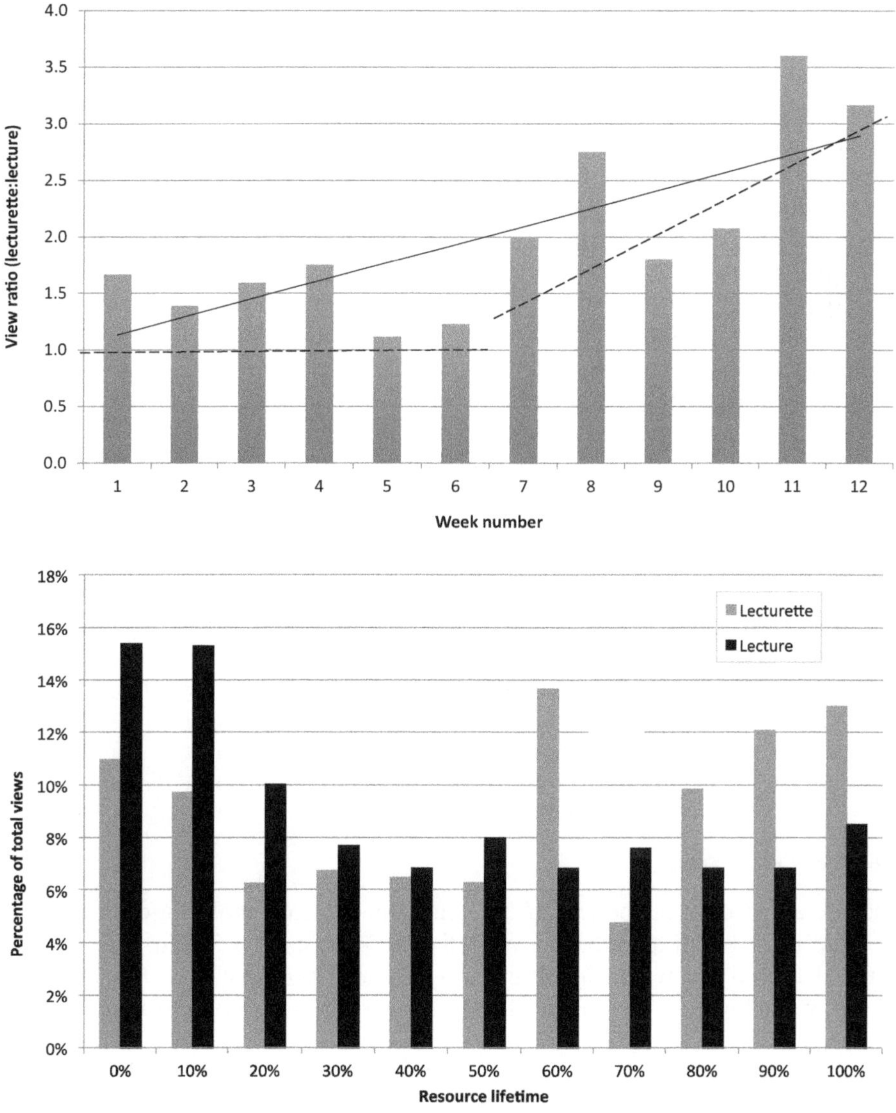

Figure 3.6 (a) Ratio of lecturette views to lecture video views each week of semester, and (b) averaged proportion of total views across lifetime of resource, comparing viewing trends for lecturettes to lectures

to lecture videos. Again, as this data represents the total number of views across the semester, this may not necessarily indicate that as the semester progressed, the students turned their preference to the lecturettes, but may be explained by decreased course engagement towards the end of the semester and a reliance on more time-efficient means to catch up on material prior to the exam.

By taking into account the proportion of total views across the lifetime of the resource, student preferences become more apparent (Fig. 3.6(b)).

3.5.1.5 Transitioning F2F tutorials and workshops

As described in Table 3.2, a weekly timetabled F2F tutorial and a workshop was offered each year of delivery. F2F tutorials provide a regular 2-week cycle of "quiz and conceptual questions" one week, followed by "quiz feedback and problems" the next week. An abridged list of conceptual questions and numerical problems are selected from the prescribed text and constitute a week-by-week exercise list, made available to students from week 1 of the course, that provides direction for the tutorials and student self-study expectations. The fortnightly tutorial quizzes consist of typically seven or eight multiple choice conceptual questions and two short answer type questions. The quiz is to take 20–25 minutes of the 50-minute tutorial class, no more. Similarly, the feedback on the quiz is not to consume more than 15–20 minutes of the following tutorial and focuses on the questions that the majority of the students had difficulty with. In 2016, short online review quizzes were developed for the interleaving weeks to guide student self-study and prepare for summative quizzes. Online versions of the summative quizzes were also prepared for the first external offering of the course. In 2016, tutorial problem-solving activities were changed from table-based group problem solving to a standing at whiteboard problem-solving format. Over the 4 years (2013–2016), no significant change in student participation in tutorial classes was seen.

The workshops are designed to provide learning support and skill development in areas that may not be sufficiently developed before entering tertiary study. The three key areas are developing a problem-solving strategy, development of computer literacy focused on professional practices and expectations, and group problem solving (how to use study groups), all of which culminate in an exam preparation session. The workshops are voluntary except for on three occasions when the timeslot is used for the summative numerical problem-solving tests. Over the 5 years (2012–2016), no statistically significant change in student participation in workshops was seen, possibly since the basic activities remained the same.

3.5.1.6 Transition result

Students are asked towards the end of each study period to complete a course evaluation consisting of a number of questions to which they respond using a 5-point Likert scale. The "overall satisfaction" score for the Engineering Physics N course over the period (2013–2015) is shown in Figure 3.7, with a relative comparison to the average result for the school, the division and the university as a whole. The shift from F2F to blended mode in 2014 did not appear to change student satisfaction; however, the flexibility of the hybrid modified flipped classroom with lecturettes in 2015 was resoundingly well received.

Success of a teaching initiative may be an improvement in student performance, in assessment. Since all other aspects of course delivery were constant, the impact of a partially flipped classroom in 2015 may be appreciated by looking at grade outcomes. Attempts were

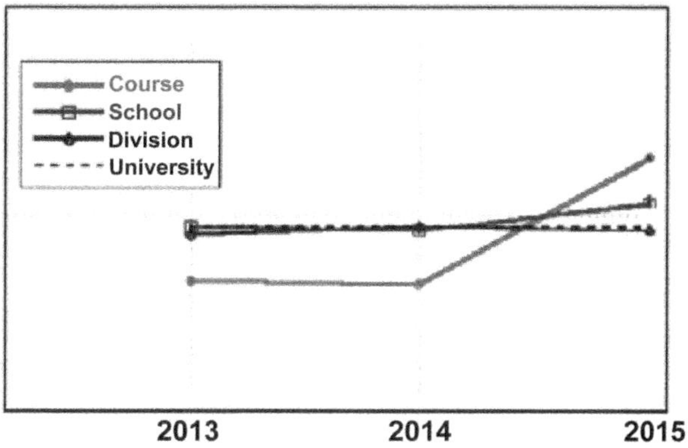

Figure 3.7 Relative course "overall satisfaction" change in relation to the average for all courses in the school, division and university over 3 years

made to discern a change in the overall grade or mark distribution year by year, and variations at the individual student level of their overall course grade GPA from their average GPA coming into the course. The overall percentage pass rate stayed the same. While students are happy with the change, clearly flipping lectures works for some and not others. No statistically meaningful trends or correlations were found within grade levels, however there was an overall improvement when considering the percentage of students whose course grade improved relative to their average GPA grade (i.e. for all their courses), from 32% in 2014 to 43% in 2015. This is significant considering that the Engineering Physics course is considered one of the more challenging in their first-year experience. Making incremental changes in delivery results in harder-to-quantify incremental changes in overall course marks. Further student feedback from focus groups and student interviews has been incorporated for the 2017 delivery.

Engineering students are familiar learning through online videos (Khoo *et al.*, 2015), but data in that study and in ours found that students do not fully engage with the required video resources to truly appreciate the body of knowledge which is expected to be covered. Average or struggling students benefit greatly from the viewing of lecturettes in our hybrid flipped classrooms; however, credit-level students performed less well, perhaps due to over-confidence, i.e. they may feel that they have already had success in university learning and with all the resources available to them so they can relax and still be successful in the course. Better preparation for this change in learning approach may improve outcomes (Quinn *et al.*, 2012).

3.5.2 *Mathematical Methods for Engineers I*

This case study explains how blended learning was successfully implemented from a F2F math course, but only after the creation of a fully online offering produced by a development team which was subsequently delivered by a sessional tutor, under the guidance of the course coordinator. It highlights the importance of F2F activities that continue in the blended course and the optimization of teacher and student meetings to enhance student engagement and learning outcomes.

3.5.2.1 Background

The course, Mathematical Methods for Engineers 1, is a first-semester first-year course offered as a part of all bachelor of engineering programs as well as our associate degree in engineering. The F2F version of this course was first introduced in 1992 with the aim of providing an introduction to mathematical concepts relevant to engineering disciplines using both analytic and software approaches. The prerequisites for this course are Mathematics 1 or Mathematical Studies at Year 12, or equivalent. While the intake is predominantly local students from the high school system, approximately 15% of students are international.

The F2F course has been taught with 3 hours of lectures for 13 weeks, 1.5-hour tutorial for 12 weeks and 1-hour computer practical for 13 weeks. Assessment since 2007 has been via a computer modeling project, continuous assessment using paper-based quizzes, two assignments and a final examination.

The course had a history of poor performance in student learning outcomes with high failure and withdrawal rates when compared to other first-year courses in the engineering degree. Interventions to detect students at risk via performance in early assessment were in place in 2008 (Johnston *et al.*, 2008) and had some impact at reducing the rate of withdrawal, but poor grades and comparatively low student satisfaction scores continued.

3.5.2.2 Transitioning materials

With the help of a development team, a fully online version of this course was offered in partnership with Open Universities Australia as part of an associate degree in engineering program from 2012. The course materials underwent some change to allow this new delivery including the development of topic notes, lecturettes and tutorial solution videos. For lecture materials, full-length in-class video recordings (1 hr × 3 lectures per week × 13 weeks = 39 hours) were initially offered, but with a change in course coordinator in 2013, these recordings were replaced with in-class lecture recordings using a document camera. These new lecture recordings were subsequently broken down into 328 titled and grouped videos of ~7 minutes length that allowed for easier navigation of the content for external online students. The F2F cohort of students did not have access to any of these lecture recordings unless there was an unexpected absence of the 2013–2015 course coordinator. Instead, using a document camera and pen and paper, the lecture capture recordings and a scanned version of the written notes were made available to F2F students after each lecture.

From 2016, the formal shift to blended learning was made for the F2F version of the course. This was driven by a change in course coordinator and the timetabling of an external (fully online) version of the course. The lecture recordings were progressively replaced with the 2016 coordinator's smartboard recordings, which were edited and titled, this time into 69 smaller videos from the original 13 weeks. The lectures were integrated into a weekly study plan for internal and external students based on a time budget (see Fig. 3.8). F2F lectures continue to be offered and are well attended.

The blended website was restructured to mirror the time budget, creating numbered checklists of activities that students needed to complete each week. The students studying externally access the same website as the internal students.

Learning activities were also changed. In 2016, six automatically marked two-attempt online quizzes were used to replace the 10 pen-and-paper quizzes that had been previously held at the beginning of tutorial classes to encourage student attendance. Topics assessed

Week	HOURS OF STUDY 1	2	3	4	5	6	7	8	9	10
1 **26/2**	Orientation Activities			On-board Quiz (OBQ) Module 4	Introduction to Modelling		App C, Chapter 1	Key terms and Q&A Forum	Install software	Computer Practical 1
2 **5/3**	OBQ Mod 5	Polynomials for Modelling; The marriage of Algebra and Geometry; Parametric equations for modelling		Chapter 1.3 and 9.4	Key terms Q&A Forum	Computer Practical 2	IMPs	Board Tutorial 1	Independent Problems Tutorial 1	Work on PSEs 1
3 **12/3**	Introducing Calculus; Limits			Chapter 2	Key terms Q&A Forum	Computer Practical 3	IMPs	Board Tutorial 2	Independent Problems Tutorial 2	**SUBMIT** Quiz 1 (2%; 2 attempts)
4 **19/3**	OBQ Mod 6	Rules of Differentiation		Chapter 3.1-3.6	Key terms Q&A Forum	Computer Practical 4	IMPs	Board Tutorial 3	Work on Problem Solving Exercises 1	
5 **26/3**	Derivatives: What could go wrong?; More on differentiation rules and notation; Applications of differentiation - Optimisation			Chapter 3.1-3.6	Key terms Q&A Forum	Computer Practical 5	IMPs	Board Tutorial 4	Independent Problems Tutorial 3/4	**SUBMIT** Quiz 2 (2%; 2 attempts)
6 **2/4**	OBQ Mod 6	Vectors		Chapter 11 or eReading	Key terms Q&A Forum	Team work form Groups	IMPs	Board Tutorial 5	**SUBMIT** Problem Solving Exercises 1 (15%)	
	MID-SEMESTER BREAK									
7 **23/4**	Implicit differentiation; Related rates; L'Hôpital's Rule			Chapter 3.9; 4.8	Key terms Q&A Forum	Project Work	IMPs	Board Tutorial 6	Independent Problems Tutorial 5/6	**SUBMIT** Quiz 3 (2%; 2 attempts)
8 **30/4**	OBQ Mod 8	Antiderivatives; Sum and Areas		Chapter 5.1-5.5	Key terms Q&A Forum	Project Work	IMPs	Board Tutorial 7	Work on Problem Solving Exercises 2	
9 **7/5**	Substitution; Natural logs			Chapter 5.7; 6.7	Key terms Q&A Forum	**SUBMIT** Group Project (15%)	Board Tutorial 8		Independent Problems Tutorial 7/8	**SUBMIT** Quiz 4 (2%; 2 attempts)
10 **14/5**	OBQ Mod 8	Applications of Integration – Volume and length of curve		Chapter 6.7-6.8	Key terms Q&A Forum	Computer Practical 6	IMPs	Board Tutorial 9	Work on Problem Solving Exercises 2	
11 **21/5**	Fundamental Theorem of Calculus; Differential equations; Other applications			Chapter 6.1, 6.2, 6.4	Key terms Q&A Forum	Computer Practical 7	Board Tutorial 10		**SUBMIT** PSEs 2 (15%)	**SUBMIT** Quiz 5 (2%; 2 attempts)
12 **28/6**	OBQ Mod 8	Complex numbers		Chapter 8.1, 5.8, 3.10	Key terms Q&A Forum	Computer Practical 8	Board Tutorial 11		Independent Problems Tutorials 9/10/11	
13 **4/6**	Review of the practice examination			Q&A Forum	Past examination papers			Discuss past exam papers with Tutor		**SUBMIT** Quiz 6 (2%; 2 attempts)
	EXAMINATION (3 hours; partial closed book with 1 x A4 double-sided hand written notes; **45%**)									

Key
Book Work Usually can be done offline
Lectures Suitable for mobile viewing
Online Activities Various learning activities that are done online
Virtual Classroom Requires online participation at a set time
Computer Work Requires access to special software not available on most computers
Online Quiz Requires online participation for assessment purposes
Assessment Work on, or **SUBMIT** larger assessments
Face to Face Tutorial, Examination

Figure 3.8 A time budget for the blended delivery of Mathematical Methods for Engineers 1 (internal students version)

in these quizzes were confined to the absolute basic foundation knowledge, leaving more advanced concepts to the other assessment tools (assignments and project). Rich feedback (principle being assessed, answer and then full solution) was embedded into the feedback for each question. This changed quizzes from being simple assessment tools to tools that also supported the teaching and student learning of the foundational core mathematics in this course. Equivalent quiz questions were created for each principle being assessed to allow randomization. Typically, students were encouraged to take the weekly quiz (approximately five items) at the beginning of the week to identify those topics that they understood and those that they did not and to use the automatic feedback to guide their learning. They then take the quiz again at the end of the week, with slightly different questions.

Figure 3.9 compares the average grade of students' first attempt at each of the six quizzes with the average grade of their highest graded attempt (data from Moodle Quiz Statistics for 2016 study period 2 cohort). This demonstrates that on average there is a 17% improvement in performance from students' first attempt to their second attempt.

In response to the standard end-of-course online survey question, "What are the strengths of this course?"

(SP2 2016) MME1 students wrote:

- "Quizzes"
- "I enjoyed the instant individual answer feedback in the quizzes. It allowed me to rethink any problems I had so that I could re-engage with the problem."
- "The quizzes were well set up"
- "The structure was great. All the learning items flowed into the assignments and quizzes."

3.5.2.3 Transitioning students

In recognition of the diverse student backgrounds and to prepare the students for online learning of mathematics, a "Maths Onboarding" website was created in 2014. This website allowed students to self-assess 46 distinct topic areas and, if necessary, develop their

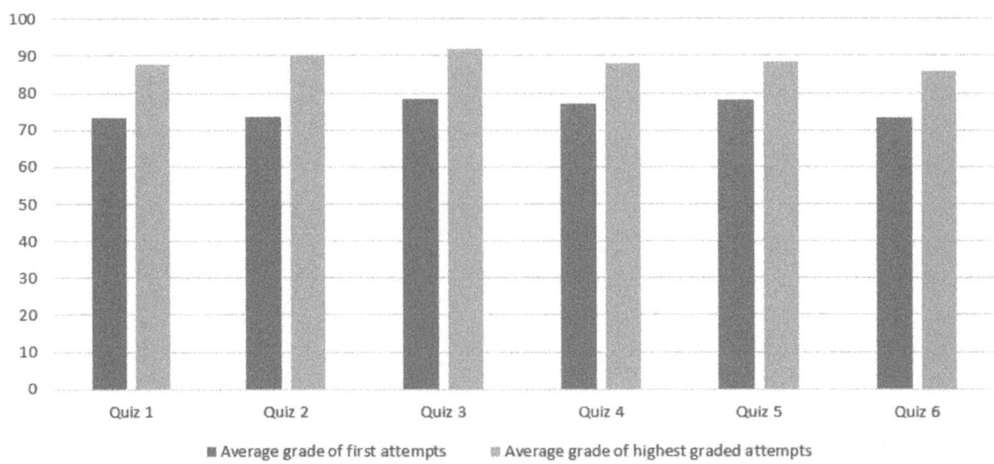

Figure 3.9 Average quiz scores from first and highest attempts in 2016 in MME1 quizzes

background knowledge in preparation for the topic that was to be introduced that week. Students are given three random questions related to the assumed knowledge for the topic using a quiz, which were also linked to study notes with videos from high-school-level courses. This pre-learning opportunity allows students to save face and independently "diagnose and treat" any areas found to be wanting.

Although not all students access these resources, those that do find it useful to revisit the website on multiple occasions. The data shown in Table 3.3 compares the aggregated webpage hits for only the first module of the Maths Onboarding online resource. Since their inception, the quizzes have been found to be used more than the independent learning resources. Both the quiz and the learning resources for assumed knowledge are repeatedly used with quizzes being revisited by users between 20 and 28 times, while learning resources are being revisited by users 8–11 times (Table 3.3).

To address potential math anxiety issues, a math mindset quiz, based on the work of Carol Dweck (Dweck, 2012) and Jo Boaler (Boaler, 2013, 2015), was developed in 2016. This educational quiz introduced students to the concepts of neuroplasticity, growth and fixed mindset, debunked "giftedness" in relation to mathematics and emphasized the importance of learning from mistakes. This awareness can help students become more robust as learners and develop a "growth mindset" in relation to studying mathematics. In 2016, the voluntary quiz was completed by only 17% of the cohort, however 89% of these participants received final grades greater than 75% (distinction level or above). This could be a positive selection bias; that is, only those students who are interested in concepts such as math mindset would also have a positive attitude to their studies, but it will be interesting to explore this learning activity longitudinally and follow individual student progress to see if the positive impact of mindset/neuroplasticity awareness that has been found for junior- and high-school-level students (Boaler, 2013) also holds for engineering students.

In addition, a standard GRIT quiz was introduced in 2016 which showed that, on average, students starting out in their engineering studies are mostly gritty. A subsection of these students will go onto Engineering Physics N and resit the GRIT quiz. There is an argument that GRIT is unlikely to change over time (Duckworth *et al.*, 2007). The grittiness of a student has been linked to academic success, but it is not clear if the grit level of students reflects how they engage with blended, flipped and online learning opportunities. Further studies in this area will be incorporated as this course continues in its various modes.

Table 3.3 Maths Onboarding usage for Module 1 topics (August 2017)

Topic	*Quiz*		*Learning Resource (notes, videos)*	
	Views	*Users*	*Views*	*Users*
1.1 Mathematical Preliminaries	20,396	719	2008	235
1.2 Polynomials	14,180	580	1553	150
1.3 Exponents	9821	474	1048	89
1.4 Rationals, Radicals and Complex Numbers	10,263	434	825	84

3.5.2.4 Transitioning F2F tutorials

The tutorial classes were also rethought to allow blended delivery. Traditionally, F2F students were provided with a list of problems to complete at home. They then went to their tutorial class where they could ask their tutor to work through solutions on the board for the problems that had proved the most troublesome. In many ways, these tutorials were more like student-directed mini-lectures. Tutorial problems had, over the years, become simplified to be more like drill-and-practice mathematical procedures than more challenging problem solving that would be relevant to an engineer. As the foundation knowledge was already being supported by the online quizzes, we took the opportunity to incorporate more challenging applied engineering mathematics problems into the tutorials as well as the computer modeling project.

Board tutorials (Liljedahl, 2016; Seaton *et al.*, 2014) were implemented to transform the otherwise passive tutorials into active peer-to-peer learning opportunities. After researching several options, a simple problem-solving strategy (Mason *et al.*, 2010) was adopted with slight modification as an initial framework to support students working with applied problems (Fig. 3.10).

To reduce the cognitive load associated with solving applied mathematical problems (Ambrose *et al.*, 2010), interactive math problems (IMPs) were designed and developed to scaffold the problem-solving process using the Lesson tool in Moodle. These online resources provide students the opportunity to select how much support they require to find solutions to board tutorial-like problems and encourage them to broaden their skill set through alternative methods and further practice.

In addition, further textbook-based extension problems with solutions were identified for students to practice. Collectively, this meant that the blended form of the tutorial was comprised of three components – IMPs, board tutorials and take-home problems.

All these changes have been well received with a 43% improvement in the number of students passing the course and a 59% improvement in overall student satisfaction rating as measured by an anonymous end-of-semester student survey from 2015 to 2016. The course currently enjoys the highest student satisfaction rating of all the first-year first-semester engineering courses.

3.5.2.5 Transitioning staff

The development of the course for online delivery in 2011 included technical and pedagogical development of staff through a series of workshops and regular meetings during the 6-month development phase. Incentives in the form of stipends, workload adjustments and equipment were obtained as well as hiring of teaching assistants to develop lecturettes. As with all courses in the associate degree in engineering, a time budget was negotiated to support staff and students with the transition to studying mathematics online.

There were three subsequent changes in course coordinators before the change to blended delivery. In 2015, 6 months before the blended offering commenced, the new course coordinator was transitioned to technological and pedagogical approaches to blended delivery through weekly meetings with development staff. Incentives were provided in the form of equipment and the hiring of assistants to create quizzes and IMPs. During this time, the course structure was refocused and new online elements designed. Tutors were also retrained for board tutorials using a series of demonstrations and workshops. A new time budget was again negotiated to represent the integration of physical and virtual experiences available to students (Fig. 3.8).

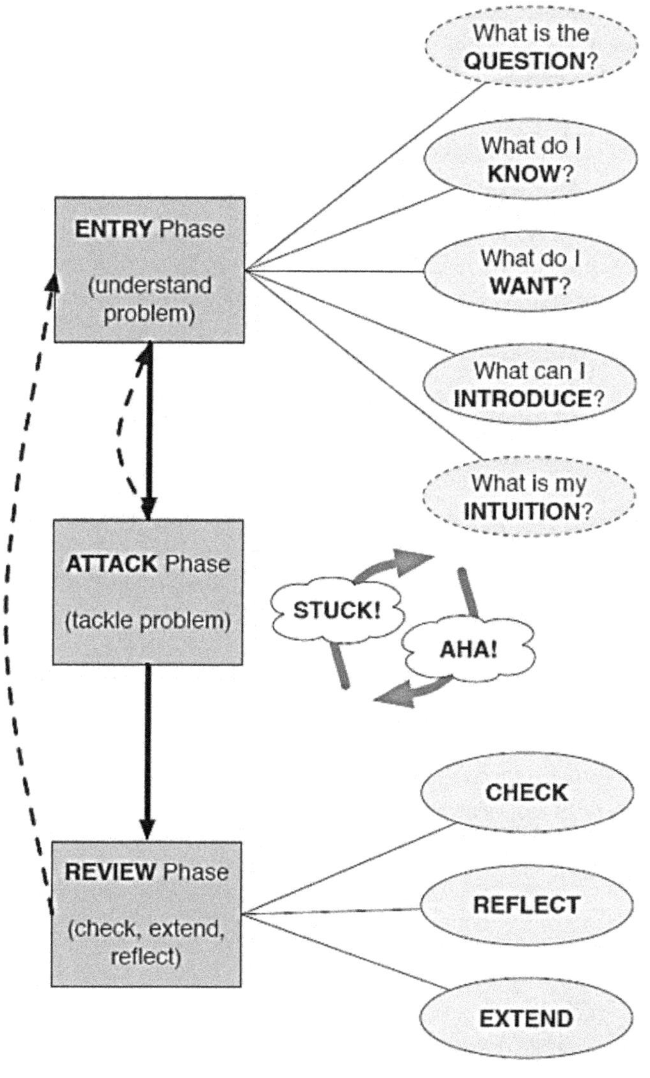

Figure 3.10 A simple problem solving technique styled after and modified from Mason *et al.* (2010)

3.6 Summary

In this chapter we have proposed a model for transitioning the delivery of first-year tertiary engineering courses from F2F environments to variable delivery modes employed in tertiary education to meet changing demand. We looked in detail at two case studies in which blended engineering courses have been developed to support first-year engineers. A key part of this process is the role that the time budget plays in first negotiating what is involved in the course and how it can be envisaged in a new modality, be it blended, flipped or online. A time budget provides a readily realized holistic view of a course for all involved in the transitioning process, before, during and on an ongoing basis, for the delivery and further refinement of the course.

Student learning has been described as three interrelated components: presage, process and product (Biggs, 1989). When the mode of delivery changes the context of learning (presage), this has flow-on effects to how students perceive the context of their learning and the approach that they take to their learning (process). In the case studies, we have discussed strategies to improve student learning outcomes by addressing presage – by helping students to self-evaluate and refocus their incoming knowledge and attitudes using online tools. Time budgets address the process of learning, by shaping students' perceptions of the teaching content, demonstrating an overview of the course's structure and demonstrating that it is fair and respectful of the amount of work required by the student. In turn the student can decide how they will approach their learning – better aware of the time commitments involved. The creation of online tools such as two-attempt quizzes, lecturettes and the implementation of high-value active F2F activities such as board tutorials, tech talks, lecture-workshops and practicals promote students taking a deeper approach to their learning leading to better outcomes (Fig. 3.11).

When transitioning the whole system – either independently as a teaching team over a series of years or in collaboration with others – it is necessary to not just add material to the course website, but to take a systems approach, coupling the best of what online has to offer with the key active F2F experiences. Only in this way can you create engaging learning environments in which the student experience has been optimized for success.

Moving to blended or online delivery, we found that for one student wanting "lectures to be retained," another wants "everything on the Web." There are no cost or time efficiencies with flipped lectures or blended/online delivery, as you cannot compensate for the personal interaction and active learning. Blended and online delivery works for some but not for others. As with anyone with responsibility for the well-being of others, the goal must always be "*primum non nocere*."

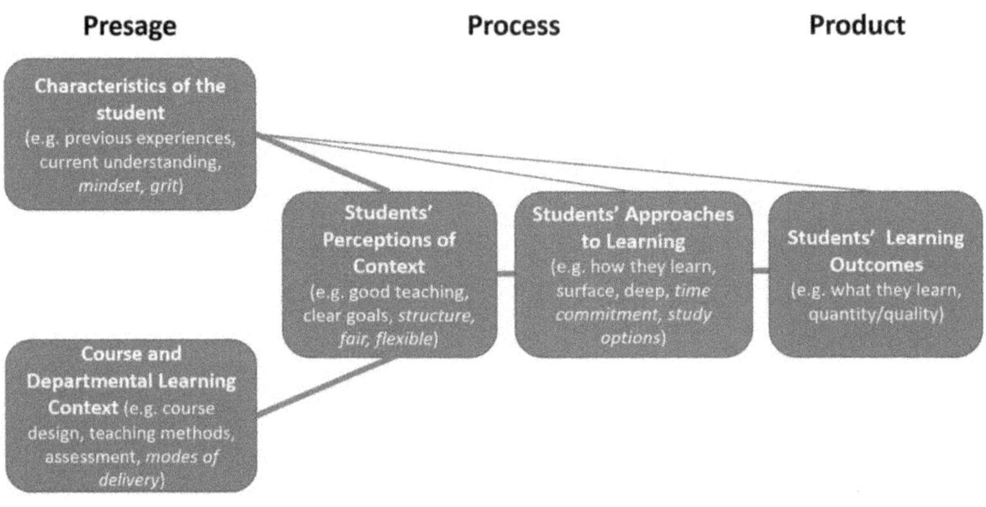

Figure 3.11 The 3P model of student learning (modified from J.B. Biggs (1989) and Trigwell and Prosser (1997))

Acknowledgements

The authors acknowledge Dr Jorge Aarao for his work in preparing the Mathematical Methods for Engineers 1 case study and Dr Kerry Smith and Dr Jane Kehrwald for their help creating the Maths Mindset Quiz.

References

Ambrose, S.A., Bridges, M.W., DiPietro, M., Lovett, M.C. & Norman, M.K. (2010) *How Learning Works: Seven Research-Based Principles for Smart Teaching*. John Wiley & Sons, San Francisco, CA, USA.

Anderson, B., Lee, S.W., Simpson, M.G. & Stein, S.J. (2011) Study orchestrations in distance learning: Identifying dissonance and its implications for distance educators. *The International Review of Research in Open and Distributed Learning*, 12(5), 1–17.

Balamuralithara, B. & Woods, P. (2009) Virtual laboratories in engineering education: The simulation lab and remote lab. *Computer Applications in Engineering Education*, 17(1), 108–118.

Biggs, J. (1999) What the student does: Teaching for enhanced learning. *Higher Education Research & Development*, 18(1), 57–75.

Biggs, J. & Tang, C. (2011) *Teaching for Quality Learning at University*, 4th ed. McGraw-Hill Education, Maidenhead.

Biggs, J.B. (1989) Approaches to the enhancement of tertiary teaching. *Higher Education Research and Development*, 8(1), 7–25.

Bishop, J.L. & Verleger, M.A. (2013) The flipped classroom: A survey of the research. *Paper Presented at the ASEE National Conference Proceedings*, Atlanta, GA, USA

Blackmore, K., Compston, P., Kane, L., Quinn, D. & Cropley, D. (2010) *The Engineering Hubs and Spokes Project-institutional Cooperation in Educational Design and Delivery*. University of Queensland, Brisbane, Australia.

Boaler, J. (2013) Ability and mathematics: The mindset revolution that is reshaping education. *FORUM*, 55(1), 143–152.

Boaler, J. (2015) Fluency without fear: Research evidence on the best ways to learn math facts. *Reflections*, 40(2), 7–12.

Borko, H., Whitcomb, J. & Liston, D. (2009) Wicked problems and other thoughts on issues of technology and teacher learning. *Journal of Teacher Education*, 60(1), 3–7.

Bourne, J., Harris, D. & Mayadas, F. (2005) Online engineering education: Learning anywhere, anytime. *Journal of Engineering Education*, 94(1), 131–146.

Brown, T., Zoghi, M., Williams, B., Jaberzadeh, S., Roller, L., Palermo, C., McKenna, L., Wright, C., Baird, M. & Schneider-Kolsky, M. (2009) Are learning style preferences of health science students predictive of their attitudes towards e-learning? *Australasian Journal of Educational Technology*, 25(4), 524–543.

Chubb, I. (2013) *Science, Technology, Engineering and Mathematics in the National Interest: A Strategic Approach*. Australian Government, Canberra, Australia.

Coates, H. & McCormick, A.C. (2014) *Engaging University Students*. Springer, Singapore.

Dawson, S. (2015) Discussion paper: Developing UniSA's Digital learning strategy. Available from: www.unisa.edu.au/PageFiles/171878/Background%20paper%20-%20Developing%20UniSA's%20Digital%20Learning%20Strategy.pdf

Dekkers, A., Adams, N. & Elliott, S. (2011) Using technology to provide a supportive mathematical pathway into university. *8th Delta Conference on the Teaching and Learning of Undergraduate Mathematics and Statistics*, Nov 2011, Rotorua, New Zealand. pp. 382–388.

Duckworth, A.L., Peterson, C., Matthews, M.D. & Kelly, D.R. (2007) Grit: Perseverance and passion for long-term goals. *Journal of Personality and Social Psychology*, 92(6), 1087.

Duckworth, A.L. & Quinn, P.D. (2009) Development and validation of the short grit scale (Grit – S). *Journal of Personality Assessment*, 91(2), 166–174. doi:10.1080/00223890802634290

Dweck, C. (2012) *Mindset: How You Can Fulfil Your Potential*. Hachette, UK.

Ellis, R. & Goodyear, P. (2013) *Students' Experiences of E-learning in Higher Education: The Ecology of Sustainable Innovation*. Routledge, London, UK.

Ernst & Young (2012) *University of the Future: A Thousand Year Old Industry on the Cusp of Profound Change*. Available from: www.ey.com/Publication/vwLUAssets/University_of_the_future/$FILE/University_of_the_future_2012.pdf

Frender, G. (1990) *Learning to Learn: Strengthening Study Skills and Brain Power*. Incentive Publications, Nashville, TN, USA.

Goodyear, P. (2013) Instructional design environments: Methods and tools for the design of complex instructional systems. *Instructional Design: International Perspective*, 2, 83–111.

Hachey, A.C., Wladis, C. & Conway, K. (2015) Prior online course experience and G.P.A. as predictors of subsequent online STEM course outcomes. *The Internet and Higher Education*, 25, 11–17. doi:10.1016/j.iheduc.2014.10.003

Hamdan, N., McKnight, P., McKnight, K. & Arfstrom, K. M. (2013) *The Flipped Learning Model: A White Paper based on the Literature Review Titled*. A Review of Flipped Learning, Elsevier Science, Amsterdam.

Harris, D., Black, L., Hernandez-Martinez, P., Pepin, B., Williams, J. & with the TransMaths Team. (2014) Mathematics and its value for engineering students: What are the implications for teaching? *International Journal of Mathematical Education in Science and Technology*, 46(3), 321–336. doi:10.1080/0020739x.2014.979893

Hedberg, J. (2004) Designing multimedia: Seven discourses. *Cambridge Journal of Education*, 34(2), 241–256.

Henry, J. & Knight, C. (2003) Modern engineering laboratories at a distance. *International Journal of Engineering Education*, 19(3), 403–408.

Hernandez-Martinez, P., Williams, J., Black, L., Davis, P., Pampaka, M. & Wake, G. (2011) Students' views on their transition from school to college mathematics: Rethinking 'transition' as an issue of identity. *Research in Mathematics Education*, 13(2), 119–130. doi:10.1080/14794802.2011.585824

Herrington, J., Reeves, T.C. & Oliver, R. (2014) Authentic learning environments. In: *Handbook of Research on Educational Communications and Technology*. Springer, New York, NY, USA. pp. 401–412.

James, P., Quinn, D. & Dansie, B. (2011) Re-engineering for Australia's engineering skill shortage. *ASCILITE-Australian Society for Computers in Learning in Tertiary Education Annual Conference*, Dec 2011, Hobart, Tasmania.

Johnston, H., Aziz, S.M., Kaya, C.Y. & Quinn, D. (2008) Engaging students: Encouraging success. *Paper Presented at the ATN Assessment Conference Adelaide*, 20–21 Nov, South Australia, Australia.

Khoo, E., Scott, J., Peter, M. & Round, H. (2015) Evaluating flipped classrooms with respect to threshold concepts learning in undergraduate engineering. *Paper Presented at the Frontiers in Education Conference (FIE)*, Oct 2015, El Paso, TX, IEEE.

Kift, S. (2009) Articulating a transition pedagogy to scaffold and to enhance the first year student learning experience in Australian higher education. *Final Report for ALTC Senior Fellowship Program*, Australian Learning and Teaching Council Strawberry Hills, NSW, Australia.

Liljedahl, P. (2016) Building thinking classrooms: Conditions for problem-solving. In: *Posing and Solving Mathematical Problems*. Springer, New York, NY, USA. pp. 361–386.

Liljedahl, P., Santos-Trigo, M., Malaspina, U. & Bruder, R. (2016) Problem solving in mathematics education. In: *Problem Solving in Mathematics Education*. Springer, New York, NY, USA. pp. 1–39.

Marshall, E.M., Staddon, R.V., Wilson, D.A. & Mann, V.E. (2017) Addressing maths anxiety within the curriculum. *MSOR Connections*, 15(3), 28–35.

Mason, J., Burton, L. & Stacey, K. (2010) *Thinking Mathematically*. Pearson, Essex, England.

Mayadas, A.F., Bourne, J. & Bacsich, P. (2009) Online education today. *Science*, 323(5910), 85–89.

McLaughlin, J.E., White, P.J., Khanova, J. & Yuriev, E. (2016) Flipped classroom implementation: A case report of two higher education institutions in the United States and Australia. *Computers in the Schools*, 33(1), 24–37.

McQuiggan, C.A. (2007) The role of faculty development in online teaching's potential to question teaching beliefs and assumptions. *Online Journal of Distance Learning Administration*, 10(3). Available from: www.westga.edu/~distance/ojdla/fall103/mcquiggan103.htm

Merriam, S.B. (1998) *Qualitative Research and Case Study Applications in Education*, 2nd ed. Jossey-Bass Publishers, San Francisco, CA, USA.

O'Flaherty, J. & Phillips, C. (2015) The use of flipped classrooms in higher education: A scoping review. *The Internet and Higher Education*, 25, 85–95.

Pollock, E., Chandler, P. & Sweller, J. (2002) Assimilating complex information. *Learning and Instruction*, 12(1), 61–86.

Porter, W.W., Graham, C.R., Spring, K.A. & Welch, K.R. (2014) Blended learning in higher education: Institutional adoption and implementation. *Computers & Education*, 75, 185–195.

Quinn, D., Albrecht, A., Webby, B. & White, K. (2015) Learning from experience: The realities of developing mathematics courses for an online engineering programme. *International Journal of Mathematical Education in Science and Technology*, 46(7), 991–1003.

Quinn, D., Amer, Y., Lonie, A., Blackmore, K., Thompson, L. & Pettigrove, M. (2012) Leading change: Applying change management approaches to engage students in blended learning. *Australasian Journal of Educational Technology*, 28(1), 16–29.

Quinn, D. & Wedding, B. (2012) Responding to diversification: Preparing naïve learners for university study using time budgets. *Paper Presented at the ASCILITE-Australian Society for Computers in Learning in Tertiary Education Annual Conference*, 25–28 Nov 2012, Wellington, New Zealand.

Seaton, K.A., King, D.M. & Sandison, C.E. (2014) Flipping the maths tutorial: A tale of n departments. *Gazette of the Australian Mathematical Society*, 41(2), 99–113.

Strayer, J.F. (2012) How learning in an inverted classroom influences cooperation, innovation and task orientation. *Learning Environments Research*, 15(2), 171–193.

Thiel, T., Peterman, S. & Brown, M. (2008) Addressing the crisis in college mathematics: Designing courses for student success. *Change: The Magazine of Higher Learning*, 40(4), 44–49.

Tobias, S. (1994) Interest, prior knowledge, and learning. *Review of Educational Research*, 64(1), 37–54.

Trigwell, K. & Prosser, M. (1997) Towards an understanding of individual acts of teaching and learning. *Higher Education Research & Development*, 16(2), 241–252.

UniSA (2017) *Assessment Policy and Procedure Manual*. University of South Australia, University of South Australia. Available from: http://w3.unisa.edu.au/policies/manual/

Wiggins, G.P. & McTighe, J. (2005) *Understanding by Design*, 2nd ed. Pearson, NJ, USA.

Wingard, R.G. (2004) Classroom teaching changes in web-enhanced courses: A multi-institutional study. *Educause Quarterly*, 27(1), 26–35.

Wlodkowski, R.J. (2011) *Enhancing Adult Motivation to Learn: A Comprehensive Guide for Teaching All Adults*. John Wiley & Son, San Francisco, CA, USA.

Chapter 4

Impact of online quizzes on students' results in a blended learning system of an Engineering subject

D. Hagare[1] *and M.M. Rahman*[2]

[1]*School of Computing, Engineering and Mathematics, Western Sydney University, Sydney, Australia*

[2]*Department of Civil and Environmental Engineering, King Faisal University, Al-Ahsa, Saudi Arabia*

ABSTRACT: Due to rapid changes in the communication technologies, teaching and learning have changed dramatically over the last two decades. Nevertheless, the face-to-face (classroom-based) teaching remains one of the most predominant ways of imparting knowledge to students. On the other hand, different online activities are increasingly being used to supplement traditional face-to-face engagement. In this chapter, engagement strategies adopted while delivering Infrastructure Engineering (one of the civil engineering subjects) of the Bachelor of Engineering program offered at Western Sydney University (WSU) are presented. The subject includes mathematical calculations and completion of a design project using computer-aided design (CAD) software. In addition, students participate in an online in-class engagement tool and after-lecture weekly quizzes, tutorial sessions and a final exam. While some of the activities in the subject, such as in-class engagement tool and weekly quizzes, are online, tutorial sessions are delivered via face-to-face mode. It was found that the failure rate was significantly lower (17%–24%) for students who passed online quizzes compared to the failure rate (31%–63%) for students who failed in the quizzes. This indicates that online quizzes have led to deeper understanding of the lecture materials covered in the subject.

4.1 Introduction

Despite the increasing popularity of online delivery and learning of subjects and courses, face-to-face (classroom-based) teaching remains one of the most predominant ways of imparting knowledge to students. In the face-to-face teaching system, the main contact point between teacher and the students is the classroom, where students listen to lectures, receive lecture materials, take notes on lectures, and attempt class tests and final exams. On the other hand, online learning is defined as learning that takes place entirely over the internet (Bourne *et al.*, 2005; Revere & Kovach, 2011). A combination of face-to-face and online delivery systems is called the blended learning approach (BLA) and is expected to enhance the learning ability of students (Bhathal, 2016; Rahman, 2017). In this chapter, effectiveness of using online quizzes in an engineering subject of Western Sydney

University, namely, Infrastructure Engineering is assessed. The assessment is based on students' performance in different activities of the subject. The chapter is organized by first discussing the background of face-to-face and online teaching techniques, which is followed by a discussion on the blended learning approach. The usefulness of online quizzes as a learning tool, compared to other conventional face-to-face systems, including tutorial, group-based major project and final exam, has been assessed through the discussion of a case study.

4.2 Face-to-face *vs.* online teaching approach

There is a considerable debate on which of the two methods of learning, face-to-face or online, provides maximum benefits to students. According to Budin (1991), teaching in face-to-face classrooms has evolved over the centuries to include a common set of skills and competencies agreed upon by most of the disciplines. A face-to-face classroom teacher possesses academic excellence, lifelong learning and personal commitment and spends considerable time both inside and outside the classroom to produce successful learning outcomes (Kerr, 1989; Tomei, 2006). The quality of learning environment in a face-to-face system is preferred by some students over an online system (Johnson *et al.*, 2000). However, the face-to-face teaching environment has been criticized by some researchers for encouraging passive learning, ignoring individual differences and needs of the learners and not giving proper attention to problem solving, critical thinking and other higher-order thinking skills (Hannum and Briggs, 1982).

On the other hand, in an online learning environment, students interact with their instructor through communication media. The instructor generally sends emails to their students within the first weeks of a semester regarding the course expectations, learning assignments or lesson objectives. Students and the teacher share their ideas using online chatrooms, which provide a near-real-time learning environment (Tomei, 2006). However, some researchers distrust that online instructions actually solve difficult teaching and learning problems and raise concerns about some barriers (i.e. changing nature of technology, complexity of networked system and the lack of stability in online learning environment) to the implementation of such a system (Brandt, 1996; Conlon, 1997).

In a comparative study conducted by Johnson *et al.* (2000), it was shown that in the face-to-face system, students have more chances to enhance social relationship among each other, as they can easily get together at least once a week (depending on course scheduling) for an extended period of time to discuss class projects and work out differences in opinion. This type of social dimension is not present in an online environment. The same authors also found differences in the way students get feedback from the instructor in the two different systems. In the face-to-face system, students get live and dynamic forms of support from the instructor, while the online students receive feedback in a form of one-way static communication mainly through emails, uploaded files and periodic telephone conversations. In terms of departmental support, students in the face-to-face system need less support as the students already have direct contact with the instructor and sometimes a part-time teaching assistant. In contrast, due to the complexities associated with online technologies, students need technical support provided by the department (Johnson *et al.*, 2000). Further, in the case of complete online delivery, it may be difficult to ensure integrity of the assessment for a given subject or course.

4.3 Blended learning approach

A blended learning approach could be a good compromise between face-to-face only or online only learning systems. It can be structured in such a way that it can include advantages of the both approaches and, at the same time, overcome the limitations of both the systems. According to Garrison and Kanuka (2004), blended learning is the thoughtful integration of classroom face-to-face learning experiences with online learning experiences; this is, not just the addition of one system to another, but an effective integration in terms of fundamental reconceptualization and reorganization of face-to-face and online systems. Garrison and Kanuka (2004) also emphasized that BLA is a multiple form of communication to meet specific learning requirements. In BLA, various learning materials are made available to the students, including online availability of recorded lectures and tutorials, hand-written tutorial solutions, discussion board and online practice quizzes (Rahman, 2017; Rahman and Al-Amin, 2014). Typically, in BLA, lectures, tutorials and final exams are classroom-centered activities; however, quizzes are online, which quite often form a small percentage of overall assessment. This ensures the overall integrity of the assessment in BLA. Depending on the purpose, the BLA can be categorized as below (Graham, 2006).

1 Enabling Blends: blends that are intended to focus on addressing issues related to access online technology and convenience to the learners;
2 Enhancing Blends: blends that are intended to allow some changes in the way of teaching, such as inclusion of some online supplementary materials in a traditional face-to-face system;
3 Transforming Blends: blends that are intended for a radical transformation of the way of teaching, such as a change from a model where learners are just receivers of information to a model where learners actively construct knowledge through dynamic interactions.

Many education researchers try to implement the transforming blend type of BLA in their curriculum (Halverson *et al.*, 2012), because they believe BLA should be transformative rather than just enhancing or enabling.

4.3.1 Enhanced in-class engagement

Student activities in the class are core to student's learning (Biggs, 1999), although it is difficult to identify the types of activity that will keep the students fully engaged in the class and thus help in learning. Close associations exist between conception of learning through discussions with approaches to face-to-face and online activities and learning outcomes (Ellis *et al.*, 2008). According to Smith et al. (2005), it would be interesting to know if there is any impact of teacher efficiency in delivering the lecture, underlying complexity of the course content, group-based activities, and connection of curriculum with the industry-related problem on student engagement in the class. However, Smith *et al.* (2005) observed that breaking up lectures with short discussion times may have positive impacts on re-engaging the students to the lectures. This type of breaking up of lectures can be achieved by implementing various online activities. These activities were implemented by many educators (Carroll et al., 2014; Dabbour, 2016; Donohue, 2014; Schultz, 2013) in many engineering as well as non-engineering courses. One such online tool is GoSoapBox™, which is discussed later in this chapter.

4.3.2 Online quizzes

Online quizzes that are usually conducted on a weekly basis were found by some researchers to be an effective method for learning in mathematics and physics for engineering students (Bhathal, 2016; Blanco *et al.*, 2009; Lim *et al.*, 2012; Martins, 2016; Siew, 2003). One of the advantages of the online quizzes is that they can save precious class time; they also allow a large number of students to be involved in the learning activities with little effort from the teacher as far as providing feedback is concerned (Broughton *et al.*, 2013; Martins, 2016). Generally, online quizzes are an out-of-class activity (Butler and Zerr, 2005; Zerr, 2007), where students get enough time to read and understand the lecture before answering questions. However, an in-class online quiz as a learning as well as engagement tool is investigated by Arteaga and Vinken (2013) for a mechanical engineering course. According to Arteaga and Vinken (2013), students got the opportunity to discuss with other students before answering the quiz questions. It is claimed by the authors that in-class discussion as well as the feedback from online quizzes helped the students to identify gaps in their knowledge.

Dobson (2008) introduced online quizzes into an undergraduate exercise physiology course to investigate if the online quizzes had any impact on the final exam performance. Three different groups were investigated in which the first group completed the original version of the course, the second group completed the course with more vigorous exam questions, and the third group completed the course similar to the second group in addition to 10 online quizzes. It was observed that the third group scored higher marks in the final exam than did the other two groups, which is because of the introduction of online quizzes in the course. Similar investigation was conducted by DeSouza and Fleming (2003), in which the impact of online quizzes on four in-class exams was evaluated. The authors compared in-class exam results of 297 undergraduate students who took online quizzes with 291 undergraduates who took traditional paper and pencil quizzes. The online quizzes were generated by an assessment program called Mallard™, which allows students to take customized quizzes on a secure server at their convenience. Results suggested that the students who were in the online quizzes group performed better in the exams. Salas-Morera *et al.* (2012) reported positive impacts of online quizzes on the students' performance when used as an assessment tool. The investigation was carried out during a period of 5 years for a course in which course components were group tutoring, discussion forums, collaborative learning, peer review and online quizzes. Tselios *et al.* (2001) suggested that the software platform used to deliver online quizzes may affect test performance. The authors reported that students performed significantly better in completing online quizzes in WebCT compared to another platform called IDLE. The content in the two platforms were the same; however, the user interface was different.

Finally, Hillman (2012) recommended some steps to make online quizzes more effective, which include (i) the students should be allowed to attend a quiz more than once and get immediate feedback on the grade; (ii) a generous time limit (e.g. 60 minutes) may encourage students to read the material ahead of time before taking the quiz; (iii) the order of quiz questions as well as their multiple choice answers should be randomized to avoid student cheating and (iv) some online quiz questions can be included in the final exam; this will encourage students to engage in long-term learning.

The main drawback of online quizzes can be said to be the difficulties associated in ensuring the integrity of the quizzes. The main issue is to what extent one can attribute the answers received for each of the questions to the individual student. There are always some possibilities for cheating. However, as long as the weighting for the online quizzes is kept at lower levels, the temptations for cheating may be reduced.

4.4 Case study on Infrastructure Engineering subject

The Infrastructure Engineering subject, delivered as part of Bachelor of Engineering course at Western Sydney University, provides students with material to assist them with transportation engineering, civil engineering, urban development and town planning projects. The subject mainly focuses on the planning, design and construction of transportation facilities using a case of subdivision (urban center) development. As per the learning outcomes of the subject, upon successfully completing this subject, the students will be able to (i) apply principles involved in the design, construction and maintenance of both small and large transportation networks comprising both roadways and railway tracks; (ii) analyze and design transportation hubs and intersections for allowing efficient traffic flow; (iii) analyze sustainable transport systems and facilities for both rural and urban areas; (iii) apply available design tools and guidelines for transportation network design and (iv) create and contribute to productive and efficient teams for designing and evaluating efficient transportation systems.

The teaching and learning activities of the subject are shown in Figure 4.1 and discussed in subsequent sections. The delivery of this subject is structured on the principles of "problem-based learning." The assessment mainly included four components as shown in Table 4.1. Students need to achieve an overall mark of 50% to pass the subject. As discussed earlier, to maintain the integrity of the assessment, relatively low weighting of 5% is assigned to online quizzes. Also, relatively low weighting of 5% is assigned for tutorial participation as well. Tutorials are particularly useful, as students are encouraged to discuss among themselves while solving tutorial questions.

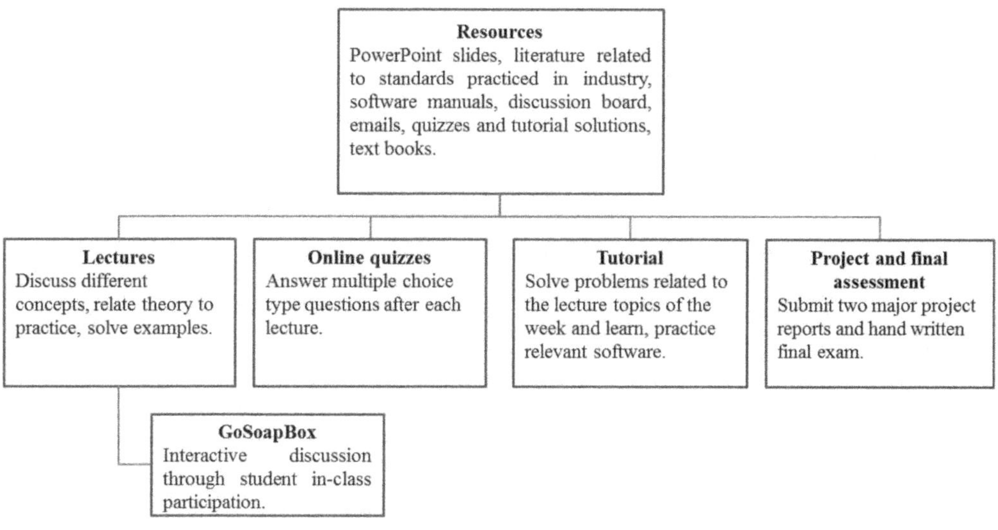

Figure 4.1 Learning and assessment components of Infrastructure Engineering subject (Hagare and Rahman, 2016)

Table 4.1 Assessment components and their weightings in Infrastructure Engineering subject

Assessment component	*Weighting*
Online quiz participation	5%
Tutorial participation	5%
Major project (group of four or less)	30%
Final exam	60%
Total	100%

This chapter presents extended analysis of data in relation to the one carried out in an earlier study (Hagare and Rahman, 2016). In Hagare and Rahman (2016), only 2015 data was analyzed. On the other hand, in this chapter, both 2014 and 2015 data are analyzed.

4.4.1 Assessment components

Lecture

In this subject, the lectures are designed to introduce students to the concepts relevant to infrastructure and to brief students on assessment requirements. Lectures are mainly delivered using PowerPoint slides and document camera. While the concepts are explained using PowerPoint slides, step-by-step solutions to examples are shown using a document camera. To check whether the students understood the concepts, an engagement tool, namely GoSoapBox™, was used.

GoSoapBox™

In the Infrastructure Engineering subject, three to four questions are asked using GoSoapBox™ during the lecture time, depending on the content of the lecture. The questions are generally simple and straightforward. These questions are mainly intended to test the understanding of the concepts introduced in the lecture by the students. When asked, the students log in to GoSoapBox™ with an access code given by the lecturer and complete the activity. Through using this tool, students are able to convey whether they have understood the concepts introduced in the lecture. It is also an instant feedback mechanism for the lecturer on the students' understanding of the lecture materials. Based on the feedback received, the lecturer decides whether to explain the concept further or move on to the next topic. There is no mark allocated for participating in GoSoapBox™ question and answer sessions.

Tutorials

In tutorial classes, the students perform activities to reinforce the concepts introduced in lectures. Tutorials are held in tutorial rooms or computer labs where students get chance to work together on their projects and to learn the relevant software. Students are encouraged to participate in tutorial classes by allocating 5% of the total marks for attendance and participation (Table 4.1). Each student must answer all the tutorial questions and/or follow the steps to use the software. The tutor, if required, assists individual students in solving the tutorial questions or the use of particular software. Each student is expected to work in collaboration with other students and, at the end of the tutorial class, to submit the answers to all the questions or show the output from the software. Based on the answers provided by the student, a mark is assigned by the tutor. This mark varies between 0 and 100%.

Online quizzes

Online quizzes are in addition to GoSoapBox™ quiz questions, where students are asked to answer five to ten questions. These questions aim at testing the understanding of underlying concepts of the topic discussed in the lecture. The students need to attempt all 13 online quizzes based on the weekly lectures. The students are allocated 6 days to answer the questions in an online quiz. Students must take these quizzes in order to gain the participation marks for this assessment component. The overall mark for this assessment component depends on the performance of each of the students in each of the online quizzes. For example, if a student receives 100% in each of the quizzes over all 13 quizzes, the student will be credited with five marks towards his/her final mark as per the weighting shown in Table 4.1. The students are given feedback on their answers in the following week of each lecture.

MAJOR PROJECT AND FINAL EXAM

Students need to submit two reports on a major project, which carries a weighting of 30%, and attend the final exam. The final exam carries 60% of total marks, as shown in Table 4.1.

4.4.2 Assessment strategy

Data related to students' performance under different activities of the subject were extracted from results of Autumn 2015 and 2014 sessions (March–June). Regression analysis was performed to find the effect of students' performance in online quizzes on the final marks. Similar analyses were carried out to find the effect of tutorials, major projects and final exams on the final marks. For this purpose, linear regression technique was used, which attempts to model the relationship between dependent and independent variables by fitting a linear equation to the observed data. In this study, the relationship between the dependent variable and the independent variables are assumed to be linear. The following represents a multiple linear regression equation (Montgomery *et al.*, 2012):

$$Y = \alpha + \beta_1 X_1 + \beta_2 X_2 + \cdots + \beta_k X_k \tag{4.1}$$

where, α is the model intercept, β (1, 2, 3, . . ., k) are the slope coefficients and k is the number of independent variables. Statistical analyses were carried out using Minitab™ statistical software.

4.4.3 Assessment outcome

In the Infrastructure Engineering subject, the marking criteria were set as fail or unsatisfactory (F) when student received 0–49%, pass (P) for 50%–64%, credit (C) for 65%–74%, distinction (D) for 75%–84% and high distinction (HD) for 85%–100% of the total marks. Distribution of students' marks in individual sections of the subject is shown in Figure 4.2 for years 2015 and 2014. It can be seen from Figure 4.2 that during the Autumn 2015 session, students performed well in tutorial and in major projects compared to online quizzes

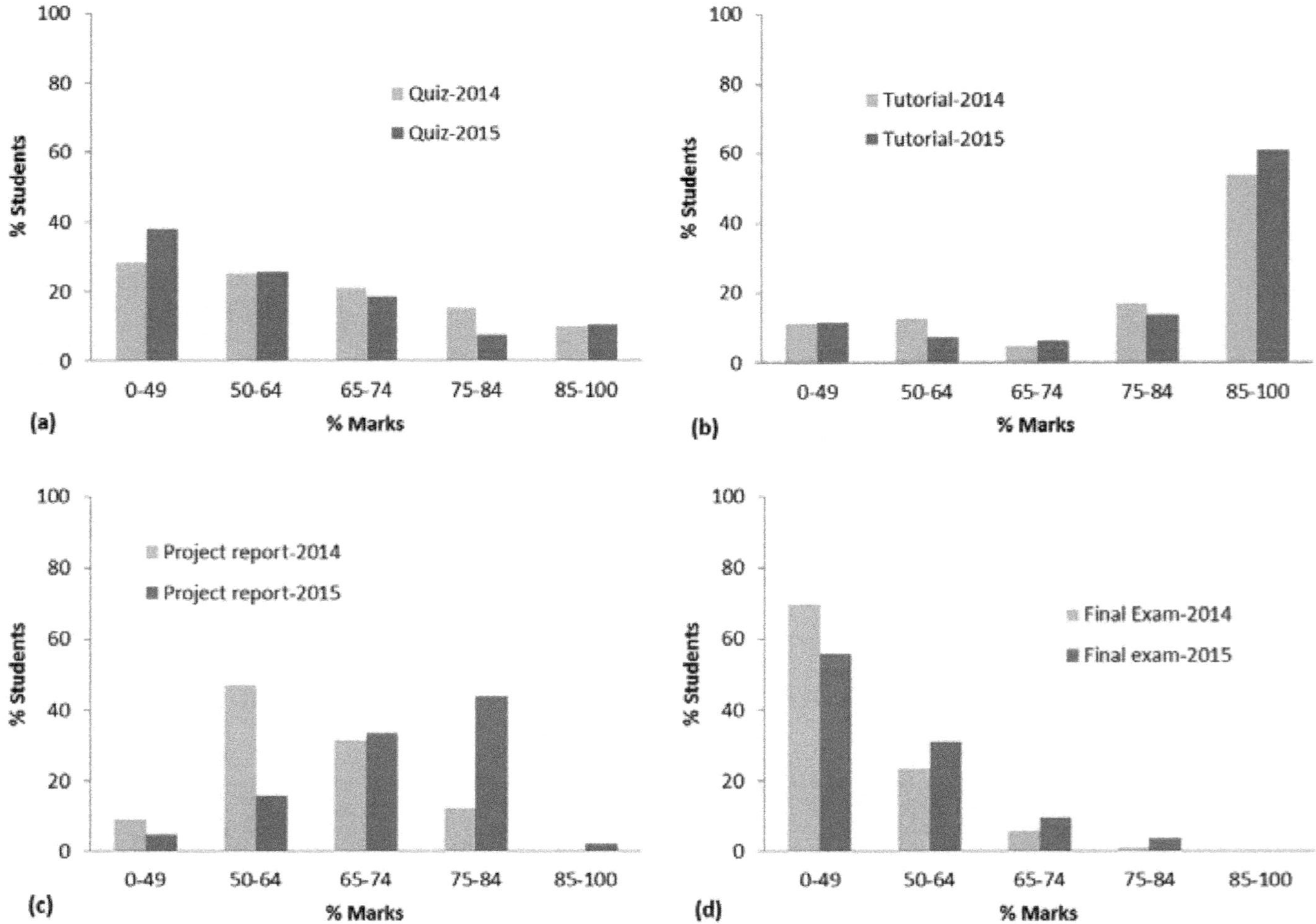

Figure 4.2 Histogram of students' score in (a) quiz, (b) tutorial, (c) major project and (d) final exam in Autumn 2014 (n = 123) and Autumn 2015 (n = 145) sessions

and final exams. In online quizzes, 38% of students failed, which is 56% for the final exam. On the other hand, in tutorials and major projects, only 12% and 7% students failed, respectively. Overall, 22% of the students failed in this subject during Autumn 2015. Similar results were observed during Autumn 2014 session. In terms of distribution of marks, although the magnitude may be different, there is a similarity between the online quiz and the final exam marks. On the contrary, the distribution of marks for tutorial participation and project reports are significantly different from that of final exam. Thus, it can be said that online quizzes have been helpful for student learning, as they reflect the understanding of the subject material by an individual student.

Investigation was carried out to see the impact of the students' performance in each assessment component on the final overall marks (Fig. 4.3). As shown in Figure 4.3, in Autumn 2015, both quizzes ($R^2 = 0.25$) and tutorial marks ($R^2 = 0.13$) are positively correlated with the final score; however, quizzes show higher correlation compared to tutorial. In both cases, the predictor variable was significant as the p-value was less than 0.001. The low p-value suggests the predictor as a meaningful entity in the regression equations. As can be seen in the figure, online quizzes have significantly better correlation than do tutorial marks. In fact, they have as good a correlation as that of project marks, which have 30% weighting

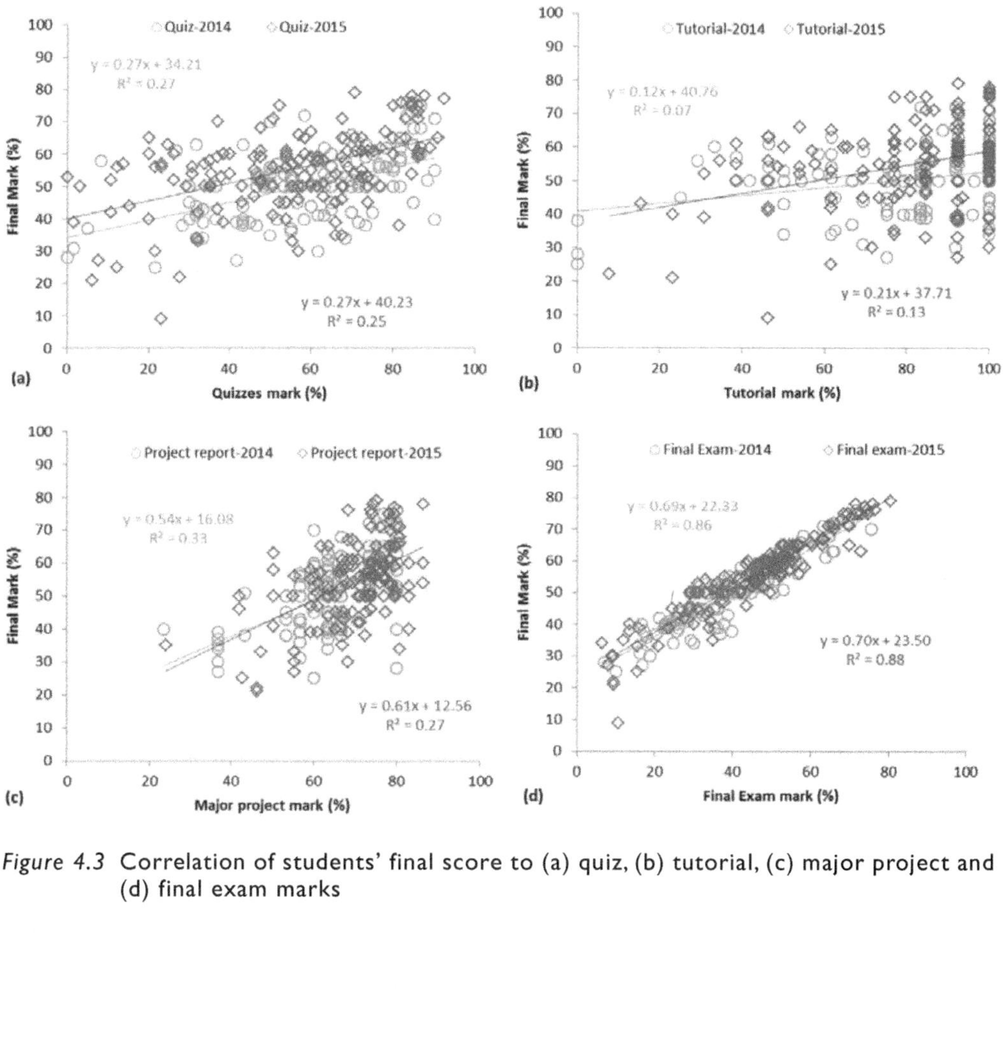

Figure 4.3 Correlation of students' final score to (a) quiz, (b) tutorial, (c) major project and (d) final exam marks

(weighting for online quizzes is only 5%). Given that the final exam has the highest correlation with the final mark of 0.88 (Fig. 4.3(d)), it can be said that the correlations of online quizzes and project marks are closer to final exam correlation than the correlation of tutorial mark to the final exam. These results again indicate that the online quizzes have contributed to students' understanding.

Weak correlation between the tutorial mark and the final mark may be partly attributed to several reasons. Some possible reasons for this include:

- Often the tutorials are very prescriptive, and students fail to pay attention;
- Not much opportunity for students to think;
- Blindly follow the tutor in solving the design problem and
- Marking may not be rigorous.

In relation to the marking of tutorials, the participation mark was given to students as long as they demonstrated participation in the tutorial activities by showing the completed solution to the particular question. While marking, the tutor may not check whether the student has understood the solution method and the basic concepts. This is mainly due to the time limitation that the tutor will have for marking the tutorial papers.

Besides analyzing correlation coefficients, more in-depth analysis was conducted to find the impact of students' performance in quizzes on the final grade (Fig. 4.4). As seen in this figure, the distribution of grades for those who failed online quizzes and tutorials was compared with that of those who passed them. It is evident from the figure that those who failed in both online quizzes and tutorials, mostly either failed or passed the unit. On the other

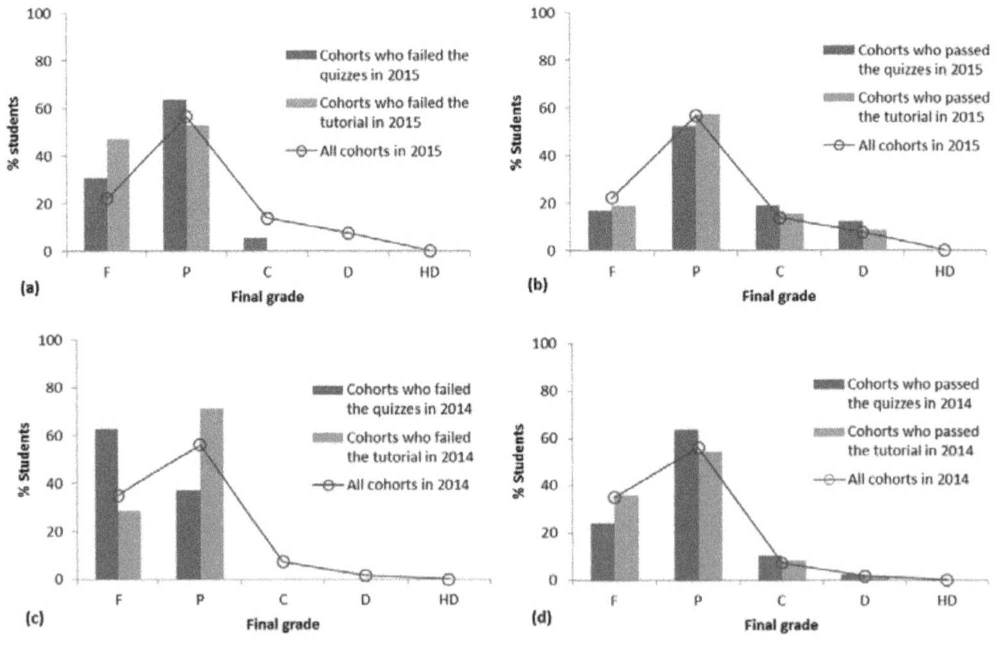

Figure 4.4 Impact of students' performance in online quizzes and tutorial on the final grade of the subject in Autumn 2015 (a, b) and in Autumn 2014 (c, d)

hand, the distribution of grades for those who passed both quizzes and tutorials has more or less similar distribution as that of the final grade. However, a closer look at the distribution of students who passed both online quizzes and tutorials (Figs. 4.4(b) and (d)) reveal that the failure rate is about 17%–24% for online quizzes compared to 19%–36% for tutorials. This again indicates slightly better student learning for those who passed online quizzes.

Overall, the above results highlight the effectiveness of online quizzes for student learning. Similar observations were made by Aziz (2003) in an engineering subject (*Computer Systems*), in which students found online quizzes to be an interesting way of learning. According to Aziz (2003), 70% of the students found the online quizzes useful for testing their knowledge and for focusing on important topics in the course; 58% of the students found the quizzes a more interesting way of learning rather than just attending lectures and tutorials. The data presented in Figures 4.2–4.4 appear to reinforce the above findings of Aziz (2003). The students may find the online quiz an interesting way to learn the subject because of its flexibility. The online quizzes can be done anytime and anywhere. The system provides opportunities to undertake the activity with greater flexibility by students who live long distances from the campus, have children or lead busy lives. Many of the students in Infrastructure Engineering work part-time and took advantage of accessing the quizzes in their preferred time over the week. It is possible that some of the students may have collaborated with their peers before answering the quiz questions. Similar observation was made by Martins (2016) regarding online quizzes. However, in the current study, attempts were made to minimize the possibility of collaboration between peers by presenting the quiz questions and possible solutions to the questions in random order. Also, in some cases the questions were presented from a bank of questions, which minimized the possibility of presenting the same question between two students.

In this case study, although it was found that online quizzes may lead to better student learning than tutorials, tutorial classes are an integral part of providing a good learning experience for engineering students. However, to be effective, the tutorial classes need to be more engaging. To provide more engaging tutorial classes, it is necessary to develop more hands-on sessions, such as those using interactive multimedia software (Ghasem, 2016; Hagare *et al*., 2000). Hagare *et al*. (2000) implemented a multimedia approach for teaching fundamental concepts in environmental engineering. In the multimedia approach, initially the lecturer introduces the concepts and then the students can use a multimedia software package to reinforce the concepts. According to Hagare *et al*. (2000), the students found the animations and simulations included in the package very useful for understanding the underlying engineering concepts. However, students must be willing to access and practice such interactive multimedia enhanced software in their own time. This will reduce the cost of course delivery as well as yield better learning outcome for the students.

5 Conclusion

This study investigated the role of online quizzes as tools for students' deeper understanding of an engineering subject (Infrastructure Engineering) in Western Sydney University and compared with other face-to-face learning activities. Online quizzes showed relatively higher correlation between the mark obtained in the quiz and the final exam mark. In addition, the distribution of marks for online quizzes was found to be similar to the overall distribution of marks for the subject. On the other hand, the distributions of marks for tutorial and project report were found to be quite different from the overall distribution of marks. These observations indicate that online quizzes have led to deeper understanding of the lecture materials covered in the subject investigated as part of this study. These results, however, need to be verified by further investigation and student evaluation. Also, an online in-class engagement tool (e.g. GoSoapBox™) practiced in the subject may have impacted on the student learning, which was not investigated as part of this study and it needs further investigation.

Acknowledgment

The authors acknowledge School of Computing Engineering and Mathematics (SCEM) and Centre for Infrastructure Engineering (CIE), Western Sydney University, Australia, for providing support in conducting this research. We also express our sincere thanks to the reviewers of this chapter for their helpful suggestions and edits.

References

Arteaga, I. L. & Vinken, E. (2013) Example of good practice of a learning environment with a classroom response system in a mechanical engineering bachelor course. *European Journal of Engineering Education*, 38(6), 652–660.

Aziz, M. (2003) Online quizzes for enhancing student learning in a first year engineering course. *Proceedings of International Conference on Engineering Education, ICEE 2003*, 21–25 Jul, 2003, Valencia, Spain.

Bhathal, R. (2016) An appraisal of an online tutorial system for the teaching and learning of engineering physics in conjunction with contextual physics and mathematics, and relevant mathematics. *European Journal of Engineering Education*, 41(5), 504–511.

Biggs, J.B. (1999) *Teaching for Quality Learning at University: What the Student does*. Open University Press, Buckingham and McGraw-Hill Education, UK.

Blanco, M., Estela, M.R., Ginovart, M. & Saa, J. (2009) Computer assisted assessment through Moodle quizzes for calculus in an engineering undergraduate course. *Scienze Matematiche*, 9, 78–84.

Bourne, J., Harris, D. & Mayadas, F. (2005) Online engineering education: Learning anywhere, anytime. *Journal of Engineering Education*, 94(1), 131–146.

Brandt, D.S. (1996) Teaching the net: Innovative techniques in Internet training. *Proceedings of the 11th Annual Computers in Libraries Conference*, Washington, DC, 27 Feb 1996. Available from: https://files.eric.ed.gov/fulltext/ED412975.pdf

Broughton, S., Robinson, C. & Hernandez-Martinez, P. (2013) Lecturers' perspectives on the use of a mathematics-based computer-aided assessment system. *Teaching Mathematics and its Application*, 23, 1–14.

Budin, H.R. (1991) Technology and the teacher's role. *Computers in the Schools*, 8(1–3), 15–26.

Butler, M.B. & Zerr, R.J. (2005) The use of online homework systems to enhance out-of-class student engagement. *International Journal for Technology in Mathematics Education*, 12(2), 51–58.

Carroll, J., Rodgers, J., Sankupellay, M., Newcomb, M. & Cook, R. (2014) Systematic evaluation of GoSoapBox in tertiary education: A student response system for improving learning experiences and outcomes. In: Chova, L.G., Martínez, A.L. & Torres, I.C. (eds) *Proceedings of 8th International Technology, Education and Development Conference, INTED2014*, 10–12 Mar 2014, Valencia, Spain.

Conlon, T. (1997) The Internet is not a panacea. *Scottish Educational Review*, 29(1), 30–38.

Dabbour, E. (2016) Assessing the effects of implementing an online student-response system in a transportation engineering course. *Journal of Professional Issues in Engineering Education and Practice*, 143(1). doi:10.1061/(ASCE)EI.1943–5541.0000293

Desouza, E. & Fleming, M. (2003) A comparison of in-class and online quizzes on student exam performance. *Journal of Computing in Higher Education*, 14(2), 121–134.

Dobson, J.L. (2008) The use of formative online quizzes to enhance class preparation and scores on summative exams. *Advances in Physiology Education*, 32(4), 297–302.

Donohue, S. (2014) Supporting active learning in an undergraduate geotechnical engineering course using group-based audience response systems quizzes. *European Journal of Engineering Education*, 39(1), 45–54.

Ellis, R.A., Goodyear, P., Calvo, R.A. & Prosser, M. (2008) Engineering students' conceptions of and approaches to learning through discussions in face-to-face and online contexts. *Learning and Instruction*, 18(3), 267–282.

Garrison, D.R. & Kanuka, H. (2004) Blended learning: Uncovering its transformative potential in higher education. *The Internet and Higher Education*, 7(2), 95–105.

Ghasem, N. (2016) Enhanced teaching and student learning through a simulator-based course in chemical unit operations design. *European Journal of Engineering Education*, 41(4), 455–467.

Graham, C.R. (2006) Blended learning systems: Definition, current trends, and future directions. In: Bonk, C.J. & Graham, C.R. (eds) *Handbook of Blended Learning: Global Perspectives, Local Designs*. Pfeiffer Publishing, San Francisco, CA, USA.

Hagare, D. & Rahman, M.M. (2016) Online quizzes to increase student learning in an engineering unit. In: Rahman, A. & Illic, V. (eds) *Proceedings of International Conference on Engineering Education & Research, iCEER2016*, 21–24 Nov, Western Sydney University, Parramatta campus, Sydney, Australia.

Hagare, D., Corderoy, R.M. & Hagare, P. (2000) Developing an interactive multimedia software package to enhance understanding of and learning outcomes in water treatment processes. *Journal of Cleaner Production*, 8(5), 407–411.

Halverson, L.R., Graham, C.R., Spring, K.J. & Drysdale, J.S. (2012) An analysis of high impact scholarship and publication trends in blended learning. *Distance Education*, 33(3), 381–413.

Hannum, W.H. & Briggs, L.J. (1982) How does instructional systems design differ from traditional instruction? *Educational Technology*, 22(1), 9–14.

Hillman, J. (2012) The impact of online quizzes on student engagement and learning. Available from: www.bk. psu.edu/Documents/IT/Hillman_TLI_report.pdf [Accessed 20 February 2018].

Johnson, S.D., Aragon, S.R., Shaik, N. & Palma-Rivas, N. (2000) Comparative analysis of learner satisfaction and learning outcomes in online and face-to-face learning environments. *Journal of Interactive Learning Research*, 11(1), 29–49.

Kerr, S.T. (1989) Technology, teachers, and the search for school reform. *Educational Technology Research and Development*, 37(4), 5–17.

Lim, L.L., Thiel, D.V. & Searles, D.J. (2012) Fine tuning the teaching methods used for second year university mathematics. *International Journal of Mathematical Education in Science and Technology*, 43(1), 1–9.

Martins, S.G. (2016) Weekly online quizzes to a mathematics course for engineering students. *Teaching Mathematics and its Applications*, 36(1), 56–63. doi:10.1093/teamat/hrw011.

Montgomery, D.C., Peck, E.A. & Vining, G.G. (2012) *Introduction to Linear Regression Analysis*, 5th ed. John Wiley & Sons, New Jersey, USA.

Rahman, A. & Al-Amin, M. (2014) Teaching of fluid mechanics in engineering course: A student-centered blended learning approach, In: Alam, F. (ed) *Using Technology Tools to Innovate Assessment, Reporting, and Teaching Practices in Engineering Education*. IGI Global, Hershey, USA, pp. 12–20.

Rahman, A. (2017) A blended learning approach to teach fluid mechanics in engineering. *European Journal of Engineering Education*, 42(3), 252–259.

Revere, L. & Kovach, J.V. (2011) Online technologies for engaged learning-A meaningful synthesis for educators. *The Quarterly Review of Distance Education*, 12(2), 113–124.

Salas-Morera, L., Arauzo-Azofra, A. & García-Hernández, L. (2012) Analysis of online quizzes as a teaching and assessment tool. *Journal of Technology and Science Education*, 2(1), 39–45.

Schultz, M. (2013) Use of an on-line student response system: An analysis of adoption and continuance. *International Journal of Innovation in Science and Mathematics Education*, 21(5), 12–26.

Siew, P.F. (2003) Flexible on-line assessment and feedback for teaching linear algebra. *International Journal of Mathematical Education in Science and Technology*, 34(1), 43–51.

Smith, K.A., Sheppard, S.D., Johnson, D.W. & Johnson, R.T. (2005) Pedagogies of engagement: Classroom-based practices. *Journal of Engineering Education*, 94(1), 87–101.

Tomei, L.A. (2006) The impact of online teaching on faculty load: Computing the ideal class size for online courses. *Journal of Technology and Teacher Education*, 14(3), 531–541.

Tselios, N.K., Avouris, N.M., Dimitracopoulou, A. & Daskalaki, S. (2001) Evaluation of distance-learning environments: Impact of usability on student performance. *International Journal of Educational Telecommunications*, 7(4), 355–378.

Zerr, R. (2007) A quantitative and qualitative analysis of the effectiveness of online homework in first-semester calculus. *The Journal of Computers in Mathematics and Science Teaching*, 26(1), 55–73.

Chapter 5

A connected e-learning framework for engineering education

E. Edilson Arenas

School of Engineering and Technology, CQUniversity, Australia

ABSTRACT: The adoption of technologies to facilitate learning and teaching has become widely accepted in higher education. At the same time, recent developments in emerging technologies have created a window of opportunity to offer affordable, high-quality education for everyone in ways never seen before and considered as disruptive to traditional education. Universities are under pressure to rethink the way they leverage technology to structure their academic programs in accordance with contemporary students' needs, personal interests, social connections and industry demands. This chapter discusses these disruptive technologies, the action possibilities of these technologies, and their implementations in an evidence-based e-learning framework that responds to the challenges and new realities surrounding higher education.

5.1 Introduction

The adoption of technologies to facilitate learning and teaching has become widely accepted in higher education. For example, today most tertiary institutions use a learning management system (LMS) as the core learning environment to support learners and teacher activities (Brown *et al.*, 2015). Emerging technologies are now being used to explore transformative and innovative ways of creating, compiling and disseminating knowledge. These developments in technology have also created window opportunities to expand informal educational programs and explore alternative ways to achieve society's educational goals (Arenas and Barr, 2013). New learning and teaching paradigms have emerged as a result, i.e. massive open online courses (MOOCs), Khan Academy, edX, Coursera and open educational resources that in conjunction with micro-learning, micro-credentials, and continuing professional development training are increasingly gaining acceptance in the industry and business sectors as genuine ways of developing skills. These alternatives to formal education are having profound implications for the normal operation of higher education, which is further exacerbated by the rapidly changing needs of contemporary learners. From the learning point of view, one of the most significant challenges for higher education is to offer affordable, high-quality education to people from diverse backgrounds whose needs vary according to geographical location, cultural background and socioeconomic circumstances and who may have varying levels of computer or digital literacy.

Engineering faculties strive to create programs that are consistent with individual learning interests, industry demands and employability; however, in such a volatile climate surrounding higher education, the task is becoming increasingly more challenging. From the institutional perspective, take for example a 4-year engineering degree compared to programs that can be completed in less time. Further, engineering faculties also invest huge amounts of money to build complex and expensive lab spaces that require lots of maintenance and continuous modernization. Contributing to this climate of uneasiness is the rapidly changing nature of the industry and knowledge economies of developed nations. These factors, in conjunction with the rapidly growing freelancing market, has created a sense of uncertainty on the employability (Bennett *et al.*, 2015) of future engineering graduates. There is a need for faculty to face these new challenges in a world of globalization, emergent technologies and disruptive education. Engineering faculties cannot afford to continue their operations guided by underlying principles from the past.

In response to these challenges, this chapter introduces a connected e-learning framework that leverages technology for transformative and meaningful learning and teaching, in accordance with the current and future generation of engineering learners. The framework uses evidence-based practices and leverages recent developments in online and mobile technology and the new realities of learning and teaching to offer high-quality engineering education under the current unfavorable circumstances.

The chapter is structured as follows. The first section sets the scene by discussing the new realities of engineering education and the need for institutional change. The next section elaborates on the affordances of blended learning, along with its transformative and meaningful impact on learning and teaching. The blended learning section is used as a preamble to discuss the connected learning framework principles and empirical evidence to demonstrate its learning values. The chapter ends with the description of a practical implementation of the framework.

5.2 The new realities of engineering education

Engineering is a complex discipline often described by academics as the mix of three things: applied science and mathematics, problem-solving, and making things (Pawley, 2009). In terms of engineering education, academics make great efforts to include those three things in their designs of learning outcomes, learning activities and assessment. The primary goal is to design an appropriate educational environment to prepare engineering students for research, make them ready for employment, prepare them to be skilled citizens for society, and provide them an intellectually stimulating education (Goodhew, 2010). Over the years, engineering faculties have successfully followed these principles to form quality engineers in accordance with the needs of society and industry demands. However, today engineering faculties are facing a disruption in the normal course of their educational activities, particularly in attracting young prospective students to embrace engineering as their preferred degree of study. Some factors contributing to this disruption are the uncertainty and complexity of the higher education landscape, the perceptions of new generation of engineering students and students' learning needs for employability readiness. These factors are discussed next.

5.2.1 The landscape of uncertainty and complexity in higher education

One of the most challenging realities affecting higher education today is the competition from new models of education facilitated by recent developments in online and mobile emergent technologies. MOOCs, Khan Academy, edX, Coursera, Udemy, OpenCourseWare and open educational resources (OERs) are examples of these new forms of education. In this regard, educational practitioners (Jordan, 2015) are expressing concern about these paradigms because of the demonstrated low levels of completion rates, the lack of student guidance, the need to be a self-regulated and independent learner, amongst others. However, when complemented with micro-credentials, competency-based education and task registries (micro-learning and learning on the job), these modes of education can be seen as disruptive to formal higher education. This is problematic, because some industries are starting to accept these micro-credentials and tasks as equivalent to university degrees, and even critical, when some governments like Singapore are supporting students financially to pay for those accreditations (Ho, 2016). In this vein, institutional bodies like the Australian Computer Society (ACS), IEEE (Institute of Electrical and Electronics Engineering), Skills Framework for the Information Age (SFIA), and the Accreditation Board for Engineering and Technology (ABET) are also contributing to the complexity in higher education by offering advanced accreditation programs to individuals who want to progress in their discipline and gain skills to escalate positions within their companies. All these are happening at the learner's own pace and abilities without the pressure of having to attend the rigorous scheme of formal education of a typical university course. There is a possibility that a four-year traditional engineering program might not be attractive anymore for the future generation of engineering students. Engineering faculties need to rethink the way they are structuring their courses and programs not only to counter these competitive forces but also to match contemporary engineering learning needs.

The current climate of uncertainty and complexity in higher education (Bennett *et al.*, 2015) might play a role in students' decisions to study in an engineering education program. The crucial point is that this could be aggravated as a result of the new realities of current and future generation engineering students.

5.2.2 Current and future generation engineering students

The livelihood of universities is affected by the perceptions, attitudes and rapidly changing needs of their students. As said, engineering is a discipline in which students are to apply science and mathematics, solve problems and make things. To this end, engineering teachers expect their students to develop and gain the traditional critical thinking, research and problem-solving skills, in addition to a number of higher-order cognitive and affective skills that are unique to engineering (Heywood, 2005) and that for some individuals are perceived as difficult to attain. However, research about the perceptions of contemporary prospective learners suggests something further. A recent study aimed at understanding the reasons why students with an interest in science, technology and math choose not to study engineering as a higher education degree found that, from the point of view of the students, engineering subjects do not support student self-development, i.e. there are no opportunities to develop students' identities in a constructive and meaningful manner (Holmegaard *et al.*, 2010). This study also found that engineering students with cross-disciplinary interests and who choose to study engineering are not coping well, and that they perceive engineering education to be structured around classical mono-disciplinary subjects (Holmegaard *et al.*, 2010). There is a concern that engineering faculties are still "attempting to educate 21st century engineers with a 20th century curriculum taught in 19th century institutions" (Duderstadt, 2010, p. 19).

From the perspective of higher education, techno-literacy is a student reality that is not well understood by many. This is important to consider because part of the context of the field of contemporary higher education is the labeling of students as the Net Generation or digital natives – students who have grown up with digital technologies (Prensky, 2001, 2007). One of the assumptions behind such labels is that our students will be predisposed to work effectively with online technologies and that they bring with them the skills and understandings required to work both effectively and powerfully in this way. However, some studies suggest that this is not the case. For example, a study conducted by Arenas and Lynch (2015) found that the new generation of students may be operationally prepared to use online technologies, and it might be assumed that they will increasingly have cultural understandings of online tool use for recreational and social purposes; however, the recontextualizing of these tools and practices to support formal learning cannot be assumed as straightforward, and students need support in seeing the relevance of social networking and web-based applications to their higher education learning and in developing the cultural and critical techno-literacies. The study points to a need for faculty engineering programs that prepare and encourage students to go beyond the operational literacies involved in using course tools and to move into the cultural and critical literacies required for the effective use of online tools for learning in higher education (Merchant, 2011). There is a place for the explicit exploration of requisite cultural and critical techno-literacies within programs that make use of online tools or blended pedagogies. By bringing students into this conversation, those students who already possess high levels of operational, cultural and critical techno-literacy in relation to online course tools are well placed to critique the ways in which these tools are embedded into formal higher education learning and to contribute to the building of effective learning environments and communities of learners.

Contemporary students are engaged in many extracurricular activities. The reality is that contemporary students are very busy individuals. Based on the findings of research exploring students' perceptions of learning environments, Arenas (2012) reports on the impact that extracurricular activities might have on students' performance. This research found that many of the students had competing demands on their time, such as to raise money for the tuition fees, in addition to daily basic activities like cooking, leisure and home maintenance.

Employability is a new reality currently exerting great influence on the decision of prospective students to study engineering. In this regard, universities have an enormous responsibility for helping students gain the skills and knowledge required that make them appealing to potential employers (Bennett *et al.*, 2015), but this is becoming more difficult given the changing labor market and the economic pressures experienced in developed countries like Australia, where the options for manufacturing and engineering jobs, for instance, have been compromised in favor of others which are more economically tangible. Under these circumstances, research suggests that faculty should adopt strategies that go beyond traditional classroom-based activities to facilitate the development of graduate employability skills (Mason *et al.*, 2009). Such strategies include providing multiple resources and new approaches to teaching, like integrated learning and workplace practice, clearer links between subjects and areas of knowledge, and real-world examples and experience (Bennett *et al.*, 2015).

In response to these realities, universities have implemented a number of strategies and invested significant time and resources to continue providing quality education to their students in difficult times. An example of these strategies is blended learning, which in its many forms has been used and successfully implemented in both training and education across all sectors (Bonk and Graham, 2006).

In the following section, blended learning and its underlying principles are used to frame the discussion around the connected e-learning framework for engineering education, the focus of this chapter.

5.3 Why blended learning

In the early 1990s, the explosive growth of the Web and emerging technologies introduced the concept of e-learning (electronic learning) (Downes, 2005; Laurillard, 2006). Since then, higher education institutions have adopted virtual learning environments to support and complement the traditional face-to-face learning and teaching paradigms (Garrison, 2000; Kanuka and Rourke, 2008). Some educational innovators have gone further to use the blended approach to teaching and learning to design, implement and deliver educational programs that support transformative deep and meaningful learning (Boyle *et al.*, 2003; Garrison and Kanuka, 2004; Garrison and Vaughan, 2008).

Blended learning has been defined in a variety of ways (Stacey and Gerbic, 2009), but most commonly the term is used to describe a wide range of forms of learning and teaching that integrate the use of information and communication technologies (ICT) with face-to-face learning (Driscoll, 2002). The term was initially used in corporate training and distance education and later in higher education (Bonk and Graham, 2006; Oliver and Trigwell, 2005; Stacey and Gerbic, 2009). Oliver and Trigwell (2005) identified the following forms of blended learning based on the notion that the term encompasses the combination of two or more different modes:

- traditional face-to-face learning with e-learning, where e-learning is the use of any form of ICT in the learning process;
- online learning mixed with traditional face-to-face learning, where online learning is mostly web-based learning;
- mixed media, including any form of narrative, communicative or iterative media;
- mixing contexts like work and study, appropriate for flexible learning;
- mixing theories of learning;
- mixing learning outcomes and
- mixing teaching approaches like distance- and campus-based (Oliver and Trigwell, 2005).

For a contemporary engineering student, all the identified modes of blended learning are likely to apply. For institutional leaders, the intention is to leverage any mode that best helps in addressing the changing needs of higher education under the current climate of uncertainty and complexity.

There is a considerable body of literature about the visions of the potential of blended learning compared to either just face-to-face or just ICT-based learning environments (Aspden and Helm, 2004; Chen and Looi, 2007; Jelfs *et al.*, 2004; Osgurthorpe and Graham, 2003). These benefits as reported by Osgurthorpe and Graham (2003) include:

- pedagogical richness, allowing teachers to change the way they use class time;
- increased access to multiple resource materials for students on a timely and flexible manner;
- social interaction;
- personal agency, enabling students to take ownership of the learning with less dependence from teachers;
- increased cost effectiveness and
- ease of revision and update, through a learning environment that is flexible, responsive and spontaneous.

All these benefits have worked in favor of blended learning designs at all educational levels with the concept being widely adopted and continuing to expand in the future (Adams Becker *et al.*, 2017). In the context of higher education, research in the field suggests a positive impact of the paradigm in student performance with an increase in creative thinking, independent learning and personalization of learning experiences (Adams Becker *et al.*, 2017). Today, the blended approach is integral to higher education to the extent that those institutions that have not embraced it yet in a robust and strong manner will struggle to survive (Adams Becker *et al.*, 2017). However, regardless of the extent of blended learning implementation, the integration of any technology in learning and teaching practices is superfluous if that integration is not informed by sound learning science theories and evidence-based practices. The next section reflects upon these theories, particularly connected learning, that appears to be congruent with new approaches to learning where technology is integral to the learning process.

5.4 Connected learning theory

The literature about learning and teaching is influenced by broader perspectives from a number of disciplines, particularly philosophy, psychology and sociology. For many years, the understanding and investigation of learning and teaching in higher education has been framed within competing approaches from the positivist or scientific paradigm to the interpretive and critical paradigms. According to Foley (2000), since the late 1960s, the positivist or scientific approach to studying education in Western education has changed to an interpretive paradigm that emphasizes the individual's construction of meaning, shaped by culture and social structures. This interpretive paradigm is known as constructivism, and since its inception it has been described from a psychology (Biggs, 2003; Laurillard, 2006) to a cognitive (Piaget, 1969) and social perspective (Vygotsky, 1978).

In terms of educational technologies, much of the research into learning uses socio-cultural theory as the foundation of communication and interaction in online collaborative learning environments (Kanuka and Anderson, 1998; McLoughlin and Oliver, 1998). More recently, a learning theory that builds on principles of constructivism in ways consistent with emerging technologies is connectivism (Siemens, 2004). According to Siemens (2004), traditional learning theories were developed at a time when technology was not seen as fundamental for learning. Today, influenced by ubiquitous technology, we live, communicate and learn in a rapidly changing world. Accordingly, there is a need for applying theories in ways that utilize the affordances of new technologies. In connectivism, Siemens (2004) explains learning as a process that occurs within nebulous environments of shifting core elements – not entirely under the control of the individual. He defines learning as actionable and explicit knowledge that can only be acquired through the established connections amongst interested parties. Modern learning theories like connectivism coupled with new developments in mobile learning technologies have given rise to various digital learning frameworks with student success at the center. An example of this is connected learning, a framework that has captured the attention of educational, technical and thought leaders.

Connected learning is an educational approach that advocates for broadened access to learning that is socially embedded, interest-driven and oriented towards educational, economic or political opportunity (Ito *et al.*, 2013). Learning is socially embedded when the learner is in constant exchange of dialogical conversations with his or her peers. These social connections are inclusive and highly engaging in a peer-supported environment where the learner contributes, shares and provides feedback. In an interest-driven learning environment, the learner is responsible for his/her own learning and achieves much higher-order learning outcomes. Connected learning is academically oriented, i.e. it fosters a learning environment where the learner can flourish educationally, economically and politically. This can be achieved by connecting learners' personal interests and social engagement to academic studies, civic engagement and career opportunity. Further, the student-centered approach of the model leverages the opportunities of new media and emergent technologies to support individual's personal and social interests to fulfill their aspirations and skills recognition.

In light of the connectivism theory by Siemens (2004) and recent research on blended learning, there is a need for the construction of connected learning frameworks, those educational ecosystems capable of supporting connections for interactive and dialogical conversations amongst learners, and hence the changing learning and teaching needs of higher education. In the next section, an e-learning framework is presented that is based on connectivism and connected learning theory, particularly applied to engineering education.

5.5 Connected e-learning framework for engineering education

Engineering education practitioners have traditionally been at the forefront of promoting innovative ways of transformative learning and excellence in learning outcomes. Examples of these forms are project-based learning, problem-based learning, work-integrated learning, flip classrooms and internships, described and shown to be effective (Felder *et al.*, 2000). At the center of these learning and teaching approaches is the concept of experience that Kolb and Kolb (2005) suggest as the source of learning and development and good balance between abstract and concrete material (theory and practice). When the connected learning theme is embedded into this engineering context, the chances of student success increases, given that the added values of individual's personal and social interests are enhanced through the affordances of innovative learning technologies.

Figure 5.1 shows how the connected learning framework fits into a broader educational context. At the center of this technology-enabled learning landscape is the learner who uses a personal assistant for lifelong learning (PAL) to explore the world around and engage with authentic learning tasks that support the need to learn and be competent in a specific skill. Depending on the learning context (formal learning, informal learning, experiential learning, project-based learning, problem-based learning, inquiry-learning, collaborative learning and the like), there is support for learners' cognitive and emotional skills. In this ecosystem, learners are responsible for their own learning, and learning analytics dashboards check that both learner's progress and performance are consistent with the agreed learning outcomes.

Learners are surrounded by a diverse range of entities that have a marked influence on how they approach learning to develop the desired competencies. Governments' qualification frameworks regulate national educational systems for formal learning. Competency-based frameworks provide guidance for professional grades, accreditation and programs. These are used by accreditation bodies to inform on the proper alignment of the workforce with industry demands. Institutionally, educational providers have a number of administrative processes and procedures to attest the learner's acquisition of knowledge. There is a learning and teaching relationship where the ultimate goal is for seamless learning opportunities created specifically to nurture individual's learning needs. This ecosystem recognizes that individuals learn differently and that they see and interact with the world in many different ways. The

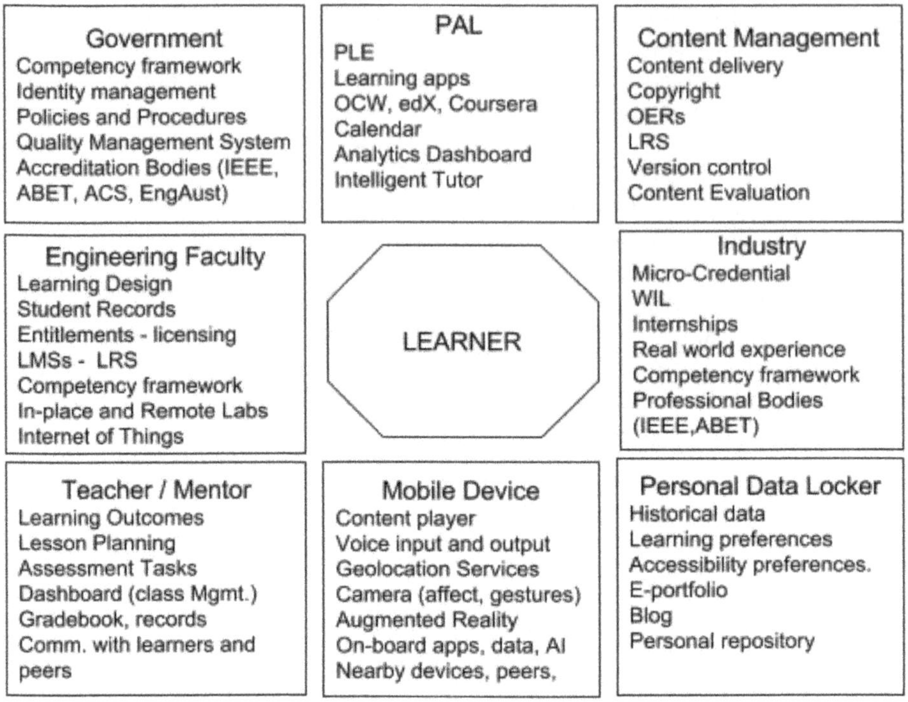

Figure 5.1 Connected e-learning framework for engineering education

capturing and analysis of those interactions lay the foundations of a personal data locker, a true e-portfolio to show and tell others about an individual's personal learning achievements.

Central to the learner is the ability to perform professionally via industry connections that foster real-world and authentic learning experiences. Work-integrated learning, internships and learn-on-the job strategies are industry-based efforts to make sure what the student learns matches industry needs.

The following sections elaborate on the applicability of the connected e-learning framework in engineering education, in terms of curricula, assessment and practice.

5.6 Curriculum design

Today, the creation and delivery of learning activities are increasingly becoming technologically based, which coupled with the uncertainties and complexities surrounding higher education makes the conceptualization of learning design more problematic. Connected learning acknowledges this complexity and considers a holistic interdisciplinary approach towards the conceptualization of learning design from a science perspective, including fields of cognitive science, neuroscience, educational technology, instructional technology, social sciences, epistemology, artificial intelligence, computer science, educational psychology and anthropology (Sawyer, 2014). A well-known learning design framework that borrows from all these fields to promote the delicate balance between learning and teaching using technology is Biggs's 3Ps model. In this model, Biggs (2003) argues that learning is constructed as a result of learners' activities, in conjunction with the distinction between learners' surface and deep approaches to learning (Biggs, 2003, p. 18). The 3P model proposes a set of learning-related factors localized in three points in time: presage, process and product. The presage is the phase that happens before the actual learning occurs. There are two kinds of presage factors: student-based and teaching context-based. The student-based factors relate to what students bring to learning, such as prior learning, previous experience, interest, motivation, culture, language, techno-literacy and curiosity, whereas the teaching context-based factors relate to the teacher's intentions, learning outcomes and assessment and the teacher's beliefs and attitudes. At the process level of the model, student-based and teaching context-based factors interact to produce the learning activities in accord with the approaches to learning. Students adopt either deep or surface approaches in response to the climate created around them. The product learning-related factors relate to students' learning outcomes which are successfully achieved with appropriate learning activities at the process level of the model. For Biggs (2003), the challenge is to bring the 3P model to a state of stability for meaningful and transformative learning. It is suggested that "learning is the result of students' learning-focused activities which are engaged by students as a result both of their own perceptions and inputs, and of the total teaching context" (Biggs, 2003, p. 20). Biggs (2003) takes further this principle and introduces the constructive alignment principle to explain the alignment that should exist between intended learning outcomes, learning activities and assessment. The next three sections provide some insights on how this can be achieved in the context of engineering education and connected learning.

5.6.1 Intended learning outcomes

Learning outcomes are crucial components in engineering learning design to state what the student is to do, the conditions under which the learning behavior is to be shown and the expected level of achievement (Wankat and Oreovicz, 2015). The whole process of mapping and collecting data for learning outcomes may become a daunting task. For engineering education, there are international supporting frameworks like the Accreditation Board for Engineering and Technology (ABET) and the Skills Framework for the Information Age (SFIA) aimed at guiding the teaching practitioner into the practice of designing good learning outcomes consistent with the demands of the industry. Nationally, for many decades, Engineers Australia has been doing this and continues being the trusted voice of the profession for most engineering programs in the country.

In addition to professional bodies, there are some technology resources available to support the engineering educators in the difficult task of designing learning outcomes. In Australia, the Commonwealth Department of Education and Training (CDET) has previously funded initiatives that focus on strategies for addressing learning outcomes. An example, technologically supported, is the whole-of-course approach (Lawson, 2017). This learning design resource gets guidance from Biggs's constructive alignment principle to ensure each element of the curricula, i.e. learning outcomes, learning activities and assessment tasks, is completely evaluated in assuring learning (Lawson, 2017). The tool supports the learning designer to embed learning outcomes in their courses that encompass course context, university themes and external requirements from themes like AQF (Australian Qualifications Framework) levels, discipline thresholds and professional body requirements. There is also support to unpack learning outcomes using rubrics, recommendations for authentic assessment tasks designed in a scaffolded manner, and ways for engaging all stakeholders in the process (Lawson, 2017).

A very important learning outcome that learning designers strive to embed into the engineering curriculum is research skill development; however, this is particularly challenging in coursework-based undergraduate courses where the main expectation is to equip students with the skills to perform professionally in the workforce. Unfortunately, resources to support student research skills development are scarce. To fill this gap, some efforts, supported by technology, have been put in place. As an example, it is worth mentioning a project conducted at Adelaide University aimed to address the conceptual difficulties faced by academics in facilitating student research-skill development (Willison, 2017). For engineering assessment and curriculum design, the resource provides some guidance in electronic, software and mechanical engineering (Willison, 2017).

5.6.2 Technology-enhanced learning activities

Like learning outcomes, the design of a learning activity is a delicate process. According to Biggs (2003), a learning activity should be built with elements of higher-order skills embedded into it. This is consistent with the constructive alignment principle to elicit the intended learning outcomes. In engineering education, however, this might not be easy owing to the nature of some learning tasks that, like lecturing, still continue being the preferred approach to teach by many academics. Lecturing is a teacher-directed learning activity and characteristic component of traditional teaching in higher education, yet the debate continues about its philosophical value (Biggs, 2003; Laurillard, 2006; Ramsden, 2003). Wankat and Oreovicz (2015) suggest that to some extent lectures might satisfy learning principles and are conducive to learning experiences of lower levels of Bloom's taxonomy. Effective teaching strategies can overcome these limitations and elevate the quality of a class with well-structured content, performance (preparation and presentation) and interpersonal rapport. Acknowledging the numerous problems associated with lecturing – such as poor attendance, focus on transmission of information and limited opportunities for interaction – connected learning complemented by the blended approach to teaching potentially offers a more engaging alternative. For example, the lecturer could pose a question or raise an issue and ask students to encapsulate their responses in a Twitter comment. Alternatively, the lecturer could provide an oral summary of a concept or process that has been described in an earlier part of the lecture and ask students to post a Twitter comment to raise areas of which they are not certain or have some confusion. The lecturer could project the Twitter site onto the lecture theater projection screen, so that the online interaction can be meaningfully integrated into his/her oral commentary and response. Such connected-blended activities, when used in large classes, might produce more effective results in comparison to the more traditional lecture with fewer opportunities for interaction. Activities like the one described here use the affordances of the technology to overcome the limitations of more conventional media.

Like teacher-directed learning activities, peer-directed learning activities are used to ensure students achieve the learning outcomes, particularly those that are conducive to student learning at higher-order thinking. In peer-directed activities, students equally engage and participate in both spontaneous and formally structured student-student learning interactions (Johnson and Johnson, 1985). They may be conducted by the students after being initiated by the teacher or totally conducted by the students outside the classroom. Regardless, students report to the teacher for guidance or clarification (Biggs, 2003). The following is an example of connecting and blending face-to-face interactions with ICT-mediated tools within a peer-directed activity that capitalizes on the affordances of a particular technology.

As with Twitter in previous example, a tool that exploits students' familiarity with social software and Web 2.0 is PeerWise™. According to Purchase *et al.* (2010), PeerWise™ is a tool in which students engage collaboratively in creating, sharing, evaluating, answering and debating a repository of assessment questions about the course content in ways that appear to promote deep learning and improve students' performance. There is a cluster of research about the benefits of giving students assessment responsibilities that traditionally are the domain of teachers and instructional designers (Dearing, 1997; Dochy *et al.* 1999; Falchikov, 2005). The following is a description of how a tool like PeerWise™ (Luxton-Reilly and Denny, 2010) could be integrated into a connected and blended learning environment such that the affordances of blending face-to-face and online communications are capitalized upon. Individually, students are required to author a question linked to the learning outcomes of the course and to provide what they believe is the correct answer to the question. Both the authored questions and the answers are stored in a repository which is available to all students. When a student accesses the repository and attempts one of the stored questions, the student can see other students' responses to the question as well as the students' normal distribution of the responses. This process enables the students to reflect on those answers and compare them with their own responses. Students are given the opportunity to review and edit their own questions based on the global feedback and engage in discussions (Luxton-Reilly and Denny, 2010). Once the questions have been authored and reviewed by the students, they may be evaluated by others in class with the teacher providing guidance and clarification.

Learning designers can also construct self-directed learning activities embedded with elements of higher-order skills to elicit the intended learning outcomes. This is becoming increasingly important these days to support student locus of control, learning ownership and self-regulation. In a self-directed learning environment, the learner is personally responsible for the management and control of the learning activities (Biggs, 2003). Self-direction is at the center of the connected and blended learning approach and essential in the context of engineering education. There are many ways of exploring the avenues through which technology may be used to develop and support self-management skills that add value to independent learning in engineering education. A traditional approach in engineering education is the use of laboratories, but as pointed out by Wankat and Oreovicz (2015), there is a tendency to minimize lab-based work because of their cost and space requirements that are always found to be problematic in engineering faculties. One solution is the use of simulations or virtual labs.

5.6.3 Virtual and remote labs in engineering education

According to Wankat and Oreovicz (2015), the use of specialized simulation programs is beneficial, because practicing engineers use simulations. This is consistent with a study conducted by Arenas and Lynch (2013), in which the blend of online classes and self-directed exploration using virtual learning spaces was a good example of how technology-mediated and face-to-face activities can be blended in meaningful and transformative ways. The virtual lab was a learning resource integrated with a computing subject. Using the tool, students were required to complete a set of self-directed activities at their own time and pace. Every week, they had to give evidence of the completion of the tasks by submitting a weekly progress report. The use of the virtual lab was meaningful in terms of its practicality, particularly in the design of courses in which it was not possible for the students to use a live environment where they could practice what they learned in theory. The tool provided students with prompt and accurate feedback on tasks and supported students' development of explorative, deductive and investigating skills that empowered them to achieve deep learning through interactive simulations of real-world problems typical of complex and technical configurations of secured networks. Acknowledging the cost and expertise required to set them up, in the context of engineering education, interactive tools like virtual labs can enhance learning by supporting a blend of face-to-face and online activities that facilitate activities not otherwise easily accomplished, by overcoming the constraints of face-to-face activities through a capitalization on the affordances of the technology.

Despite the action possibilities of these virtual learning spaces, simulations will never replace the real-world learning experience of in-person physical equipment (Wankat and Oreovicz, 2015). Research shows that the intended learning outcomes of practical work in engineering education are achieved much better through the use of physical laboratories (Finkelstein *et al.*, 2005; Nickerson *et al.*, 2007; Sauter *et al.*, 2013). In that vein, there is a trend in higher education to implement practical and viable solutions through remote labs. In this scenario, students access the labs via the internet to remotely monitor and control physical laboratory apparatus (Lowe, 2014).

Since the inception of remote labs in the mid-90s, there has been a growing interest amongst engineering learning designers to implement and configure remote labs that meet the concrete and practical needs of distance education learners. More recently, the push is for remote labs that can scale to large number of students, like in MOOCs (Tawfik *et al.*, 2014). Of particular interest is the Smart Device specification, a set of well-defined metadata interfaces used to decouple the client from the server functionalities. This promotes interoperability and broader sharing of remote labs (Salzmann *et al.*, 2015). This specification is the foundation of a collaborative effort currently under development: the IEEE P1876 metadata standard for online laboratories. The P1876 standard aims to offer any online laboratory as a service (LaaS) and as an interactive open educational resource (laboratory as an OER, or LaaO). Some examples of large-scale implementations of virtual and remote physical labs include Labster, Open Science Lab, Labshare, Pirate, MIT and GOLC.

Labster uses a combination of sophisticated mathematical algorithms and gamification elements to build fully interactive advanced virtual labs for science. The virtual labs are used worldwide by many universities, including Harvard, MIT-iLab, California State, Stanford, Berkeley, Hong-Kong and New England.

The Open Science Lab (The Open University, UK) is an initiative of the Open University and the Wolfson Foundation aimed to improve science education globally. The lab uses real data for the construction of virtual scenarios, experiments and learning activities, with some of them available to all at no cost.

The Physics Innovations Robotic Astronomical Telescope Explorer is also an Open University initiative. Pirate is a pilot project that explores the use of a remotely operable telescope in university distance education. Currently, Pirate is part of an Open Science Laboratories project to support Open University courses in astronomy and planetary sciences.

Labshare (University of Technology Sydney) is a consortium offering a range of centralized services for the use of remotely accessible laboratories within the engineering educational sector. Labshare has a number of engineering rigs equipped with sensors, and data acquisition systems that can be remotely accessed and operated by engineering students to run experiments anytime and anywhere.

MIT-iLab is a Massachusetts Institute of Technology initiative aimed to create a rich set of experiment resources that facilitate students and academics around the world to share their physical labs over the internet. Currently, the MIT-iLab facility offers case studies in microelectronics, chemical engineering, polymer crystallization, structural engineering and signal processing.

The Global Online Laboratory Consortium (GOLC) is an international institution focused on promoting the research, development, interoperability and sharing of remotely accessible laboratories within the higher educational community. Many universities around the world are part of the consortium, including UTS, MIT, University of Stuttgart and Nanyang Technological University.

Emergent technologies like virtual and remote labs are transforming the traditional way engineering faculties support the experimental and practical curricula of their engineering programs. These online experimental environments are currently being technologically enhanced and augmented further through the introduction of augmented reality-enabled interfaces.

5.6.4 Leveraging augmented reality in engineering education

Augmented reality (AR) technologies are gaining popularity to support connected personalized self-directed learning activities. AR is the use of technology to create a reality that is enhanced and augmented through the overlaying of virtual objects into a real-world environment (Dunleavy and Dede, 2014). In terms of cost, accessibility and usability, the use of mobile devices is becoming the *de facto* standard to delivery AR. In a recent systematic review of the literature on AR in education conducted by Akçayır and Akçayır (2017), the most cited advantages of AR in educational settings are the enhancement of learning achievement, learning motivation and learning enjoyment. Practical examples portraying the use of AR in education can be found at the Augmented Reality for Enterprise Alliance (AREA). AREA is an entity focusing on supporting organizations (including educational institutions) to achieve greater operational efficiency through the smooth introduction and widespread adoption of interoperable AR-assisted enterprise systems. Of particular importance are those examples related to on-the-job training on systems that are hard to access or dangerous for humans to deal with, like nuclear reactors, power generators and the like. This is particularly relevant to engineering education approaches like work-integrated learning (WIL). The educational settings of WIL, including the integration of theory and workplace experience and the preparation of learners to be work-ready, can be enhanced through AR. Take for example the current situation of manufacturing companies leveraging advances in automated control systems, predictive maintenance software and sensor networks to improve their production lines. As a result, there is an employment shift with a number of traditional jobs slated to disappear, and the creation of new opportunities demanding more specialized sets of skills. The need to learn on the job is becoming critical, particularly in retraining staff. The author of this chapter is currently contributing to a research project seeking the standardization of an AR specification, named ARLEM to model learning activities and working environment. The P1589 – IEEE standard for an augmented reality learning experience model (ARLEM) – aims to define a learning activity in which any type of learning content can be experienced in a workplace through the use of elements extracted from AR and the Internet of Things. The specification comprises two modeling languages: activity modeling and workplace modeling (activityML and workplaceML). The activityML is used to describe the augmented learner's experience and interactions with the learning objects. The workplaceML describes the tangibles (people, places, things), configurables (devices, apps) and triggers (overlays, detectables) of a workplace. The system is integrated into the xAPI learning experience for the analytics of learner's performance tracking and experience (xAPI will be discussed later in a separate section).

5.6.5 Artificial intelligence in education – adaptive learning

For many years, researchers in the field of cognitive science have dedicated much effort to developing technologies that monitor learners' progress and uses these learning data to modify instruction at any time (O'Connell, 2016). This pedagogical approach is referred to as adaptive learning and is enabled by research in artificial intelligence, particularly machine learning. According to Pugliese (2016), adaptive learning systems can be used to dynamically adjust the type of content in accordance with learner's ability attainment. This content brokering aspect of adaptive learning systems allows for both automated and teacher interventions, with timely feedback that may accelerate learners' performance. Adaptive learning systems have been criticized by their instruction-centric aspect; however, recent research conducted by VanLEHN (2011) suggests that within their limited area of expertise, currently available adaptive learning systems seem to be just as good as human tutors or instructors. According to Moore (2016), adaptive learning is one of Gartner's top strategic technologies for higher education with the potential to actualize the promise of scalable personalized learning and granular predictive learning analytics. In this regard, National University, a private institution of higher education in California (Fain, 2017), is the latest higher education institution combining adaptive learning, predictive analytics and competency-based learning to support the personal learning needs and interests of their students.

5.6.6 Content authoring in engineering education

The authoring of educational resources for engineering education demands lots of time and effort. Over the years, academics have trusted publishing companies as the main source of educational resources in the form of traditional textbooks and more recently e-books. Since their inception, e-books have been used merely as the duplication of the printed page on a tablet or smartphone and more recently, thanks to the affordances of emergent technologies, the trend is for the authoring of educational resources as e-books that incorporate cloud-based authentic learning activities and resources. This trend is consistent with learners' expectations of being able to work, learn and study anything, anywhere and at any time in a challenged-based and active learning environment (Arenas and Barr, 2013). For academics in general, the vast majority of these educational services are currently authored and delivered via LMSs which, as mentioned, are integral to the core learning environment of universities. Regarding the use of LMSs, faculty sentiment is mixed. Recent research on the next generation digital environment conducted by EDUCAUSE highlights the teacher-centric affordance of the LMS by enabling the administration of learning but very little in terms of enabling learning itself (Brown *et al.*, 2015). As a result, the trend is for the use of the LMS as a repository and backend system to support storage and administrative tasks typical of learning and teaching and the use component-based digital systems to enable transformative and meaningful learning (Brown *et al.*, 2015). Later in the chapter, this issue is elaborated and examples are provided that particularly resonate with the theme of connected and personalized learning.

The traditional process of sourcing educational materials from publishing companies is currently being challenged with the strong emergence of open educational resources, discussed next.

5.6.7 Open educational resources

Open access education and open educational resources are concepts that usually pop up in addressing issues of knowledge hoarding and social inequality (Peter and Deimann, 2013). The philosophy behind OERs is encapsulated in David Wiley's 5Rs framework (Wiley, 2014):

1 Retain – the permission to make, own and control copies of the content.
2 Reuse – the permission to use the content in a wide range of ways and contexts.
3 Revise – the permission to adapt, adjust, modify or alter the content itself for different purposes.
4 Remix – the permission to combine the original or revised content with other open content to create something new.
5 Redistribute – the permission to share copies of the original content, your revisions or your remixes with others (Wiley, 2014).

Some leading universities have capitalized on the concept by adopting the approach in partnership with MOOC organizations such as Coursera, Khan Academy and edX for the provision of universal access to high-quality education resources at no cost (Arenas *et al.*, 2016). Recent developments in e-learning standard models have also promoted the use of high-quality open access content in higher education (Arenas *et al.*, 2016). For example, universities like Washington State University and New York State University are actively engaged in the process of forming task forces to recommend the increased use of OERs in their faculties aimed to reduce students' financial burden (Anderson *et al.*, 2017). This has not been without issues, particularly in disciplines like engineering, in which OERs are not readily available and the selection of good materials is difficult. Anderson *et al.* (2017) looked into this issue by studying engineering faculty's perceptions and needs about OERs. Based on that study, they suggest best practices for outreach and implementation plans of OERs in the engineering curricula:

- Conduct a review of available OERs in engineering;
- Pursue teaching opportunities if knowledge regarding OERs is limited;
- Understand engineering faculty practices for the adoption of course materials;
- Develop strong partnerships with instructional designers, curriculum coordinators and experts in copyright/intellectual property;
- Speak to peers about course material needs, maintaining focus on student success and
- Recommend OERs as supplemental resources if course material needs cannot be otherwise filled with open content (Anderson *et al.*, 2017).

The following are some examples of high-quality OER systems currently widely used, including OER-Commons, Merlot and OpenCourseWare.

5.6.7.1 OER-Commons

OER Commons (open educational resources Commons) are learning and teaching resources often under the Creative Commons license that provides guidance on how to retain, reuse, revise, remix and redistribute the resources. OER Commons offers a wide range of curated collections from full university courses to interactive mini-lessons and simulations, adaptations of existing open work, open textbooks and K–12 lesson plans.

5.6.7.2 MERLOT

MERLOT (multimedia educational resource for learning and online teaching) is an online educational collaborative effort that started in 1997 at California State University intended to support the formation of online communities of learning for the creation and distribution of shared knowledge bases of learning resources. Strategically, MERLOT aims to support learning and teaching through a corpus of peer-reviewed educational resources that can be easily integrated into faculty learning activities. Structurally, MERLOT has only the learning object metadata (LOM) that allows access to educational resources hosted in the connected educational repositories.

Today, MERLOT is partnering with many higher education institutions, professional societies and the industry. MERLOT members of the discipline communities contribute materials of their own and recommend peer-reviewed resources to be added to the MERLOT collection. At the center of MERLOT communities are editorial boards in charge of controlling the quality of the contributed materials. Editorial board members are academics with extensive teaching and industry experience. In partnership with the IEEE Computer Society, the author of this chapter currently contributes to two MERLOT editorial boards: Computer Science and Information Technology. MERLOT uses cutting-edge technology to deliver the materials ranging from web services (API) to LMS integration including Blackboard™, Canvas™, Desire2Learn™, Moodle™ and metadata batch import.

5.6.7.3 OpenCourseWare (OCW)

OCW (Massachusetts Institute of Technology) is a free and open online initiative started by MIT in 1999 and used by millions of learners and educators around the world. OCW publishes collections of quality educational content from thousands of MIT courses, videos and interactive resources. In 2004, the adoption of the Creative Commons license positioned OCW as the OER leader provider amongst higher education teachers, students and self-regulated learners. OCW is being successfully used for a wide range of purposes: from educators claiming improvement in personal knowledge to students claiming enhancement of their personal knowledge, and self-learners saying that the platform allows them to explore areas outside of their professional field.

5.7 Learning and academic analytics

A concept that has captivated the imagination and attention of worldwide educational technologists and innovators is learning and academic data analytics. Data analytics originated in business intelligence models used for years by business organizations to leverage massive amounts of data sets that traditional database models have not been able to. The term big data emerged in the form of 3Vs – volume, velocity and variety – to describe the nature of amount of data, speed of data in and out and range of data types and sources, respectively. The idea is that computational analysis of big data has the potential to uncover hidden patterns for decision-making (Long and Siemens, 2011). Higher education institutions are complex environments in which mountains of data are generated and stored. Like business organizations, it makes sense for higher education to leverage data analytics.

In terms of education, Long and Siemens (2011) divide the education analytics concept into two broad branches – learning analytics and academic analytics – to separate the analytics that are largely concerned with learner's improvement performance and success from the analytics that are concerned with improving organizational effectiveness, respectively. Through the lens of the connected learning principles, in my opinion, this holistic approach towards the improvement of education leveraging data analytics is balanced and recognizes that today's approaches to learning and teaching should be learner-centric and aligned with learners' preferences and personal interests.

In the last 5 years, we have seen the development of analytics tools aimed to harvest education data. The following is a short description of two competing learning analytics systems currently used: IMS Caliper and Experience xAPI (xAPI).

5.7.1 IMS Caliper analytics

The Caliper Analytics framework (IMS Global Learning Consortium) builds upon existing IMS Global standards, methods and technologies. Its implementation aims to capture and present measures of learning during learner's engagement with the learning activities. To that end, Caliper:

- defines a common language for learning data marking;
- provides standard mechanisms to measure the level of learner's performance through metric profiles and
- creates the IMS learning sensor API to define basic learning events and standardize and simplify the gathering of learning metrics across diverse learning environments.

This will enable faculty interventions to measure, compare and improve learning outcomes. As mentioned, Caliper leverages and extends IMS Global standards and technologies like learning tools interoperability (LTI) and question and test interoperability (QTI). However, to date, the degree of interoperability is limited, since IMS has full control of the development of new metric profiles (Serrano-Laguna *et al.*, 2017).

5.7.2 The experience xAPI

The xAPI (advanced distributed learning) is an application program interface that allows the use of human and machine-readable "activity streams" to track data; and provides mechanisms to access and store information about state and content. The tracking of activities can be from any platform, including LMSs, mobile devices, virtual worlds, wearable computers and the Internet of Things. Unlike Caliper that uses events to measure the level of learner's engagement, the xAPI uses statements. In terms of data models, Caliper can be considered an event scripting language, whereas xAPI can be considered an activity scripting language.

The xAPI is a free open source tool that is currently used in many domains and services. Of particular relevance is its use in education research (Bakharia *et al.*, 2016). In this respect, the most influential and active education analytics research group in the world is the Society for Learning Analytics Research (SoLAR). SoLAR aims to explore the role and impact of education data analytics on teaching, learning, training and development. To this end, SoLAR has fostered many initiatives to support collaborative and open research around learning analytics, promoting the publication and dissemination of learning analytics research, and government advising and consulting. For example, with support from the Department of Education and Training, the University of Queensland is leading a collaborative study to develop an open source toolkit for performing sophisticated analysis of learners' engagement in connected learning environments. The goal with this toolkit is to help students and teachers to harvest data about their activities in standard social media environments external to the institution's LMS, and then provide immediate feedback and reports (Kitto *et al.*, 2015).

In the last section of this chapter, a number of use-cases specifically relevant to engineering education that use the xAPI as the education analytics framework are provided.

5.8 Authentic assessment in engineering education

With reference to the constructive alignment that should exist between intended learning outcomes, learning activities and assessment, the latter is perhaps the most challenging and sensitive learning design element. From a teaching perspective, Ramsden (2003) argues that assessment can be seen as a process through which teachers may report on learners' achievements, and even as a process through which teachers' practices may be changed to improve quality learning. Similarly, assessment is influential from a learner's perspective. According to Entwistle and Entwistle (1997), learners appear to mainly focus their understanding on examination requirements, to the detriment of the actual higher order understanding needed to see the world differently. Both students' and teachers' perceptions of what constitutes assessment make it even more difficult to preserve the constructive alignment between the elements of learning. The analysis of these intricacies is beyond of the scope of this chapter; however, in terms of a connected learning framework and learner's personal interests (sometimes highly biased by social connections), if we strive for the provision of authentic learning tasks (Herrington *et al.*, 2014), then logically students should be assessed with authentic meaningful tasks.

In assessing authentically (Herrington *et al.*, 2014), institutions focus on learners using and applying knowledge and skills in real-life scenarios like simulations or role plays, completion of a real-world task or assessment in a workplace setting (Mueller, 2016). One of the strengths of authentic assessment is that learners use their theoretical knowledge to solve problems in unpredictable, ambiguous and complex environments, typical of today's workplaces. In doing so, learners demonstrate the level of capacity gained during the learning process. The implementation of authentic assessment has a number of challenges, particularly the significant workload required to develop authentic learning activities and the interpretation and grading of the assessments. These two combined issues of assessing authentically and efficiently have encouraged educational institutions to use technology. There are software solutions already in place, from Microsoft™ Word add-ins to electronic rubrics, annotation and feedback tools, audio recording grading systems, Turnitin and robo-graders, with some of them integrated within popular LMSs like Moodle™ and Blackboard™. Some of these systems constitute an improvement in the efficiency to grade learners' work; however, they are far from being practical in terms of interoperability and ability to grade learners' authentic work.

Today's trend is for the use of education analytic systems that support teachers to harvest massive amounts of unstructured information and evaluate whether the learner has achieved the desired learning outcomes. Authentic learning and assessment is expected to be transformative, and accordingly, we are starting to see learning environments capable of:

- tracking what learners do, assignments, progress, and grades across multiple institutions and multiple online learning systems;
- analyzing extensive learner's background data to deliver more personalized, culturally relevant and educationally effective learning experiences;
- maintaining students' history and preferences in an external e-portfolio updated and queried by multiple adaptive learning systems and
- integrating peer assessment to improve students' understanding of course materials as well as improve their metacognitive, critical thinking and interpersonal skills.

The reference to e-portfolios and peer-assessment listed above is quite relevant because they are becoming exemplars of authentic ways of learning and evaluating learning outcomes in technology-mediated large-scale learning environments.

5.8.1 Portfolios in engineering education

For years, portfolios have been used as learning and assessment tools in higher education. A portfolio is a collection of student's work and includes a reflective dimension on what the student has learned (Mueller, 2016; UNSW, 2017). According to Mueller (2016), a portfolio has the properties of authentic learning and assessment because of their focus on a meaningful collection of student performance and meaningful reflection and evaluation of that work. From a learning perspective, the use of portfolios has been used and advocated as a very powerful approach to help students build their profile in line with their professional interests and aspirations. In terms of authentic assessment, Cain (2013) used the constructive alignment principle to identify effective assessment criteria which enabled quick, accurate assessment of student portfolios in an introductory programming unit.

5.8.2 Peer assessment

Combined with scaffolding, peer assessment can be an effective strategy to assess students authentically (Kearney *et al.*, 2016). In peer assessment, students' work is assessed by other students of equal status and based on teacher's grading guidelines. Traditionally, the method has been used as a metacognitive tool to develop the critical and reflective skills of students (Topping, 2009). More recently, the practice of peer assessment is becoming widely adopted in large-scale learning environments like MOOCs, as a novel approach to cope with the extremely high number of students who typically take these courses (Meek *et al.*, 2017). The empirical evidence about the effectiveness, credibility and accuracy of peer assessment in MOOCs is still to be determined, and more research is needed. In that regard, to get a greater understanding of what constitutes peer assessment (and of course as matter of curiosity as well), the author decided to have a taste of the method by enrolling himself in a MOOC course of his interest.

5.8.2.1 My edX experience – peer assessment

As a musicophile, I took a first-year university course on music theory run by Berklee Music School hosted at edX. The course had a duration of six weeks. At the end of each week, I was asked to complete a quiz and a peer-reviewed assignment. Each peer-reviewed assignment (six in total: one per lesson) consisted of two components: a small project and reviewing five of my peers. The rationale was that if I did not complete the peer review portion of each assignment, then I was to receive an incomplete for that assignment. I had to remember to keep the time zone in mind when I was submitting my assignments. The course did not allow for late assignment submissions or extension of deadlines.

The expectations about the use of peer assessment for each lesson were clearly defined and articulated well before the course started. For example, it was clearly mentioned the learning goals of peer assessment, what it meant, and the learning values as a cognitive and affective tool. Students were asked to provide fellow students with specific, helpful and kind feedback. For the peer-assessment assignments, I first submitted my own projects for each lesson by the due date. After I had submitted my assignments, I needed to provide feedback for at least five of my fellow students per project (30 peer reviews in total) by the final day of instruction. As said, that was mandatory before I was able to see the feedback on my own assignment. The feedback was supposed to be relevant to the question provided.

The connected learning aspect of the course was achieved by a student-led forum (the teaching team intervened very little with timely moderation). It was a virtual learning space to discuss ideas and concepts we covered in class. It was also possible to connect with fellow classmates via the course Facebook group (I did not use this facility because I am not a Facebook user).

This is the first time I ever enrolled in a course like this. In hindsight, I enjoyed it a lot and surprisingly, I scored very well (full marks) in both types of assessments: auto graded

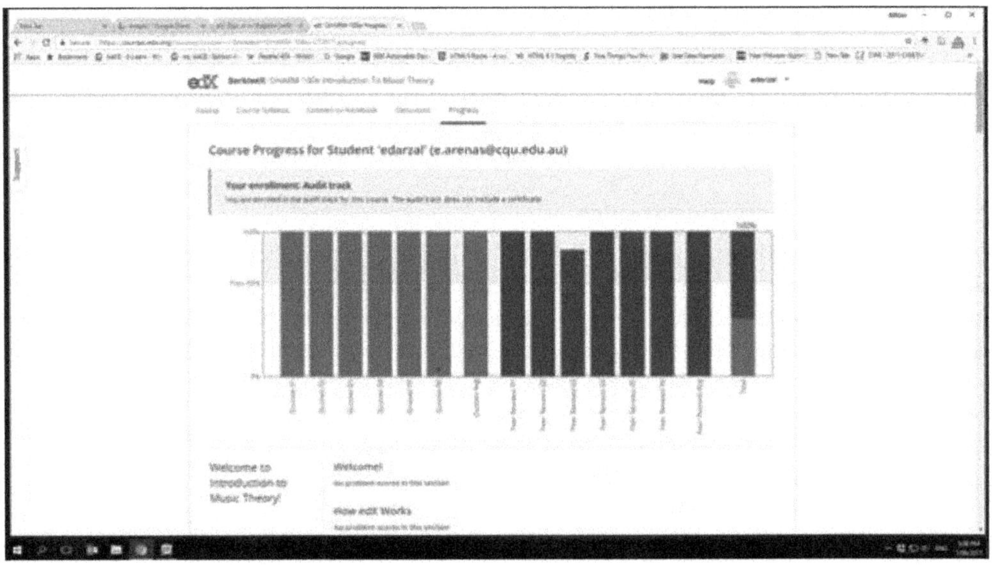

Figure 5.2 Results from the MOOC course

assessment (40%) and peer assessment (60%). The videos were of a high-quality definition (no longer than 5–6 minutes each) and performed by a highly qualified and engaging teacher. With reference to the peer-assessment task, I must admit that it was quite a remarkable activity. Consistent with the research literature, I was able to verify the potential of the method to support my cognitive, critical and reflective skills. It was an interesting exercise and highly recommended.

In the previous sections, I discussed the relevance of using technology to transform the traditional way we teach and learn in engineering. On that note, I am currently contributing to a number of educational technology projects that integrate the vast majority of connected learning principles discussed in this chapter. The following is an abstract of one of these projects, the Actionable Data Book (ADBook).

5.9 Pilots and examples of connected learning implementations

The ADBook was born as part of a research and development special project sponsored by the IEEE-LTSC (Learning Technology Standard Committee). In 2013, the IEEE-LTSC activity sponsored a special interest group (SIG) that proposed a preliminary ADBook architecture aimed to prototype the interoperable learning activity seen as a key enabler for a true sustainable and scalable strategy to a global platform for connected learning and open access education (Arenas *et al.*, 2013). Since then the architecture has been refined and new key functionalities added based on contributions from a special interest group of experts in authoring tools, xAPI code development, metadata, experiential analytics vocabulary and ontology, EPUB 3 conversion tools, EPUB 3 readers, rights and permissions, and content distribution (Arenas *et al.*, 2016)

The ultimate goal of the ADBook is to describe and, as practicable, demonstrate an actionable data book with attributes as follows (Arenas *et al.*, 2016):

- A specialized educational ecosystem that resembles a cloud-based interactive mobile application creating opportunities for rethinking educational publishing for electronic delivery;
- Based on open standards (EPUB 3, HTML5, CSS, JavaScript, xAPI);
- Tailored to support most formal and informal learning paradigms;
- Supporting learner accessibility preferences, usage preferences, learning analytics, experiential analytics and learning pathway tracking and assessment (formative, summative and ipsative);
- Containing interactive content that can exchange data with hardware and other applications within the learner's mobile platform, local devices (taking advantage of the sensors in the mobile device), and cloud resources and
- Full ability to operate in offline or online modes and provision content locally on the mobile device depending on the user's situation.

5.9.1 ADBook as a connected learning ecosystem

The overarching tenet in an improved ecosystem for connected learning is connectivity with other elements of the ecosystem. At the center of the ADBook learning ecosystem is a content consumer (learner) who uses a mobile device to launch the ADBook with the intention or need to learn (see Fig. 5.3). On the left side of the figure, the digital content producers (authors, publishers, knowledge experts and the like) store in open educational repositories their previously EPUB 3 encoded, formatted and packaged content. The EPUB 3 format is an advance over more traditional forms of connectivity, like portable document format (PDF). We believe that there is an opportunity to extend the paradigm initialized with the EPUB 3 format even further so that connectivity is leveraged to the fullest degree in an ecosystem which permits both standalone and connected use, whichever suits the user within the particular time and place of use (Arenas *et al.*, 2016).

The content can be any kind of digital media including multimedia objects (text, audio, video, animation, simulation and the like), learning modules, interactive widgets, and managed learning activities. The learning object and activities can be called in at any time to augment the learner (reader) supported closed loop control. From the learning perspective, the figure also illustrates how learners' self-awareness metacognitive skill is supported by allowing them to see themselves through the lenses of their own ADBooks. The scriptable components (also known as scripted widgets) enable the creation and inclusion of dynamic and interactive components within the EPUB 3 content, for example, the creation of intelligent content (Rockley *et al.*, 2015) to enhance the learning experience. The connectivity built into the ADBook through widgets allows the learner designer, for example, to link digital content to sophisticated online offerings including immersive simulation environments for practice and assessment, multi-player games, adaptive testing systems, robo-graders and intelligent tutoring systems. In terms of the teaching practice, this also allows academics to assign, monitor and participate in learning activities in real time (Arenas *et al.*, 2016)

The Learning Record Store (LRS) service is used to validate and store the activity statements generated as a result of the learning activities. For performance tracking and analysis of what the learner does with the content, the ADBook relies on the flexibility and adaptability of the xAPI specification. The xAPI tracking can be done regardless of the nature of the learning environment, from formal learning using traditional LMSs to informal, social

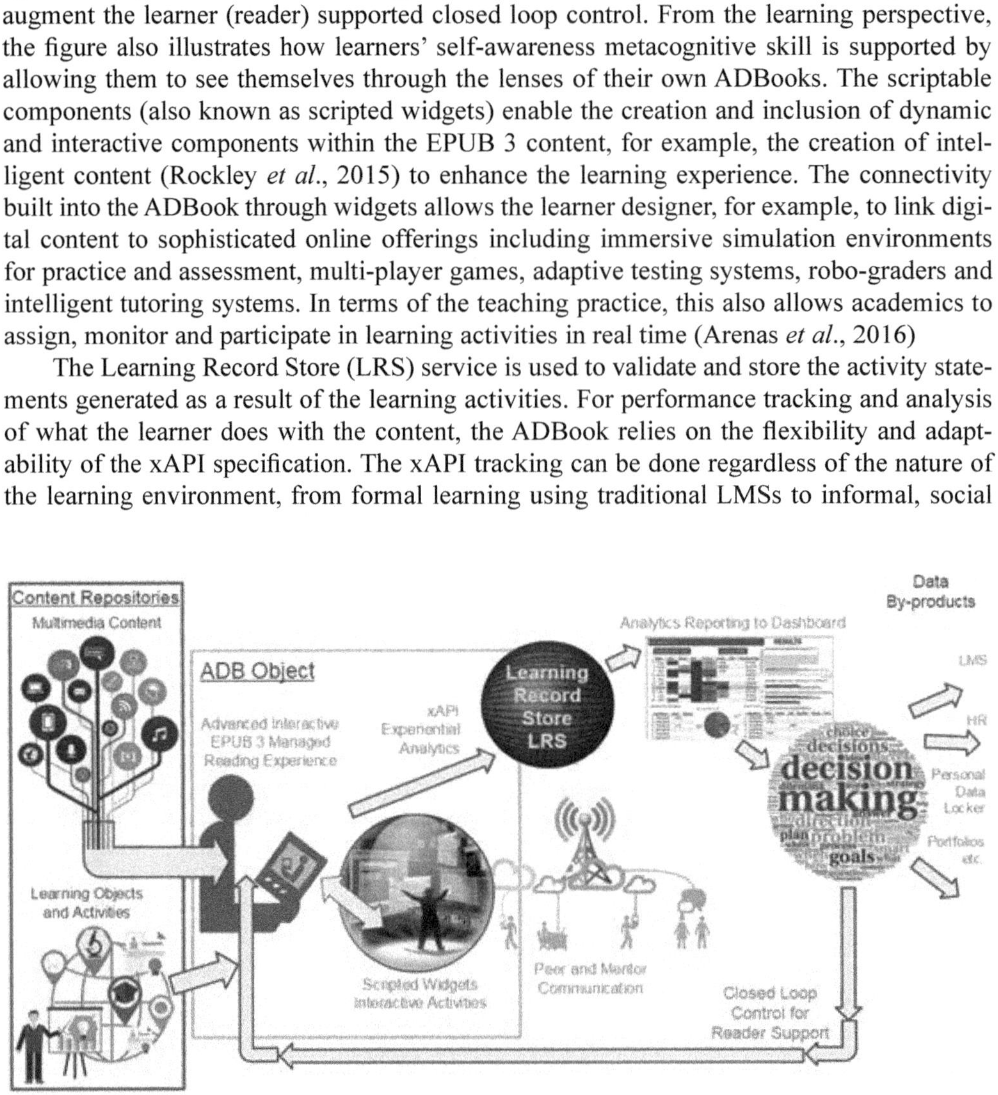

Figure 5.3 Connected learning ecosystem as supported by the ADBook

and real-world learning experiences. The fine-grained, human- and machine-readable xAPI data enables educational stakeholders like teachers, academics, learning designers and content creators to personalize interventions to support the learner. It also supports storage of performance-based learning data in badges, personal learning lockers and portfolios. Measurement of the effectiveness of interventions by the xAPI can also be used by institutions to make decisions on academic processes, policies and procedures, curriculum development, governance and the like (Arenas *et al.*, 2016). Figure 5.3 also shows the connected learning feature of the ADBook, i.e. learning which is open, equitable, social and participatory (Ito *et al.*, 2013). In this learning ecosystem, there are opportunities for internal peer and mentor communication and also, leveraging widget activity, for external Web 2.0 and social media interactions.

The e-book within the ADBook retains the metaphor and underlying linear narrative of a traditional book in addition to its adaptable hypertext. The broader ADBook concept is really a learning delivery platform (content, e-reader and hardware) where content (packaged as a. Zip archive, as required by EPUB 3) is delivered online, or preloaded for offline use (usually) on mobile tablet or smartphone devices. In this regard, the ADB acts (in addition to delivering textbook or reference content) as a learning activity manager, i.e. a central connection point for learning systems and external software (many in the form of cloud services) enabling a wide range of baseline learning-centered operational functions. This is consistent with current educational research recommendations suggesting a confederation of digital learning systems in which learners can access their learning directly with self-selected apps (Lego approach) rather than through traditional monolithic LMSs handling all of the communication and interworking between functions, internally.

To date, the ADB project has led to several functional prototypes and some pilot projects that demonstrate the integration of e-book content with xAPI-based cloud services. One example is RePubIT's SMART Skills pilots for the Catholic Relief Services nongovernmental organization (NGO): Savings and Internal Loan Communities and Marketing Basics (Fig. 5.4).

Results of the CRS ADBook testing in Ethiopia and South Africa and comments from the NGO community were very positive and clearly indicate the suitability of this approach for work in developing countries.

The communicative and interactive capabilities of the ADBook have also the potential to transform large-scale learning environments like MOOCs, particularly to address the low completion rates of these systems. To this end, the ADBook provides several types of learning activities which may include peer or institutional communications, problem-solving and (leveraging mobile device platforms) contextual data-gathering and reporting. The learner should no longer be principally a passive information receiver. Analytic audits of peer group interactions and inferences of engagement may inform interventions to remediate likely poor attainment and non-completion.

Empirical evidence on the connected learning effectiveness of the ADBook is currently being validated in a longitudinal design-based research project led by ADL (Advanced Distributed Learning) called TLA (Total Learning Architecture) (ADL, 2017). In its first stage, the TLA project used a Delphi strategy with a panel of 54 international participants (including the author of this chapter) to develop and evaluate the technical and functional adequacy of the specifications. In addition to this panel, the research conducted a live prototype and demonstration with 73 US special groups of operation soldiers. The preliminary findings of this research project provided empirical evidence on the learning potential of the TLA

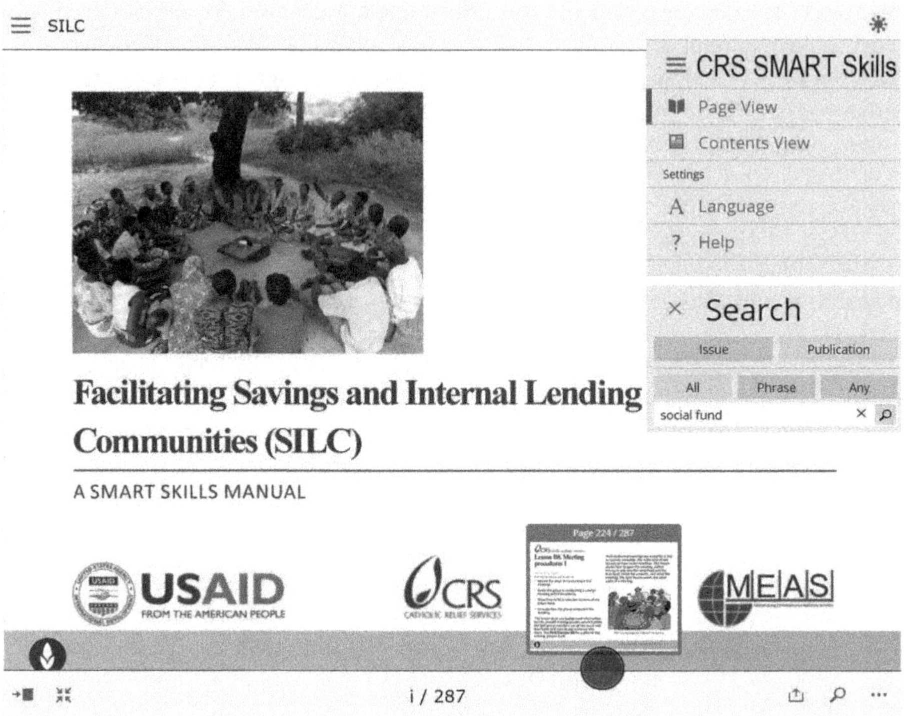

Figure 5.4 ADB example – RePubIT's CRS SMART Skills

framework, including insights on system's functionality, performance and user experience. As a longitudinal study, the final outcomes of this research are still to be seen; however, these preliminary findings show how critical is the separation between the functionalities of the system and the actual quality of learning experience. The suggestion is for instructional design aspects of the system to be a responsibility of the adopting institutions and for the instructional components to be intuitive enough for faculty and educators to develop and personalize. Technically, the preliminary findings also suggest for the creation of common standards and requirements for more refined competencies and educational learning outcomes across institutions (ADL, 2017).

To sum up, the problem TLA aims to solve is interoperability between future personalized adaptive learning systems that deliver the right learning at the right time and in the right way of new generation of learners. In the spirit of this chapter, the TLA acknowledges the connected learning tenets by enabling the actualization of next generation technologies to support personalized, data-driven and lifelong learning (ADL, 2017).

5.10 Conclusion

In this chapter, the connected e-learning framework is discussed for engineering education. The framework uses evidence-based practices and leverages the latest advancements in blended learning to support the connected learning needs, personal interests and social connections of contemporary engineering students. The chapter also addresses the issues of uncertainty and complexity currently faced by higher education as a result of the new realities of learning and teaching in a world of globalization, emergent technologies and disrupted education.

In the chapter, further evidence is provided on the positive impact of blended learning designs in engineering education, particularly in student performance with an increase in creative thinking, independent learning and personalization of learning experiences. Central to blended learning is the integration and adoption of technology. In this respect, there are key trends accelerating technology adoption that will impact learning and teaching. The chapter discussed some important developments in educational technology including adaptive learning, artificial intelligence, next generation of learning environments and education analytics. These and newer technologies will continue to be relevant in the future to support higher education in its endeavors to implement more personalized and lifelong learning experiences in times of uncertainty and complexity.

The chapter makes an important contribution to the field; however, it does not provide enough empirical data to support the application of the connected e-learning framework. There is a need for further research to address this gap.

References

Adams Becker, S., Cummins, M., Davis, A., Freeman, A., Hall, C. & Ananthanarayanan, V. (2017) *NMC Horizon Report: 2017 Higher Education Edition*. The New Media Consortium, Austin, TX, USA. Available from: www.nmc.org/publication/nmc-horizon-report-2017-higher-education-edition/

ADL. (2017) Total Learning Architecture – ADL Net. Available from: www.adlnet.gov/tla [Accessed 13 September 2017].

Akçayır, M. & Akçayır, G. (2017) Advantages and challenges associated with augmented reality for education: A systematic review of the literature. *Educational Research Review*, 20, 1–11. https://doi.org/10.1016/j.edurev.2016.11.002

Anderson, T., Gaines, A., Leachman, C. & Williamson, E.P. (2017) Faculty and instructor perceptions of open educational resources in engineering. *The Reference Librarian*, 0(0), 1–21. https://doi.org/10.1080/02763877.2017.1355768

Arenas, E. (2012) *Blended Learning in a Higher Education Multicultural Environment*. Doctoral Thesis, Deakin University, Australia.

Arenas, E. & Barr, A. (2013) The digital book in higher education: Beyond the horseless carriage. In: *30th Ascilite Conference 2013 Proceedings*. Available from: www.ascilite.org/conferences/sydney13/program/papers/Arenas.pdf

Arenas, E., Costa, J., Berking, P. & Barr, A. (2016) A technical architecture for open access education. *EDULEARN16 Proceedings*. IATED, Barcelona, Spain, 4–6 July 2016. pp. 5900–5908.

Arenas, E. & Lynch, J. (2013). Pedagogical principles for the design of virtual learning spaces in higher education. In: Frielick, S., Buissink-Smith, N., Wyse, P., Billot, J., Hallas, J. and Whitehead, E. (eds.) *Research and Development in Higher Education: The Place of Learning and Teaching*. Auckland, New Zealand, 1–4 July 2013. pp. 26–36.

Arenas, E. & Lynch, J. (2015) Techno-literacy and blogging within a formal higher education setting. *International Journal on E-Learning*, 14(4), 409–422.

Arenas, E., Richards, T. & Barr, A. (2013) The IEEE actionable data book: A platform for inclusive education. In: *IEEE Global Humanitarian Technology Conference 2013*, IEEE, San Jose, CA, USA.

Aspden, L. & Helm, P. (2004) Making the connection in a blended learning environment. *Educational Media International*, 41(3), 245–252.

Bakharia, A., Kitto, K., Pardo, A., Gašević, D. & Dawson, S. (2016) Recipe for success: Lessons learnt from using xAPI within the connected learning analytics toolkit. In: *Proceedings of the Sixth*

International Conference on Learning Analytics & Knowledge. ACM, New York, USA. pp. 378–382. https://doi.org/10.1145/2883851.2883882

Bennett, D., Richardson, S., Mahat, M., Coates, H., MacKinnon, P. & Schmidt, L. (2015) Navigating uncertainty and complexity: Higher education and the dilemma of employability. In: *Learning for Life and Work in a Complex World*. HERDSA, Melbourne, Australia.

Biggs, J.B. (2003) *Teaching for Quality Learning at University: What the Student does*, Vol. 2. Society for Research into Higher Education, Open University Press, Philadelphia, PA, USA.

Bonk, C.J. & Graham, C.R. (2006) *The Handbook of Blended Learning: Global Perspectives, Local Designs*. John Wiley & Sons, Inc., San Francisco, CA, USA.

Boyle, T., Clare, B., Chalk, P., Jones, R. & Pickard, P. (2003) Using blended learning to support student success rates in learning to program. *Journal of Educational Media*, 28(2–3), 165–178.

Brown, M., Dehoney, J. & Millichap, N. (2015) The next generation digital learning environment-A report on research (ELI Paper). EDUCAUSE Learning Initiative. pp. 1–11. Available from: http://library.educause.edu/~/media/files/library/2015/4/eli3035-pdf.pdf

Cain, A. (2013) Developing assessment criteria for portfolio assessed introductory programming. In: *Proceedings of 2013 IEEE International Conference on Teaching, Assessment and Learning for Engineering (TALE)*. IEEE, Bali, Indonesia. pp. 55–60. https://doi.org/10.1109/TALE.2013.6654399.

Chen, W. & Looi, C.-K. (2007) Incorporating online discussion in face to face classroom learning: A new blended learning approach. *Australasian Journal of Educational Technology*, 33(3), 307–326.

Dearing, R. (1997) *Higher Education in the Learning Society*. The National Committee of Inquiry into Higher Education, London, UK.

Dochy, F., Segers, M. & Sluijsmans, D. (1999) The use of self-, peer and co-assessment in higher education: A review. *Studies in Higher Education*, 24(3), 331–350. https://doi.org/10.1080/030750799 12331379935

Downes, S. (2005) E-learning 2.0. *ELearn*, 2005(10). Available from: http://portal.acm.org/citation.cfm?id=1104968

Driscoll, M. (2002) Blended learning: Let's get beyond the hype. Available from: http://elearningmag.com/ltimagazine/article/articleDetail.jsp?id=11755

Duderstadt, J.J. (2010) Engineering for a changing world. In: *Holistic Engineering Education: Beyond Technology*. Springer Science & Business Media, New York, NY, USA.

Dunleavy, M. & Dede, C. (2014) Augmented reality teaching and learning. In: Spector, J.M., Merrill, M.D., Elen, J. & Bishop, M.J. (eds) *Handbook of Research on Educational Communications and Technology*). Springer, New York, USA. pp. 735–745. https://doi.org/10.1007/978-1-4614-3185-5_59

Entwistle, N.J. & Entwistle, A. (1997) Revision and the experience of understanding. In: Marton, F., Hounsell, D.J. & Entwistle, N.J. (eds) *The Experience of Learning: Implications for Teaching and Studying in Higher Education*. Scottish Academic Press, Edinburgh. pp. 145–158.

Fain, P. (2017) National U experiment combines multiple pieces of personalized learning. Available from: www.insidehighered.com/news/2017/08/01/national-u-experiment-combines-multiple-pieces-personalized-learning [Accessed 13 September 2017].

Falchikov, N. (2005) *Improving Assessment through Student Involvement: Practical Solutions for Aiding Learning in Higher and Further Education*. RoutledgeFalmer, New York, USA.

Felder, R.M., Woods, D.R., Stice, J.E. & Rugarcia, A. (2000) The future of engineering education II. Teaching methods that work. *Chemical Engineering Education*, 34(1), 26–39.

Finkelstein, N.D., Adams, W.K., Keller, C.J., Kohl, P.B., Perkins, K.K., Podolefsky, N.S., . . . LeMaster, R. (2005) When learning about the real world is better done virtually: A study of substituting computer simulations for laboratory equipment. *Physical Review Special Topics – Physics Education Research*, 1(1), 010103. https://doi.org/10.1103/PhysRevSTPER.1.010103

Foley, G. (2000) *Understanding Adult Education and Training*. Allen & Unwin, St. Leonards.

Garrison, D.R. & Kanuka, H. (2004) Blended learning: Uncovering its transformative potential in higher education. *The Internet and Higher Education*, 7(2), 95–105.

Garrison, D.R. & Vaughan, N.D. (2008) *Blended Learning in Higher Education: Framework, Principles, and Guidelines*. John Wiley & Sons, San Francisco, CA, USA.

Garrison, R. (2000) Theoretical challenges for distance education in the 21st century: A shift from structural to transactional issues. *International Review of Research in Open and Distance Learning*, 1(1), 1–17.

Goodhew, P.J. (2010) *Teaching Engineering – All You Need to Know About Engineering Education But Were Afraid to Ask*. UK Centre for Materials Education, Liverpool.

Herrington, J., Reeves, T.C. & Oliver, R. (2014) Authentic learning environments. In: *Handbook of Research on Educational Communications and Technology*. Springer, New York, USA. pp. 401–412. https://doi.org/10.1007/978-1-4614-3185-5_32

Heywood, J. (2005) *Engineering Education: Research and Development in Curriculum and Instruction*. John Wiley & Sons, Hoboken, NJ, USA.

Ho, O. (2016) Shop for new skills at supermarket-styled roadshow for SkillsFuture Credit. *The Straits Times*. Available from: www.straitstimes.com/singapore/education/shop-for-new-skills-at-supermarket-styled-roadshow-for-skillsfuture-credit

Holmegaard, H.T., Ulriksen, U. & Madsen, L.M. (2010) Why students choose (not) to study engineering. In: *Joint International IGIP-SEFI Annual Conference 2010*. Trnava, Slovakia. Available from: www.researchgate.net/publication/268425887_Why_students_choose_not_to_study_engineering

Ito, M., Gutiérrez, K., Livingstone, S., Penuel, B., Rhodes, J., Salen, K. . . . Craig Watkins, S. (2013) *Connected Learning: An Agenda for Research and Design*. Digital Media and Learning Research Hub, Irvine, CA, USA. Available from: https://dmlhub.net/wp-content/uploads/files/Connected_Learning_report.pdf

Jelfs, A., Nathan, R. & Barrett, C. (2004) Scaffolding students: Suggestions on how to equip students with the necessary study skills for studying in a blended learning environment. *Learning Media and Technology*, 29(2), 85–96.

Johnson, R.T. & Johnson, D.W. (1985) Student-student interaction: Ignored but powerful. *Journal of Teacher Education*, 36(4), 22–26. https://doi.org/10.1177/002248718503600406

Jordan, K. (2015) Massive open online course completion rates revisited: Assessment, length and attrition. *The International Review of Research in Open and Distributed Learning*, 16(3). https://doi.org/10.19173/irrodl.v16i3.2112

Kanuka, H. & Anderson, T. (1998) Online social interchange, discord, and knowledge construction. *Journal of Distance Education*, 13(1), 57–74.

Kanuka, H. & Rourke, L. (2008) Exploring amplifications and reductions associated with e-learning: Conversations with leaders of e-learning programs. *Technology, Pedagogy and Education*, 17(1), 5–15.

Kearney, S., Perkins, T. & Kennedy-Clark, S. (2016) Using self- and peer-assessments for summative purposes: Analysing the relative validity of the AASL (Authentic Assessment for Sustainable Learning) model. *Assessment & Evaluation in Higher Education*, 41(6), 840–853. https://doi.org/10.1080/02602938.2015.1039484

Kitto, K., Cross, S., Waters, Z. & Lupton, M. (2015) Learning analytics beyond the LMS : The connected learning analytics toolkit. In: *Proceedings of the 5th International Learning Analytics and Knowledge (LAK) Conference*. ACM, Poughkeepsie, New York, USA. Available from: https://eprints.qut.edu.au/81343/

Kolb, A.Y. & Kolb, D.A. (2005) *The Kolb Learning Style Inventory – Version 3.1: 2005 Technical Specifications*. Hay Group, Boston, MA, USA. Available from: www.whitewater-rescue.com/support/pagepics/lsitechmanual.pdf

Laurillard, D. (2006) E-learning in higher education. In: Ashwin, P. (ed) *Changing Higher Education: The Development of Learning and Teaching*. Routledge, London. pp. 71–84.

Lawson, R. (2017) Assuring learning. Available from: www.assuringlearning.com/ [Accessed 12 September 2017].

Long, P. & Siemens, G. (2011) Penetrating the fog: Analytics in learning and education. *EDUCAUSE Review*, 46(5). Available from: http://er.educause.edu/articles/2011/9/penetrating-the-fog-analytics-in-learning-and-education

Lowe, D. (2014) MOOLs: Massive open online laboratories: An analysis of scale and feasibility. In *2014 11th International Conference on Remote Engineering and Virtual Instrumentation (REV)*. pp. 1–6. https://doi.org/10.1109/REV.2014.6784219

Luxton-Reilly, A. & Denny, P. (2010) Constructive evaluation: A pedagogy of student-contributed assessment. *Computer Science Education*, 20(2), 145–167.

Mason, G., Williams, G. & Cranmer, S. (2009) Employability skills initiatives in higher education: What effects do they have on graduate labour market outcomes? *Education Economics*, 17(1), 1–30. https://doi.org/10.1080/09645290802028315

McLoughlin, C. & Oliver, R. (1998) Maximising the language and learning link in computer learning environments. *British Journal of Educational Technology*, 29(2), 125–136.

Meek, S.E.M., Blakemore, L. & Marks, L. (2017) Is peer review an appropriate form of assessment in a MOOC? Student participation and performance in formative peer review. *Assessment and Evaluation in Higher Education*, 42(6), 1000–1013.

Merchant, G. (2011) Unravelling the social network: Theory and research. *Learning, Media and Technology*, 37(1), 4–19. https://doi.org/10.1080/17439884.2011.567992

Moore, S. (2016) Gartner highlights top 10 strategic Technologies for higher education in 2016. Available from: www.gartner.com/newsroom/id/3225717 [Accessed 13 September 2017].

Mueller, J. (2016) Authentic assessment toolbox home page. Available from: http://jfmueller.faculty.noctrl.edu/toolbox/index.htm [Accessed 13 September 2017].

Nickerson, J.V., Corter, J.E., Esche, S.K. & Chassapis, C. (2007) A model for evaluating the effectiveness of remote engineering laboratories and simulations in education. *Computers & Education*, 49(3), 708–725. https://doi.org/10.1016/j.compedu.2005.11.019

O'Connell, A.J. (2016) The Blurry Definitions of Adaptive vs. Personalized Learning. Campus Technology. Available from: https://campustechnology.com/articles/2016/12/20/the-blurry-definitions-of-adaptive-vs-personalized-learning.aspx

Oliver, M. & Trigwell, K. (2005) Can blended learning be redeemed? *E-Learning*, 2(1), 17–26.

Osgurthorpe, R.T. & Graham, C.R. (2003) Blended learning environments: Definitions and directions. *The Quarterly Review of Distance Education*, 4(3), 227–233.

Pawley, A.L. (2009) Universalized narratives: Patterns in how faculty members define "engineering". *Journal of Engineering Education*, 98(4), 309–319. https://doi.org/10.1002/j.2168-9830.2009.tb01029.x

Peter, S. & Deimann, M. (2013) On the role of openness in education: A historical reconstruction. *Open Praxis*, 5(1), 7–14. https://doi.org/10.5944/openpraxis.5.1.23

Piaget, J. (1969) *The Psychology of the Child*. Routledge & Kegan Paul, London, England.

Prensky, M. (2001) Digital natives, digital immigrants. *On the Horizon*, 9(5), 1–6.

Prensky, M. (2007) How to teach with technology: Keeping both teachers and students comfortable in an era of exponential change. Available from: http://publications.becta.org.uk/download.cfm?resID=25940

Pugliese, L. (2016) Adaptive learning systems: Surviving the storm. *EDUCAUSE Review*. Available from: http://er.educause.edu/articles/2016/10/adaptive-learning-systems-surviving-the-storm

Purchase, H., Hamer, J., Denny, P. & Luxton-Reilly, A. (2010) The quality of a PeerWise MCQ repository. In: *Proceedings of the Twelfth Australasian Conference on Computing Education – Volume 103*. Australian Computer Society, Inc., Darlinghurst, Australia. pp. 137–146. Available from: http://portal.acm.org.ezproxy.cqu.edu.au/citation.cfm?id=1862219.1862238

Ramsden, P. (2003) *Learning to Teach in Higher Education*, Vol. 2. RoutledgeFalmer, London and New York, USA.

Rockley, A., Cooper, C. & Abel, S. (2015) *Intelligent Content: A Primer*. XML Press, Laguna Hills, CA, USA.

Salzmann, C., Govaerts, S., Halimi, W. & Gillet, D. (2015) The Smart Device specification for remote labs. In: *Proceedings of 2015 12th International Conference on Remote Engineering and Virtual Instrumentation (REV)*. pp. 199–208. https://doi.org/10.1109/REV.2015.7087292

Sauter, M., Uttal, D.H., Rapp, D.N., Downing, M. & Jona, K. (2013) Getting real: The authenticity of remote labs and simulations for science learning. *Distance Education*, 34(1), 37–47. https://doi.org/10.1080/01587919.2013.770431

Sawyer, R.K. (2014) *The Cambridge Handbook of the Learning Sciences*. Cambridge University Press, Cambridge, UK.

Serrano-Laguna, Á., Martínez-Ortiz, I., Haag, J., Regan, D., Johnson, A. & Fernández-Manjón, B. (2017) Applying standards to systematize learning analytics in serious games. *Computer Standards & Interfaces*, 50, 116–123. https://doi.org/10.1016/j.csi.2016.09.014

Siemens, G. (2004) Connectivism: A learning theory for the digital age. Available from: www.elearnspace.org/Articles/connectivism.htm [Accessed 1 November 2016].

Stacey, E. & Gerbic, P. (eds). (2009) Effective blended learning practices. *IGI Global*. Available from: www.igi-global.com/bookstore/titledetails.aspx?titleid=301

Tawfik, M., Salzmann, C., Gillet, D., Lowe, D., Saliah-Hassane, H., Sancristobal, E. & Castro, M. (2014) Laboratory as a service (LaaS): A novel paradigm for developing and implementing modular remote laboratories. *International Journal of Online Engineering (IJOE)*, 10(4), 13–21.

Topping, K.J. (2009) Peer assessment. *Theory into Practice*, 48(1), 20–27. https://doi.org/10.1080/00405840802577569

UNSW. (2017) Assessing by portfolio. Available from: https://teaching.unsw.edu.au/assessing-portfolio [Accessed 13 September 2017].

VanLEHN, K. (2011) The relative effectiveness of human tutoring, intelligent tutoring systems, and other tutoring systems. *Educational Psychologist*, 46(4), 197–221. https://doi.org/10.1080/00461520.2011.611369

Vygotsky, L.S. (1978) *Mind in Society: The Development of Higher Psychological Processes*. Harvard University Press, Cambridge, MA, USA.

Wankat, P.C. & Oreovicz, F.S. (2015) *Teaching Engineering*, 2nd ed. Purdue University Press. Available from: http://ezproxy.cqu.edu.au/login?url=http://search.ebscohost.com/login.aspx?direct=true&db=nlebk&AN=1067069&site=ehost-live

Wiley, D. (2014) The access compromise and the 5th R. Available from: http://opencontent.org/blog/archives/3221

Willison, J. (2017) Research skill development for curriculum design and assessment. Available from: www.adelaide.edu.au/rsd/ [Accessed 12 September 2017].

Chapter 6

The gamification of education

A case study in using positive psychology and game design to increase student engagement

T.B. Nguyen

College of Engineering & Science, Victoria University, Melbourne, Victoria, Australia

ABSTRACT: This chapter describes the implementation of gamification in a first-year engineering physics course to increase students' engagement with the course. Following completion of the gamified course, student engagement, performance and understanding of the subject matters all increased.

6.1 Introduction

Today's students have grown up in the age of gaming and have not known a world without video games. It is estimated that an average young American adult can spend 10,000 hours playing games by the age of 21. According to a Neilson report, 61% of gamers are millennials between the age of 18 to 34. Today's students have come to expect sustained stimulation and short-term rewards offered by games. This vastly differs from the realities of the university experience where long hours of passive learning dominates. This leads to disengagement and boredom of the learning experience. Student engagement in engineering is of concern to engineering schools and faculties, particularly in the first year where attrition rates are historically high compared to higher years. Games incorporate elements and mechanisms that can successfully engage and motivate a player and change the player's behavior. Gamification is the application of these game elements and mechanisms to non-game contexts, such as education. Adding a layer of gamification to the curriculum can allow students to exercise more freedom to determine their own learning approaches in a way that best suits their learning styles. Game principles have the potential to address key issues relating to student learning and engagement that include:

- Mastery of learning: gamification allows students to repeat learning activities until they have mastered certain skills or comprehension before attempting levels that are more difficult. Through cycles of learning and practice, expertise and mastery are developed. This benefits both the retention and understanding of the subject matter.
- Active learning through safe exploration of ideas and reflection: gamification can encourage students to think critically by taking risks and exploring concepts in the safety of the game sandbox while understanding the consequences of these actions without suffering from those consequences. An example of this is the use of simulations of activities that are potentially dangerous or harmful if mistakes are made in the real environment.
- Flexible learning: gamified learning can be done anytime and anywhere with one or multiple players in collaborative and social environments. This affords students flexibility

to learn as individuals and as a team. By embedding the scaffolding that games offer to players, students are able to customize their own learning experience, enabling them to understand about how they learn, as well as about learning and thinking for themselves.

- Increase motivation: motivation lies in activities that feel challenging, but are achievable. Gamified learning can cater to the needs of individual students by simultaneously offering different levels of difficulty. Gamified learning can thus effectively target an individual student's zone of proximal development (Daniels, 2017). Gamified learning makes the learner the active agent, and this agency is rewarded through the achievement of learning goals, which are subsequently recognized with incentives such as rewards, status and ranking. This rewards-based system promotes active participation and engagement.

The principles behind effective gamification are the applications of positive psychology that appeals to a player's intrinsic motivations for mastery, autonomy and purpose (Pink, 2009). Gamification has been extensively applied in the business and consumer markets for many years as well as in the K–12 education sector (Gee, 2007; Zichermann and Linder, 2010). However, the uptake of gamification in higher education has been low. One reason for this is that many academics feel that games are childish and trivialize learning or are condescending to students' abilities and motivations. Others note that gamification has only short-term impact and does not drive long-term behavioral change (Farzan *et al.*, 2008). Academics in engineering disciplines have traditionally been conservative in their approach to teaching and have stuck with the lecture as the primary mode of teaching. In recent years, the benefits of gamification in education have been studied (Klopfer *et al.*, 2009), and gamification in courses has been gaining traction (Younis and Loh, 2010). A report by consulting firm Sage Road Solutions identifies game-based learning in higher education in the US as being at the innovators stage of the adoption curve but is quickly gaining momentum (Derryberry, 2012). This is further supported by the New Media Consortium's Horizon 2016 report that predicts large-scale adoption of game-based learning into mainstream education (Johnson *et al.*, 2016).

Much of the gamification of education involves game-based learning where games are used as tools for delivering knowledge and learning content, to enhance the learning experience (Killi, 2005) and to develop skills that then transfer to non-game applications (Tobias *et al.*, 2014). Others have gamified the curriculum by adding social elements or altering the content to make it more game-like. This type of gamification can be referred to as content gamification (Kapp, 2013). Examples of content gamification include learning games, simulations, role-play and case-based learning. The structure of the course can also be made more game-like in what is referred to as structural gamification by adding game mechanics such as points, levels and leaderboards to the learning experience (Kapp, 2013).

This chapter describes how structural gamification was applied to a first-year introductory engineering physics unit of study, or course. The chapter reports on the effects that the gamification had on student engagement of the course, as well as their performance in the course.

6.2 The structure of a game

The structure of most games consist of some typical elements and mechanisms, of which, the main ones are described below.

1 All games have a clear *goal* or purpose. In a competitive game, like tennis, the goal is to beat the other opponent. In a game like Monopoly, the goal is to bankrupt your

opponents. In the popular video game, Angry Birds, the goal is to knock off all of the pigs on the level. These goals are aligned with the player's own purpose and intrinsic motivations. These motivations may include recognition of their efforts or skills, earning rewards (financial or otherwise) or personal enjoyment.

2 All games have *rules*, which all players must follow to play the game in order to reach the end goal. These rules ensure fairness and consistency in how the game is played as well as defines how the player progresses through the game.
3 Most games have *levels* that define the player's ability or performance. The player must start at the lowest level and progress up to the top or final level. At each level, the activities become progressively more difficult or challenging. In the FIFA World Cup for example, football teams from different nations play in tournaments in continental zones to qualify for the final tournament. The final tournament itself consists of a number of levels, starting with round-robin matches, followed by the knock-out stage, followed by the quarter-finals, then by the semi-finals, then finally to the grand final where the winner is decided.
4 Games provide some type of recognition or *reward* to acknowledge the player's performance or achievements. This may include earning points, rankings on a leaderboard, levels of status or some other artifact that represents achievement. In professional games like sports, players are financially rewarded through prize money for their performance or achievements.
5 To play any game well, players must be proficient in the skills, knowledge and methods of the game. This requires effort. Games are thus good at training the player to *learn* a particular skill. When a player takes an action in a game, there is instant, or near instant, *feedback* provided on how the player has performed. When a tennis player hits the ball out of court, the player knows instantly that the swing of the tennis racquet was not correct.
6 The act of playing a game is a *meaningful activity* to the player who has engaged in the game to achieve some personal goal. If playing the game was not meaningful and did not help the player progress towards achieving his/her goal, the player would be unmotivated to continue playing.
7 Lastly, most games afford some level of *social connectedness*, which many players consider to be an important reason for playing the game. The competitive nature of many games, where one player is pitted against another player, is often a driving factor for motivation. This is true even in games where the players do not directly compete against each other (e.g. in the game of golf).

These elements underscore some basic principles of intrinsic motivations that increase engagement. Games are structured to provide the player with a purposeful challenge leading to the attainment of some goal, reward or recognition. The game environment provides the player with a sandbox to safely learning new skills. The player is provided with feedback and scaffolding to explore and master these skills autonomously. This combination of purpose, autonomy and mastery is key to any game's ability to engage the player with that game. Consequently, games can powerfully change player behavior.

6.3 Comparing games to education

Compare the structure and elements of a typical game to that of a typical formal learning environment as shown in Figure 6.1; the similarities become evident.

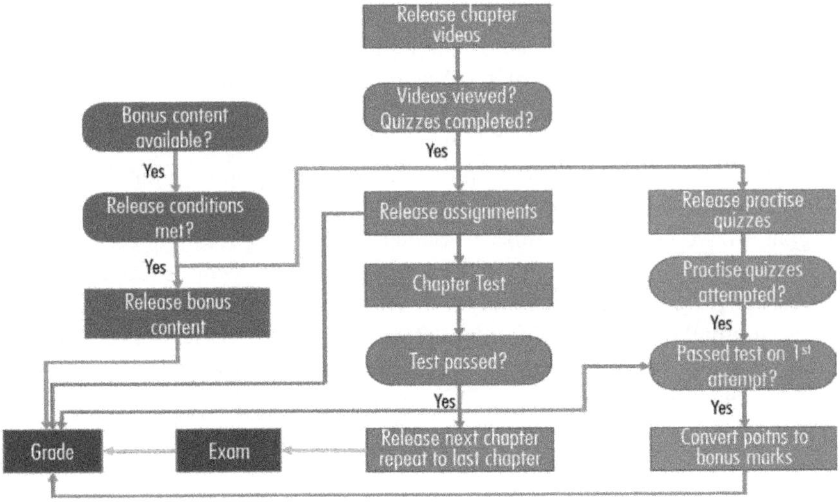

Figure 6.1 Managed progression through the course

Table 6.1 Comparison of a game structure and a course

Game Elements	*Course Elements*
Goals	Learning outcomes
Rules	Policies/assessments
Levels/progression	Prerequisites/hurdles
Rewards/recognition	Marks/grades
Feedback/scaffolding	Feedback/scaffolding
Meaningful activity	Learning activities
Social connectedness	Classroom and team-based learning

Despite such similarities, there are notable differences in the ways that these elements are constructed and implemented. This results in differences in how the user (student) interacts with those elements, thus affecting user engagement. Discussed in further detail here are three of the main differences.

6.3.1 Course progression

A course is structured much like a game in that there is a logical progression through the course from start to end. Students engage in learning activities in which they acquire the necessary knowledge and skills to complete assessments, which can earn marks that count towards their grade. The student is deemed to have satisfactorily passed the course when that student has met the assessment rules, demonstrated the attainment of the learning outcomes and earned a passing grade for that course. Yet it is possible to complete a course without having acquired a conceptual understanding of the course content and without having completed all of the necessary learning activities or assessments (Michael *et al.*, 2017). It is also possible to progress through the course without having fully engaged with the learning activities.

Studies have reported class attendance rates vary in range from 81.5% (Marburger, 2001) to 30% (Moore, 2006). A meta-analysis conducted by Credé *et al.* showed a strong correlation between attendance rates and class grades (Credé *et al.*, 2010). Participation in a course is flexible with the onus on the students to be self-regulated in their level of participation. Contrast this with a game where progression through the game is tightly controlled and the player must demonstrate a certain level of proficiency at one level before being entitled to attempt the next level. This forces the player to spend time practicing and gaining the minimum level of proficiency before progressing forward in the course, thus ensuring that the player has attained a certain level of preparedness and proficiency for the next level. On the other hand, a course will permit a student to progress to the next topic within the syllabus without having passed the assessments from the previous topic. In this situation, the student progresses to the next level with some deficiency that leaves that student less prepared for that level, thus decreasing his or her chances of successfully completing that level. This is akin to someone being allowed to apply for a heavy vehicle license without first successfully attaining a car license.

6.3.2 Course grading

In a game, players earn points as they progress through the game starting with score of zero. This is logical, as the player does not have any achievements at the beginning of the game to be rewarded with any points. Points are earned whenever the player completes or succeeds at a point-bearing task. For example, a player will earn points every time an enemy is killed in a first-person shooter video game. Points are awarded continuously, or quasi-continuous, and accumulated throughout the game. Contrast this with how marks are allocated in a course. Students are graded based on the outcomes of an assessable activity, such as an assignment or a test. There are usually a small number of tests and assignments throughout the semester, so assessments are infrequent. When a mark is allocated for the completed assessment, it is common to say that the student has "earned" that mark. This in fact is incorrect and the student has actually lost marks. The approach to grading assessments is punitive and students lose marks for areas in the assessment that they got wrong. To illustrate this, take a simplistic example of a course where there are only two tests, with each test worth 50% of the overall course grade, e.g. each test carries 50 marks. Consider a student who only receives 30 marks for the first test. This means that the 20 marks that the student did not receive is lost, thus limiting the maximum grade that can be received for the course to 80 marks subsequent to that first assessment (assuming that the student will earn full marks on the second test). If the student does not earn full marks for the second test, then that maximum possible grade is further reduced by the amount equals to that lost from the second tests. Unless the student earns full marks for each assessment, there is a progressive reduction in the maximum grade which can be earned. Each non-perfect score on an assessment takes marks away from students and their overall grades reduce from 100% downwards. Any experience of loss produces stress and anxiety caused by the release of cortisol, a stress hormone, which negatively impacts on the student's mental state, and thus their performance (Preuss *et al.*, 2010). The experience generates a negative feedback loop causing students to be more stressed and anxious at the next assessment. At best, students are relieved when they receive a high mark. Contrast this with the experience of a gamer who earns points as he/she progresses through the game. The reward of earning points creates a positive experience due to the release of dopamine, a neurotransmitter that controls the brain's reward and pleasure centers (Schultz, 1997). The release of dopamine generates a positive feedback loop that encourages the player to continue to seek out the reward by earning more points and thereby staying engaged with the game. Thus, the points system in games encourages engagement while the grade system in education discourages engagement.

6.3.3 Feedback

In a game, players often receive instant, or near instant, feedback on their performance. When a basketball player shoots the basketball at the ring, within a matter of seconds or less, the player will know if the shot was too hard, too soft, too much to the left or right, or if the shot was perfectly executed. This instant feedback allows the player to readjust his/her shot the next time to make any corrections to the shot if necessary. In nearly every game, the consequences of the player's actions are instant. In the learning environment, however, this is not the case when it comes to receiving feedback from assessments. Assignments and tests take time to be marked and students will not be notified of their marks for days or even weeks after the assignment due date. The received mark is just one form of feedback. Other feedback may include comments, corrections or solutions. Feedback is universally considered to be an important factor in learning and is essential in good teaching practices. Despite this, the level of timely feedback is often inadequate (Orrell, 2006). It is also arguable that much of the feedback is of little consequence. This is because students rarely have the opportunity to use the feedback to correct their submissions for reassessment. After one topic in the syllabus is complete and assessed for, the teacher then moves the entire class on to the next topic in the syllabus. This leaves no time for students with any deficiencies in their understanding to address those deficiencies before moving on. The deficiencies put those students at a higher risk of failing the next topic, and the next, and the next. The deficiencies and the risks of failure accumulate.

6.4 A gamified course

To address the issues raised in the previous section, a course in introductory engineering physics was gamified using structural gamification. Traditionally, physics has been taught as a series of lectures followed by tutorials and laboratory experiments. Students listened to the lectures, practiced applying the content to tutorial questions and undertook practical experiments, which provided physical applications of the theory. The students' understanding of the theory was evaluated through in-semester tests, laboratory reports and end-of-semester examinations.

The introductory physics syllabus covers content that includes measurements and errors, kinematics, forces, energy, momentum and wave motion. As students progress through the course, they are periodically assessed with two in-semester tests as well as submitting five practical assignments, one every fortnight. In total, there are seven assessments in a 12-week period. The final assessment is an exam worth 50% of the total grade. After sitting a test or submitting an assignment, students move on to the next topic. They may have a sense of their performance, and how well they are doing, but will not know for sure if they have passed or failed the course until the release of their grades after the final exam.

The gamification of the course involved breaking the syllabus into a sequence of small learning activities, with the smallest activity being instructional videos that replaced traditional lectures. Each chapter in the syllabus contains 9–13 short videos totaling about 1 hour in duration. After all of the videos within the chapter have been viewed and the in-video quizzes completed, the chapter assignment is then released. When the chapter assignment has been completed and submitted, students then attempt the chapter test as the last hurdle to passing the chapter. If the chapter test was successfully completed, the next chapter will be released and the process for progressing through that chapter repeats until the last chapter has been completed. The final assessment is the end-of-semester examination. Figure 6.1 shows the managed progression through the course.

In addition to the controlled progression, other game mechanisms have also been added to the gamified course delivery.

6.4.1 Learning for mastery

One of the key characteristics of most games is that the game can be repeated as often as required for the player to master the game. In an educational context, mastery means having a thorough understanding of the underlying concepts and knowledge and the skills to apply that knowledge to solve an array of problems. Mastery comes through practice. It has been shown that teaching for mastery can increase student learning by one standard deviation compared to traditional lecturing (Bloom, 1984). Increasing mastery of learning can be facilitated through the use of videos to replace lectures as well as increasing the number of problems that students are exposed to. The videos in this course range in duration from 2 minutes to 10 minutes. They were purposefully kept short to allow students time between videos to absorb and understand the video content. Breaking a learning sequence into a smaller number of "chunks" helps to improve memory retention of the learning materials (Miller, 1956). Student surveys have suggested that the optimum duration for such videos is between 5 and 10 minutes (Tsur, 2016). Additionally, the short duration helps maintain viewer attention and maximizes learning concentration (Shephard, 2003).

To increase student understanding of the learning materials, the videos included short multiple choice quizzes, an example of which is shown in Figure 6.2.

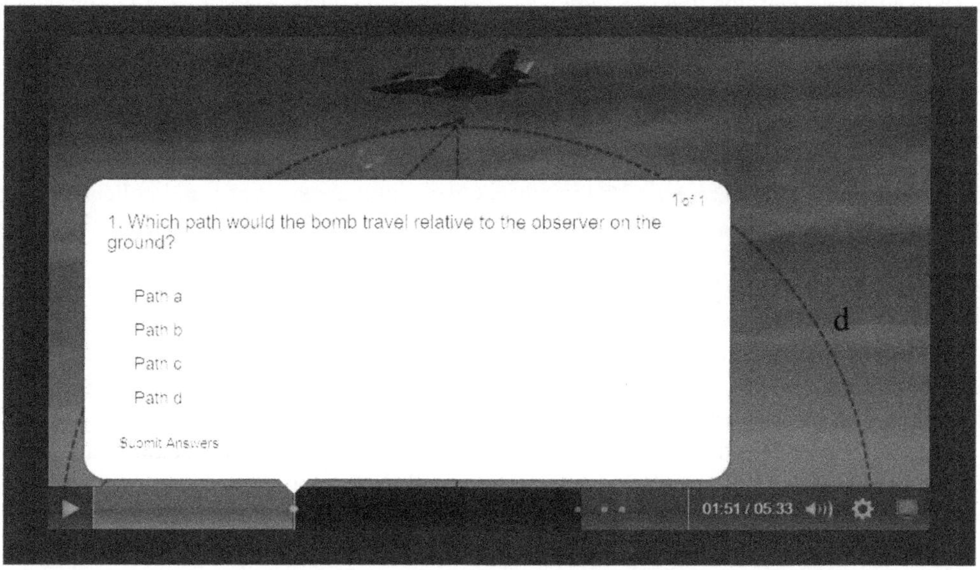

Figure 6.2 Screenshot of an online video and in-video quiz

The number of quizzes ranged from two to five per video. They acted as checkpoints to gauge student understanding. Students were required to correctly answer the in-video quizzes, demonstrating their comprehension of the learning concepts, before being allowed to progress to the next video. In total, 70 videos were created containing over 210 in-video quizzes. The short duration of the videos and the high number of in-video quizzes ensured that students were receiving frequent feedback on their understanding in small increments that helped them to improve their progress through the course. As most quizzes were multiple choice and conceptual in nature, the workload required to complete the high number of quizzes was minimal and did not pose a burden on students. Consistent with game design, students had to demonstrate their understanding of the content to be able to continue through the video sequence. This ensured that the pre-requisite knowledge needed for subsequent chapters was sufficiently acquired. Students who did not sufficiently understand one part of the syllabus would not carry this deficiency to the next part of the syllabus, thus improving their successful progression and completion of the course.

6.4.2 Purposeful learning

The short online videos were designed to break the syllabus into small chunks, enabling students to make small incremental steps towards completing the course. The videos also afforded students the flexibility and autonomy in when, and how, they interacted with the videos. The practical assignments were also designed to give students flexibility and autonomy. More importantly, they provided a more meaningful reason for learning. In a traditional laboratory practicum, students are given a certain amount of time to complete an experiment by following a prescribed set of instructions for conducting that experiment. While this may be considered as hands-on and practical learning, it is nonetheless passive, as students do not have to think about how the experiment was designed to elicit the desired outcome. On the other hand, active learning requires students to participate in the process of creating their own learning experience (Blumenfield *et al.*, 1991). This increases the ownership of the learning experience and thus increases motivation.

In this implementation, assignment briefs were provided to give an objective and context, but they did not prescribe how the objective would be met or how the solution could be found. Instead, students initiated their own scientific inquiry by analyzing, understanding and refining the problem, creating a hypothesis, designing experiments, and collecting and analyzing data that would test that hypothesis. They would then draw conclusions and communicate their findings. Throughout this process, their understanding is continuously subject to revision and therefore improvement. The inquiry-based learning affords students the capacity to fail, without consequence to their progression, and to learn from those failures. This empowers students to explore ideas at a deeper level as well as laterally without fear of failing. The autonomous, self-directed learning increases the self-agency of the student, thus helping to increase motivation.

The assignments in this course were practical assignments delivered in a problem-based learning (PBL) format in which students are required to develop their own techniques and experiments to demonstrate a physics principle or solve a practical problem (Nguyen, 2012). An example of such a problem is to measure the drag coefficient of a student's own car and evaluate how this resistance affects the car's fuel efficiency. The problems are real, non-trivial and connect with the student's own prior knowledge, thus providing them with a meaningful learning experience that they can directly relate to. The PBL assignments encourage students to take a systematic and scientific approach to formulating solutions to problems through a seven-step process, as shown below.

Students are assessed on each of the seven steps to ensure that the scientific method of inquiry is instilled in their approach to learning. Students are not only graded on the outcome, but also on the process. As each assignment follows the same assessment rubric and format, students can thus see their improvements on each criterion in the assessment rubric with each iteration of the assessments.

6.4.3 Rewarding effort and performance

In a typical learning environment, the learner is "rewarded" with marks on performance and ability. While this may seem logical (a student who understands the work most and can answer more questions correctly will earn higher grades), it can have negative psychological effects on the learner.

As explained previously, the grading system is punitive and discourages engagement. Assessments convey to the learner that if you do not do well, you will be punished with low marks. This puts the learner into the mindset of failure avoidance, a mindset that fixes the learner to studying for the assessment rather than studying to learn. This failure avoidance leads to superficial, surface learning that is both ineffective and inefficient.

In a gamified learning environment, students have opportunities to earn marks or rewards that reflect both their effort and performance. As shown in Figure 6.1, activities that students can earn marks from are released after they have completed watching the chapter videos. These activities include practical assignments, quizzes, chapter tests and additional bonus activities that can earn students extra credit. These bonus activities include games and simulations as well as more advanced assignments and more difficult quizzes. All of these activities are conditionally released to students based on their individual performance. For example, a student who scores highly on the in-video quizzes in the first attempt will show good understanding of the underlying principles described in the video and will therefore receive more advanced content to match their level of comprehension. This keeps them challenged and more engaged. Students who have completed watching their chapter videos within a certain time can unlock learning tools, such as formula sheets and online formula calculators. Students can attempt practice quizzes for which they will earn conditional points. These conditional points are converted into real marks if they pass the chapter test on the first attempt. This condition creates a positive reinforcement loop in which students are incentivized to attempt practice quizzes to better prepare them for the test, and in passing the test, are rewarded for completing the quizzes. Figure 6.2 shows the breakdown of the marks for the assessment tasks in this course.

From the bonus activities, students are able to earn more points than what is usually allowed in a traditional deficit grading scheme. This affords students with a number of benefits. Students can make up for lost marks if they have not performed so well in mandatory assessments, or they can use the bonus activities to boost their overall mark and move them to higher grade scales. The bonus marks also enable students to pass the unit before having to sit for the final examination. This reduces the reliance of passing the examination to pass the course and therefore reduces the stress and anxiety that is often felt during exam time. Students who are less stressed during the exam are more likely to do better (Culler and Holahan, 1980).

Table 6.2 Breakdown of assessment tasks

Activity	*% of Total Mark*
Chapter tests	30%
Practical assignments	20%
End-of-semester exam	50%
Bonus activities	20%
TOTAL	120%

6.4.4 Developing discipline

While effective teachers and well-developed curriculum are important in student learning, extensive research over the past few decades has shown that student success and achievement are also strongly determined by student attitudes towards themselves and their own learning abilities (Dweck *et al.*, 2014). Student attributes such as grit and the perseverance that come from having a growth mindset (Dweck, Mindset: The New Psychology of Success, 2007) are strong predictors of success (Duckworth, 2016). There is an expectation that students are disciplined and self-regulated in their learning. However, courses lack the structure that helps to develop or reinforce discipline. Students live increasingly busy lives, spending less time studying compared to the past. A common strategy for students to cope with the demands of studying is to cram prior to an assessment. Cramming is the tactic of intensive study immediately before an assessment is due. In this gamified course, the frequent number of assessments, and their reduced weighting, smooth out the level of cramming and reduce their intensity in a way that improves the management of student workloads. The regular requirement to have work completed in order to progress through the course encourages good study habits and effective study time. Both of these factors have shown to increase student performance (Nonis and Hudson, 2006).

6.5 Outcomes from a gamified course delivery

The following section describes the outcomes of the gamification of the physics course. It looks specifically at the levels of engagement, student performance and improvements in their overall understanding of the course content.

6.5.1 Increased engagement

The gamified course was delivered on the university's learning management system, Desire-2Learn (D2L). D2L has built-in analytics to monitor and analyze the performance and level of engagement of each student. The analytics reveal a high level of engagement among students with the online learning content as shown in Figure 6.4. For example, the average time spent by students watching the first video, *Intro to Numbers*, was 7:31 minutes compared to the video length of 2:56 minutes. On average, students viewed each video 2.9 times. The average time spent watching the videos was three times the video duration. In other words, students were watching each video almost three times in full.

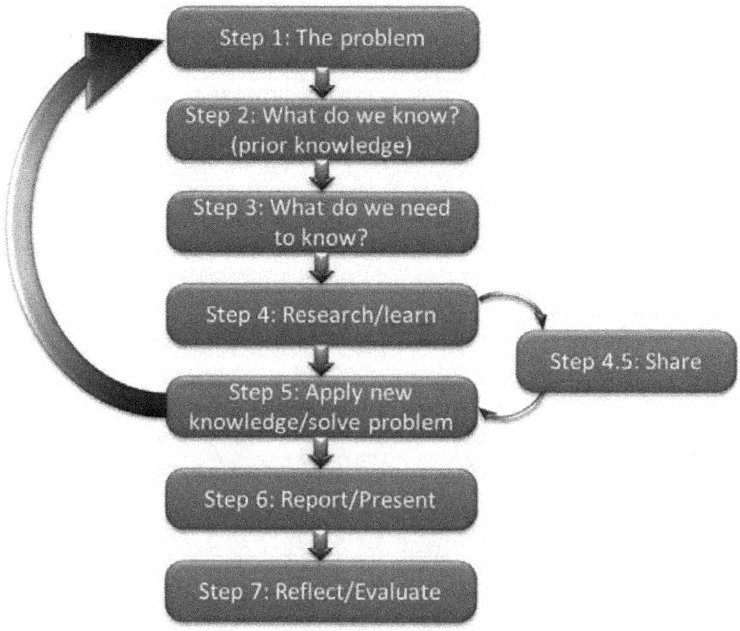

Figure 6.3 The Victoria University problem-based learning cycle

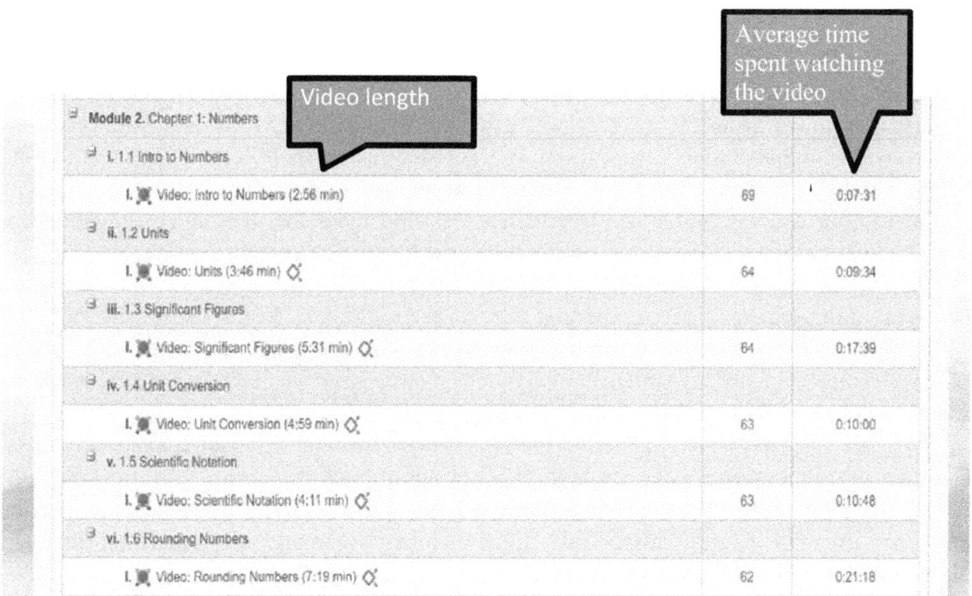

Figure 6.4 Screenshot of D2L analytics showing student engagement with the online learning resources

Table 6.3 Time spent on the student learning management system (D2L) for each course in the semester

Course	*Average time spent*
Engineering Physics I	31:12 hours
Engineering Mathematics I	4:39 hours
Intro. to Engineering Design	7:07 hours
Engineering Fundamentals	4:28 hours

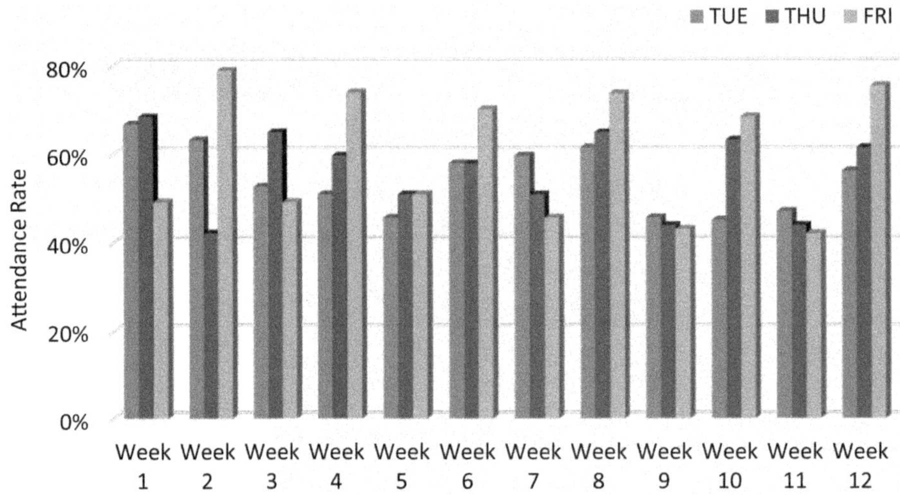

Figure 6.5 Class attendance rates

As the in-video quizzes are multiple choice, it is possible that students can guess the answers, then move on to the next video. However, the repeated views and the time spent provide evidence that students were utilizing the video resources to learn. They were going back to the videos to review concepts and ideas and repeating quizzes.

Of the four courses that students undertook during the same semester, students spent over 31 hours on average engaging online with the course compared to an average of only 5.5 hours for the other three courses, as shown in Figure 6.3. This is almost six times more time spent online on this course compared to other courses taken during the same time.

Despite the high level of online learning, students still regularly attended face-to-face classes. Further evidence of student engagement can be seen in class attendance rates, as shown in Figure 6.5.

Sixty-six students were enrolled in the course with the average attendance being 57%. Of the 66 enrolled students, 14 students did not commence the course but did not withdraw. Excluding these 14 non-commencing students, the average attendance would be 72%. This attendance rate is historically high for this course and was sustained throughout the semester. Of the remaining 52 students, 46 (88%) completed all learning activities (viewed all videos and completed all practice quizzes, tests and assignments). All 52 students completed at least one bonus activity, with 30 students completing all available bonus activities. These results showed students' willingness to put in more effort to earn higher grades if given the opportunity. Indeed, 13 students completed the semester with marks that were greater than 50% and have thus passed the course, even before sitting for the exam.

6.5.2 Increased performance

The extra marks awarded for extra work have enabled students to increase their overall grade, as shown in Figure 6.6. There were increases in the percentage of students earning grades of credit and above, with more than a doubling of students earning high distinctions. The shift towards higher grades may be perceived as grade inflation. However, grade inflation typically refers to awarding higher grades for the same standard of work that would have previously received lower grades (Pattison *et al.*, 2013). In this case, students were being awarded more marks for more work rather than for the same work.

6.5.3 Increased comprehension

In the gamified delivery, where student grades can be increased by completing more tasks, grades are thus not the best indicator of students' comprehension of the course content. Instead, this was measured using an instrument called the Force Concept Inventory (FCI) (Hestenes *et al.*, 1992). The FCI is a list of 30 multiple choice questions that test students' understanding of Newtonian physics concepts. In particular, it evaluates students' misconceptions and commonsense beliefs about motion and forces. The FCI can show if the student has done well because that student has learned by rote for the exam or if the student actually understood the subject matter. Students take the FCI twice in the course, once at the beginning

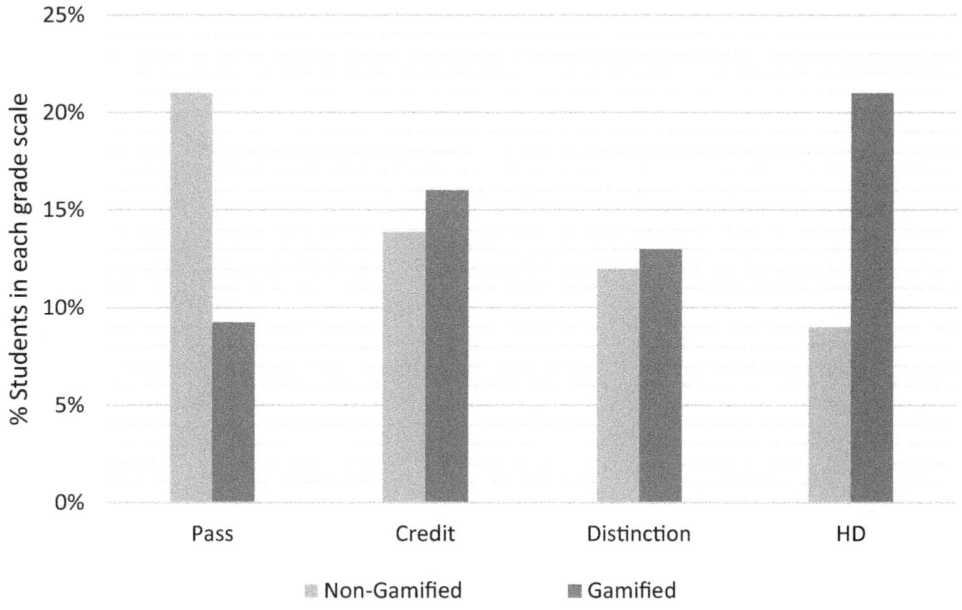

Figure 6.6 Comparison of the normalized grade distribution between a gamified and non-gamified delivery for the same course

Table 6.4 Comparison of FCI gains in a gamified and non-gamified course delivery

Delivery	*Pre-Sem. Average*	*Post-Sem. Average*	*Gain*
Lecture-based	30%	42%	12%
Gamified	29%	52%	23%

of the course, then again at the end of the course. The difference between the post-semester and pre-semester results provides information about whether the student has developed *Newtonian thinking* or whether the basic threshold concepts are not yet understood. While the FCI can be used to measure improvements in student learning, it is more often used as a tool to evaluate the effectiveness of the teaching (Hestenes & Halloun, Interpreting the Force Concept Inventory, 1996).

The results of the FCI showed significant improvements between the pre-semester and post-semester results when the course was taught using gamification compared to the more traditional delivery, as shown in Figure 6.4. The improvement represents almost a doubling of the students' understanding. The FCI results show that student understanding of the fundamental concepts in forces and mechanical physics have improved with the gamified delivery.

6.6 Conclusion

In this introductory engineering physics course, students were presented with a course that was delivered using structural gamification. Students interacted with the course in a way that was much like a game. They progressed through the course by attempting learning activities and earned points for different activities. Each level in the course must be completed before the student can progress to the next level, thus ensuring a minimum level of competency was achieved. With clearly defined rules for progression and rewards, students navigated through the course to earn as many points as they could. The more activities that they did, the more points they earned. Difficult challenges earned higher points. These points were converted into marks and then subsequently a grade. The clear progression rules enabled students to better manage their progress, thus setting the expectation of self-regulated learning.

The gamified delivery has shown to have increased engagement with online learning as well as face-to-face learning. On average, students spent six times more time online in this course compared to the other courses taken during the same semester. Nearly 90% of commencing students completed all of the course levels with the same percentage of students attempting bonus activities to earn extra points. Students earned points incrementally, starting from zero, enabling them to earn higher marks than otherwise possible in a traditional marking scheme. This had the effect of increasing the overall pass rate and improving the average performance. Through the Force Concept Inventory, it was also observed that students' understanding of force concepts and Newtonian physics was improved.

References

Bloom, B. (1984) The 2 Sigma Problem: The search for methods of group instruction as effective as one-to-one tutoring. *Educational Researcher*, 13(6), 4–16.

Blumenfield, P.C., Soloway, E., Marx, R.W., Krajcik, J.S., Guzdial, M. & Palincsar, A. (1991) Motivating project-based learning: Sustaining the doing, supporting the learning. *Educational Psychologist*, 26(3 & 4), 369–398.

Credé, M., Roch, S.G. & Kieszczynka, U.M. (2010) Review of the relationship of class attendance. *Review of Educational Research*, 80(2), 272–295.

Culler, R.E. & Holahan, C.J. (1980) Test anxiety and academic performance: The effects of study-related behaviors. *Journal of Educational Psychology*, 72(1), 16–20.

Daniels, H. (2017) *Introduction to Vygotsky*, 3rd ed. Routledge, New York, USA.

Derryberry, A. (2012) *A Scan of Game-based Learning in U.S. Higher Education*. Sage Road Solutions, CA, USA.

Duckworth, A. (2016) *Grit: The Power of Passion and Perseverance*. Scribner, New York, USA.

Dweck, C.S. (2007) *Mindset: The New Psychology of Success*. Ballantine Books, New York, USA.

Dweck, C.S., Walton, G.M. & Cohen, G.L. (2014) *Academic Tenacity: Mindsets and Skills that Promote Long-term Learning*. Bill & Melinda Gates Foundation, Seattle, WA, USA.

Farzan, R., DiMicco, J.M., Millen, D.R., Dugan, C., Geyer, W. & Brownholtz, E.A. (2008) Results from deploying a participation incentive mechanism within the enterprise. In: *CHI '08 Proceedings of the SIGCHI Conference on Human Factors in Computing Systems*. ACM, New York and Florence, Italy. pp. 563–572.

Flamberg, M. (2017) *Games 360 2017*. Nielson, New York, USA.

Gee, J.P. (2007) *What Games have to Teach us about Learning and Literacy*. Palgrave Macmillan, New York, USA.

Hestenes, D. & Halloun, I. (1996) Interpreting the force concept inventory. *The Physics Teacher*, 33, 502–506.

Hestenes, D., Wells, M. & Swackhamer, G. (1992) Force concept inventory. *The Physics Teacher*, 30, 141–158.

Johnson, L., Brown, M. & Becker, S. A. (2016) *The NMC Horizon Report: 2016 Higher Education Edition*. New Media Consortium, Austin, TX, USA.

Kapp, K. (2013, March 25) *Two Types of Gamification*. Available from: Karl Kapp: http://karlkapp.com/two-types-of-gamification/ [Accessed 17 June 2017].

Killi, K. (2005) Digital game-based learning: Towards an experiential. *The Internet and Higher Education*, 8, 13–24.

Klopfer, E., Osterweil, S. & Salen, K. (2009) *Moving Learning Games Forward*. The Education Arcade, MIT, Cambridge.

Marburger, D.R. (2001) Absenteeism and undergraduate. *Journal of Economic Education*, 32, 99–110.

McGonigal, J. (2011) *Reality is Broken: Why Games Make Us Better and How They Can Change the World*. The Penguin Press, New York, USA.

Michael, J., Cliff, W., McFarland, J., Model, H. & Wright, I. (2017) Conceptual assessment of student learning. In: *The Core Concepts of Physiology*. Springer, New York, USA. pp. 123–131.

Miller, G.A. (1956) The magical number seven, plus or minus two: Some limits on our capacity for processing information. *Psychological Review*, 63(2), 81–97.

Moore, R. (2006) The importance of admissions scores and. *Journal of The First-Year Experience & Students in Transition*, 18(1), 105–125.

Nguyen, T. (2012) A competitive approach to laboratory practicals. *3rd International PBL Symposium*. Republic Polytechnic, Singapore. pp. 388–394.

Nonis, S.A. & Hudson, G.I. (2006) Academic performance of college students: Influence of time spent studying and working. *Journal of Education for Business*, 81, 151–159.

Orrell, J. (2006) Feedback on learning achievement: Rhetoric and reality. *Teaching in Higher Education*, 11(4), 441–456.

Pattison, E., Grodsky, E. & Muller, C. (2013) Is the sky falling? Grade inflation and the signaling. *Educational Researcher*, 42(5), 259–265.

Pink, D.H. (2009) *Drive: The Surprising Truth About What Motivates Us*. Riverhead Books, New York, USA.

Preuss, D., Schoofs, D., Schlotz, W. & Wolf, O.T. (2010) The stressed student: Influence of written examinations and oral presentations on salivary cortisol concentrations in university students. *The International Journal on the Biology of Stress*, 13(3), 221–229.

Schultz, W. (1997) Dopamine neurons and their role in reward mechanisms. *Current Opinion in Neurobiology*, 7(2), 191–197.

Shephard, K. (2003) Questioning, promoting and evaluating the use of streaming video to support student learning. *British Journal of Educational Technology*, 34(3), 295–308.

Tobias, S., Fletcher, J.D. & Wind, A.P. (2014) Game-based learning. In: Spector, J.M., Merrill, M.D., Elen, J. & Bishop, M. (eds) *Handbook of Research on Educational Communications and Technology*. Springer, New York, USA. pp. 485–503.

Tsur, M. (2016) *The State of Video in Education 2016*. Kaltura, New York, USA.

Younis, B. & Loh, C.S. (2010) Integrating serious games in higher education programs. *Academic Colloquium 2010: Building Partnership in Teaching Excellence,* Ramallah, Palestine.

Zichermann, G. & Linder, J. (2010) *Game-Based Marketing: Inspire Customer Loyalty Through Rewards, Challenges, and Contests*. John Wiley & Sons, Hoboken, NJ, USA.

Chapter 7

Engineering education online

Challenges, opportunities and solutions adopted in Australian, US and EU universities

S. Shanmuganathan

University of Tun Hussein Onn Malaysia (UTHM), Malaysia

ABSTRACT: The chapter elaborates on the innovative ways, i.e. blended learning[1] (BL), e-learning, flexible learning (also sometimes referred to as hybrid learning) or "cloud" campus, that different universities in Australia, the United States of America (USA) and the European Union (EU) have embarked on to address some critical issues specific to providing engineering education online.[2] In particular, the chapter looks at the challenges, limitations, opportunities and the major research findings relating to curriculum development, pedagogy and the delivery of engineering courses online in this 21st century.

7.1 Introduction

Without a doubt, the emergence of the Internet as the main channel of communication has led an increasing number of universities to offer or explore offering engineering courses online from the certificate/diploma level to the master's level and beyond, e.g. PhD (SREG, 2017). Presently, it is possible to acquire professional body accredited engineering qualifications online in a wide spectrum of specialties, e.g. accredited qualifications from aerospace to vehicle maintenance (EECom, 2017).

The extensive use of advanced information and communication technologies (ICT) has redefined the concept and delivery of distance (Bourne *et al.*, 2005) as well as the traditional face-to-face (F2F) learning immensely (El-Zein *et al.*, 2009; Gonzalez *et al.*, 2013; Sowells *et al.*, 2017; Sun *et al.*, 2017). Significant ICT-influenced transformations in the delivery as well as the content of engineering education began to appear since the beginning of this century. However, as of today, there are still some limitations (Bourne *et al.*, 2005; Rutledge and Sugumaran, 2017) when delivering engineering courses online, as noted below.

1 The need for the quality of the programs either to be equal to or better than those of the traditional classroom setup;
2 The availability of the programs when and where needed as well as accessibility by any number of students and
3 The availability of courses on a wide range of topics/specialties.

Furthermore, despite the fact that online education in the engineering discipline presents new opportunities, there have been some lapses when implementing the programs even in the USA (Allen and Jeff, 2008). The reasons for the lapse have been identified as lack of:

1 Online teaching experience;
2 Course development and
3 Implementation of technical aspects when delivering engineering education in an online format (Kinney, 2015).

Engineering courses require considerable hours of practical/laboratory sessions incorporated into the curricula unlike some other disciplines, such as law, management, accounting and education (Keller *et al.*, 2017; Wait and Nichols, 2015). Similar views have been reflected in McCaslin and Brown (2014) that elaborated on the planning and preparation of online course materials for mechanical engineering courses.

Yet even with the above limitations experienced and a gap observed in adapting to the contemporary trend caused by a few critical issues, the impact of online education on the future of engineering schools as "providers of engineering education" is envisaged to be significant (Berdanier *et al.*, 2017). In this context, the chapter elaborates on the innovative aspects so far adopted by different universities in Australia, New Zealand, USA and the EU countries to address the issues specific to the engineering discipline, especially relating to curriculum development and the delivery of courses via online formats. Furthermore, recent major research findings relating to teaching and learning online in general as well as specific to engineering education are outlined.

7.1.1 The opportunities

The following are the new opportunities relating to online education when compared to the traditional higher education delivery, such as F2F lectures and instructor-assisted laboratory sessions:

1 Flexibility (allows for furthering skills of working professionals while on the job);
2 "Always available" learning activities;
3 Options for broadening instructional delivery, i.e. learning anywhere, anytime, remote laboratories;
4 Access to further education for individuals in rural and remote areas and
5 Sharing of resources.

7.1.2 The challenges

The following are the challenges encountered when delivering engineering education online:

1 Development of curricula for the delivery of courses online;
2 Choosing the right online format to deliver course materials;
3 Design course material to overcome the lack of social and teacher presence in online courses that lead to student attrition and
4 Aligning learning objectives with the prevailing body of knowledge and professional practices.

The last issue in the above challenges relates to obtaining accreditation for engineering courses offered in countries that are signatory to Washington and Dublin accords. Both are international agreements though the former is between bodies responsible for accrediting undergraduate and postgraduate programs, and the latter deals with qualifications relating to engineering technicians/associate level programs, i.e. establishing the educational base for engineering technicians. The Washington Accord was originally signed in 1989 by an initial group of six signatories. It has so far grown into 19 signatories and it mainly focuses on "academic programs which deal with the practice of engineering at the professional levels."

Bacon and MacKinnon (2016) examined the issues when developing online engineering courses with a particular focus on group-based collaborative working and learning that were considered to be more difficult to address in STEM[3] subjects. The reason for this being because STEM students are expected to work with heavyweight processes that require significant sharing of expertise and resources rather than the lighter weight processes experienced by open and discursive groups in other subject areas. Thereafter, the authors looked at a few major designs in practice and found concurrent design[4] (CCD) to be an excellent tool for collaborative use in engineering-related disciplines, despite the fact that it has not been tested yet for suitability in an online environment to support large-scale learning. Nevertheless, the authors found the so-called MOOC[5] to be less challenging with committed students i.e. course fee paid, in smaller groups where a tutor could guide the groups if needed. The authors described CCD as in need of practice before being applied to solve a real engineering problem online due to CCD's high learning curve. Furthermore, it was suggested that the students had completed a short course on CCD first before enrolling in a MOOC course because of the complex and changing motivations required from the students for a study in MOOC. Increasingly, MOOC is seen as an ideal model for the future as online education is anticipated to rise due to the rapidly increasing global demand for higher education, the inability of many countries to meet their local demand (such as India and in the EU), and finally the ever-increasing cost of tertiary education.

7.2 Future of online teaching and learning

In view of the interest on how online teaching and learning would shape the future of tertiary education in general, Contact North/Nord launched an online repository called Teachonline in 2010 for post-secondary educators in Ontario, Canada (Teachonline, 2016). The non-profit organization provides faculty and instructor access to profiles of innovations, tools and trends, webinars, resources for training and development and an updated list of conferences around the world on related topics. From a report under tools and trends, pertaining to the immediate future of program design, eight key changes have been identified, whereas for teaching and learning seven key changes have been listed, as presented below (Teachonline, 2016).

7.2.1 Program design

1. More flexible program designs.
2. More use of open educational resources.
3. More creative assessment processes.
4. More micro-credit and nano-degrees.
5. More co-op and experiential components within programs.
6. More international collaborative programs.
7. More transfer and international recognition agreements.
8. Blurring of lines between college and university.

7.2.2 Teaching and learning

1 Learning will no longer be defined by time, place or institutional offerings.
2 Learners will create their own learning agenda that will reflect their own career, personal and lifelong learning goals.
3 Learners will secure their learning outcomes through a combination of formal, informal, self-directed, instructor-delivered, in-class and online learning.
4 Learners will expect personalized learning services and support for their learning agenda.
5 New mechanisms for meeting personal learning agenda will appear in the marketplace as the "unbundling" of learning continues.
6 Courses will be less important than mentoring, coaching, counseling, advising and assessment.
7 Diverse and new forms of credentials will appear which reflect the varied needs of learners, employers, social agencies, innovation organizations and entrepreneurs.

With that introduction to the major aspects relating to delivering tertiary education online in general and specific to engineering education, the rest of the chapter is divided into three sections that look into the ways and means opted by Australian, US and EU universities in delivering engineering education online.

7.3 Online engineering courses in Australian universities

Australian universities seem to have opted for both fully online and blended learning (BL) options for delivering engineering and built environment courses at the diploma, undergraduate and postgraduate levels (OSA, 2017b). Currently, there are three universities in Australia in addition to the Engineering Institute of Technology (EIT) that offer online bachelor's degrees in engineering with specializations in civil, electrical, mechanical, mechatronics and environmental engineering (OSA, 2017c). The course durations of these associate and bachelor of engineering degrees are 3 and 4 years for full-time study, respectively (or equivalent for part-time study). These courses are accredited by Engineers Australia. The fully online engineering courses use a combination of remote laboratories, residential schools, workplace experience and simulation software to facilitate student learning as well as to test the knowledge gained during the study.

Among the Australian universities that have chosen BL for delivering engineering education online is Deakin University. It uses BL for four majors, namely (1) civil, (2) electrical and electronics, (3) mechanical and (4) mechatronics, all under its four-year bachelor of engineering (honors) program. All their online courses are accredited by Engineers Australia, and the university's "cloud" (online)-enrolled students are required to attend an on-campus module (in Geelong, Victoria) during "Engineering Practice Week" in week 8 of each trimester.

The fully online courses offered by EIT consist of working labs set up at various locations around the world, where students can login and proceed through the various practical sessions for completing a variety of online courses. Australia's EIT has its head office in Perth, Western Australia, and in addition, it has administrative support offices around the world, including in Canada, the United Kingdom, New Zealand, Egypt and India. Lately, there has been an additional hurdle to overcome by the institutions offering professional courses offshore. They are now required to obtain accreditation from professional institutions in both the offering as well as the offshore accreditation bodies. This is mandatory especially for educational institutions in countries that are signatory to the Washington Accord (Kootsookos *et al.*, 2017). At present, there are few universities offering online engineering courses at the master's level specializing in technical as well as general management and project management courses in Australia (OSA, 2017a).

It is interesting to note that Online Study Australia (OSA, 2017a) has ranked the Australian universities that offer the best accredited online programs based on four criteria: (i) size (number of external students), (ii) depth of courses offered online, (iii) graduate satisfaction with courses, and (iv) flexible study options (e.g. pathway courses). Based on these factors, the University of New England has topped the overall ratings in Australia's best online universities list (OSA, 2017d), and it offers civil and environmental engineering courses online (OSA, 2017e). With 16,400 external students, the University of New England is the third biggest in size among the online universities in Australia.

There has been intensive research into making engineering education more diverse and flexible in Australia. In the last decade, some far-reaching efforts (i.e. special conference sessions on "engineering education" held at international conferences) were made to look at the current and future trajectory of engineering education research in Australia and the EU (Jesiek *et al.*, 2010). The main aim of these efforts was to look into reducing the gap between engineering education and the real-world demands on engineers, which resulted in CDIO,[6] an innovative educational framework developed in the 2000s for producing the next generation of engineers. More recent research is aimed at looking at implementing CDIO. For example, in Lucke *et al.* (2016), the authors investigated the opportunities and barriers to implementing the CDIO framework for distance, online and BL educational modes when delivering engineering education. Also suggested that the CDIO framework could be successfully implemented with careful curriculum planning, consultation and engagement with key stakeholders, by making use of current technology and by applying appropriate leaning theories of active and collaborative learning.

With those issues encountered and measures opted by Australian universities when delivering engineering courses online, the next section discusses how US universities have handled the situation.

7.4 Online engineering courses in US universities

There is a major gap in the delivery of engineering education online in the USA (Kinney, 2015). A US survey conducted on online education in different areas found that, of the eight discipline areas examined for Fall 2007, the percentages were equally represented except for one discipline, which was engineering. Only 16% of the institutions were offering a fully online engineering program, whereas the other disciplines were found to be ranging from 24% to 33% (24% – psychology and 33% – business; Allen and Jeff, 2008).

Despite the above gap reported in 2007–2008, at present there are 255 accredited online schools offering 1,540 engineering courses, including engineering management as well as software engineering. Most of the programs offered are at the master's level (165), followed by certificate (112), bachelor's (89) and associate (41) levels (GOS, 2017). The GOS (2017) guide to online schools in the United States, maintained by SR Education Group, is a resource for potential online students, which provides details on online undergraduate, masters, graduate certificate and PhD degrees in all disciplines with majors/specialties within each discipline. Within engineering, there are 12 specialties listed: aerospace, biomedical, chemical, civil, computer, electrical, environmental, engineering management, industrial, mechanical, software and systems engineering. In order to help prospective students in their quest for finding high-quality engineering degrees, a separate list of top online colleges are included, along with their respective annual tuition fees and academic strength (GOS, 2017). Of the top 25 online colleges that offer engineering degrees, Georgia Institute of Technology and Brandeis University both have the highest academic strength (84%), with annual tuition fees of $16,300 and $1,624, respectively. All the other universities have at least 60% academic strength. The GOS (2017) also gives name, accreditation, annual tuition fees and recommended rate along with direct links to the schools. One of the 255 online engineering education providers in the US is City University of Seattle, and the following are details provided for this university: (i) accreditation: NWCCU (Northwest Commission on Colleges and Universities); (ii) annual tuition: $13,190 and (iii) recommended rate: 61%.

Of the 255 online engineering education providers in the US, the only institution that offers a certificate-level qualification in engineering is Pinnacle Career Institute, and the program is called "Wind Turbine Technician." The course is accredited by the Accrediting Council for Independent Colleges and Schools. The annual tuition fees are in the range of $14,615 to $20,465, with a recommendation rate of 37% (Pinnacle Career Institute, 2017).

A search for courses in civil engineering through the *Guide to Online Schools* (GOS, 2017) revealed 35 schools offering accredited degrees in different specialties within the main field civil engineering. Most of the courses offered (27) are at the master's level, followed by 10 certificate, two doctoral, two bachelor's level and just one at the associate level. The two universities that offer civil engineering at the bachelor's degree level are Old Dominion University and the University of North Dakota. Furthermore, of these 35 schools, only 24 offer academically rigorous and affordable options for an online civil engineering course. Among the 24 institutions with academic strength rated over 65%; Iowa State University is listed as the number one with an academic strength score of 72% and annual tuition fees of $8,145. Stanford University (in fourth place) and Columbia University in the City of New York (ranked 18th) have academic strength scores of 92%; these are the only two universities with higher academic strength in the top online college list that offer engineering courses.

The *Guide to Online Schools* website updated on a regular basis by SR Education Group also gives details on course admission requirements, credits and lab sessions needed to complete each and every course (GOS, 2017). Information on careers, pathways, advanced courses on specialties within different fields with respect to each and every course as well is provided by GOS (2017). For instance, an associate degree is the minimum qualification requirement to begin the entry-level work as a technician in engineering, and students who enroll in an associate degree are provided with a background in the technological and mathematical concepts that engineers use to design new systems or products. They are taught introductory theories and practices in their chosen fields. Students may choose to study aerospace, civil, electrical and electronics, environmental, or mechanical engineering or technology. Even though curriculum material varies by area of concentration, most programs consist of basic classes in computer concepts, physics, chemistry, calculus and other advanced forms of mathematics. These programs often allow students to combine classwork with hands-on instruction at local sites or laboratory settings. The students are expected to complete 60–64 credit hours over a period of 2 years. On completion of an associate degree, engineering technicians usually work under the direct supervision of more experienced workers and provide assistance to their supervisors as needed. Engineering technicians who want to advance into roles as full-fledged engineers must complete a bachelor's degree. It is also possible to become certified by the National Institute for Certification in Engineering Technologies to enhance their career options. Similarly, the *Guide to Online Schools*' (GOS, 2017) online resource provides career-related details for all online degrees/certificates.

Even though there are a number of local accreditation bodies, such as the National Committee on Accreditation (NCA) and the Higher Learning Commission (HLC), the Accreditation Board for Engineering and Technology (ABET) is the most important program accrediting board for engineering studies in the US. Most US states require students to obtain an ABET-accredited degree before practicing engineering and earning an engineering license. The majority of the ABET accredited courses are onsite with some content offered online; however, there are a few 100% online ABET accredited courses as well (ABET, 2017b). Among those, the following are the programs in engineering:

1 Arizona State University Tempe, Arizona: Electrical Engineering and Engineering Management;
2 Daytona State College Engineering Technology;
3 Eastern Kentucky University, Richmond: Fire Protection and Safety Engineering Technology;
4 Excelsior College, Albany, New York: Electrical Engineering Technology, Nuclear Engineering Technology;
5 Grantham University: Lenexa, Kansas: Electronics Engineering Technology;
6 Stony Brook University, New York, NY: Electrical Engineering;
7 The Johns Hopkins University, Baltimore, Maryland: Systems Engineering;
8 Thomas Edison State University, Trenton, New Jersey: Nuclear Energy Engineering Technology and Electronics Systems Engineering Technology and
9 University of Southern Mississippi, Hattiesburg, Mississippi: Construction Engineering Technology and Industrial Engineering Technology.

It can be observed that most of the above 100% online programs are in electrical engineering. In addition to the above online education sources, *US News* (www.usnews.com/education) as well gives information on online programs along with their rankings and career pathways. The course ranking details provided by *US News* is mostly based on pure statistics on admission selectivity, student satisfaction, engagement, completion and other related data. The US graduate schools in engineering require a minimum of an engineering degree to enroll into postgraduate programs.

Interestingly, there has been research into looking at different pedagogies for higher education in general as well as for delivering engineering education online. An online format of the "flipped classroom"[7] approach applied to an introductory engineering course at Colorado Technical University in an effort to expand the university's existing online engineering programs has been proved to be effective (Santiago *et al.*, 2017). In the piloted program, two approaches were implemented; first in the form of short videos (i.e. YouTube) that were watched by students before they worked on a short assessment (i.e. simple True or False click selections). The assessment was run soon after watching each video to simulate a conversation with e-learners. This was aimed at preparing the students for a weekly final quiz (using Google Docs) and hands-on electronic labs. In the second approach, an iteration of knowledge checks were embedded in the videos, and it seemed the checks had increased student engagement. Both approaches were described to have added an element of surprise and some anticipation thereby increasing student engagement.

Remarkably, the flipped classroom pedagogy is being implemented for a wide variety of engineering subjects and at all levels, from first- through to final-year programs reported (Kerr, 2015), who ran a survey on flipped classroom approaches in engineering education. The survey concluded with positive gains shown in problem-solving skills, conceptual understanding, student retention and satisfaction with the flipped classes based on literature reviewed for the survey. However, the authors pointed out that such studies need to provide details regarding the integration of out-of-class and in-class activities so that more information regarding good practices and guidelines for flipped classes in engineering education becomes available for future improvements.

At Oakland University in the US, the flipped classroom approach is reported to have further increased student understanding when applied to an online course in management information systems (Rutledge and Sugumaran, 2017). The paper further stated that student engagement was increased through question-embedded videos that in turn forced the students into much needed preparation prior to lecture and working on homework assignments (the main characteristic feature in the flipped classroom). The authors found the technique had ultimately increased students' understanding.

Another case study on the application of a hybrid flipped course used for 2 years in a first-year introductory engineering course at Rowan University, Glassboro, NJ, USA, is presented in Everett *et al.* (2014). It was classified as "hybrid" because the textbook as well as the majority of the homework exercises were delivered online. The course was stated as "flipped" because students had worked on online quizzes on the course e-book before the material was discussed in class. The authors stated that the approach was found to be successful as the students and teachers were satisfied with the online aspects i.e. e-book, notification of assignment due dates, online problem solving and immediate feedback (step-by-step). However, it was found that the exercises were not always rigorous enough, and some improvements were suggested to increase the success rate.

Following that discussion on the ways and means that US universities have opted to resolving some major issues relating to the delivery of online engineering education, in the next section general approaches adopted by EU universities are presented.

7.5 Online engineering courses in EU universities

Recent research and developments in the domain of online education in engineering in EU universities reveal the immense efforts undertaken to improve the delivery of engineering education, making it as attractive as in any other discipline. The more recent efforts in this research domain include a wide range of techniques, such as using mobile technology and remote laboratories to encourage creative, experiential and research-based learning and sharing of resources among universities to reduce costs. These efforts are aimed at applying the contemporary trends in delivering engineering education online as applied to other disciplines and are presented in this section.

Of the 1,243 online courses listed by Keystone Academic Solutions (KAS, established in 2002), 27 are in engineering, and the specialties within these engineering courses include many unconventional fields, such as health and safety, power engineering, landscaping, 3D printing of body parts, introduction to process mining with ProM, satellite orbit determination and big data analytics (data engineer) (KAS, 2017a). However, a search for online bachelor's degree in engineering resulted in nine; the specializations included automotive, civil, engineering (Top-up BSc of Eng. with honors), environmental, industrial, safety and systems

(KAS, 2017b). Meanwhile, there are seven programs at the master's level in engineering management. Details on mode, i.e. online, full/part-time, duration, next enrollment as well as university location, are provided in the web-based resource.

Sharing resources is an aspect EU universities have taken utmost advantage of by designing a system of interinstitutional online programs. This sharing of resources was developed and prepared for five different institutions based in four different European countries via an online-distributed platform. This new European interinstitutional online master's degree program offered online across five European institutions is oriented to meet the labor market needs for qualified graduates in EU countries (Castro *et al.*, 2015).

A conceptual framework for an individual learning environment proposed in Terkowsky *et al.* (2017) deployed mobile technology in combination with an e-portfolio system to facilitate and foster creative laboratory learning in engineering education. The aim of the application of mobile devices was described to be setting up an environment for boosting the creative thinking processes. Incorporating a mobile device into the system was intended to allow the user to make a quick note of a spontaneous idea or to record it. It as well permits the storage of any artifacts observed and activities related to the idea for further elaboration. The framework also consists of features that can be used to foster creativity through introducing personal learning environments. The features are described as enriching the formal classroom activities and laboratory work in order to attain high-level learning outcomes.

Universities in Dortmund (Germany), Palermo (Italy) and Stockholm (Sweden) implemented PeTEX (or Platform for E-Learning and Telemetric Experimentation) project that incorporated an opportunity to encourage experiential learning using real laboratory equipment remotely accessed i.e. via the Internet without being physically in the laboratory (May *et al.*, 2017). Among the many functions in PeTEX are the e-portfolios that allow students to document their own learning process with teacher guidance throughout the process. Attempts are being made to make e-portfolio software available on mobile devices so that the students can access the software from anywhere and at all times. With the combination of experiential learning along with e-portfolios, PeTEX is said to offer the potential to promote the learner's creativity.

A technical solution with guidelines to improve accessibility in e-learning platforms based on an extensive literature analysis is presented in Batanero *et al.* (2017). The goal of the study was to achieve scalable, reusable and easily manageable platforms. The method has been so far tested through a case study designed and conducted on Moodle[8] for several engineering education courses. The results are described to have shown a high degree of perceived usefulness; interestingly, even deaf students were found to have encountered fewer problems, but the blind students did not, as their issues are more complex.

7.5 Conclusions

As universities around the world are challenged to provide faster, more agile, more creative and more flexible educational solutions and to avoid becoming obsolete, the tertiary institutions are incorporating more online components into their traditionally F2F classroom teaching, in turn implementing institution-specific BL models in place of the 100% classroom formats. However, adopting any form of online education into engineering courses has limitations mainly due to the practical components traditionally provided by on-campus laboratory setups in engineering education. Literature reviewed for this study reveals that the limitations have already caused a gap in this regards. Nonetheless, despite the lapse and the challenges posed, online education provides new opportunities, and to take advantage of the situation, many institutions have lately embarked on innovative ways of delivering engineering education online. To overcome the issues discussed in the introduction of this chapter, different institutions in higher education have undertaken a range of measures: Australian universities and institutions have adopted BL (e.g. cloud campus, workplace experience and remote labs) to substitute the traditional on campus practical sessions. US universities have adopted flipped classroom, fully online with local workplace experience, labs and hybrid (online and flipped classroom) approaches as well CCD and MOOC pedagogies in place of 100% F2F programs. Meanwhile, EU institutions are more focused on developing ways and means for using modern technologies, i.e. mobile devices and more flexible online software platforms (PeTEX's), as well as sharing resources when delivering engineering education online. The EU universities have also tested open source e-learning platforms, such as Moodle, to achieve scalable, reusable and easily manageable systems for delivering engineering education online. These tertiary institutions have embarked on such innovative ways because as providers of engineering education, they certainly understand the impact of online education in the future of the engineering profession in a modern society that requires the engineers to possess multiple skills to solve real-world problems efficiently, amicably and in a timely fashion.

Notes

1 "Blended learning (BL) systems combine face-to-face (f2f) instruction with computer-mediated instruction" (Graham, 2006, p. 5). A recent study presented in Spracklin-Reid et al. (2014) also described BL as the mode that uses a combination of e-communication and F2F interactions. Contemporary BL models adopted by various educational institutions around the world can be defined as combinations of F2F instruction (ranging from 25% to 75%) with online (using various technologies) self-guided modalities (Duarte & Craven, 2017). Meanwhile, fully online is defined as 80% and over, i.e. all or most of the content delivered online, with no F2F meetings (Allen & Jeff, 2008). Today, many universities are compelled to incorporate BL formats by societies that demand faster, more agile, more creative and more flexible educational solutions. Hence, in order to avoid becoming obsolete, the universities are adopting institution-specific BL models in place of the 100% traditional F2F classroom teaching.

2 In the current context, academic institutions globally have incorporated some components of their courses online, i.e. assignments, research, and group/class projects, and some other universities have introduced fully online individual courses, i.e. 100% online. All such programs are called online, as the term "online" in a program is not well defined. Furthermore, the online component within a program is even subject to frequent changes (ABET, 2017a).

3 "STEM is an acronym, which stands for science, technology, engineering and mathematics. It also encompasses design, technology and computing" (www.stem.org.uk/press).

4 "Concurrent Design (CCD) is a method used to manage the definition, design, and implementation of complex systems and projects. It is a multidisciplinary approach that integrates all key aspects of a system in a structured way, and is applicable to different types of projects" (RHEA, 2017).

5 "Massive open online courses (MOOCs) are free online courses available for anyone to enroll. MOOCs provide an affordable and flexible way to learn new skills, advance your career and deliver quality educational experiences" (www.mooc.org).

6 "*The CDIO initiative is an innovative educational framework for producing the next generation of engineers*. The framework provides students with an education stressing engineering fundamentals set in the context of Conceiving – Designing – Implementing – Operating (CDIO) real-world systems and products. Throughout the world, CDIO initiative collaborators have adopted CDIO as the framework for their curriculum and outcome-based assessment. CDIO collaborator recognizes that an engineering education is acquired over a long period of time and in a variety of institutions and that educators in all parts of the spectrum can learn from practices elsewhere. *The CDIO network therefore welcomes members in a diverse range of institutions ranging from research-led*

internationally acclaimed universities to local colleges dedicated to providing students with initial grounding in engineering" (CDIO, 2017).

7 Flipped or inverted classroom means "events that have traditionally taken place inside the classroom now take place outside the classroom and vice versa" (Rutledge & Sugumaran, 2017, p. 2). In the flipped classroom models, typical lecture and homework components of a course are reversed. Videos and other lecture contents are posted online for students to prepare ahead of the F2F classroom time so that the class time is spent on interactive group learning activities, discussion of difficult concepts and problem solving. The concept has been shown to facilitate more interaction and engagement with students in many recent studies and research surveys.

8 Moodle is an open source learning management system that can be adapted to online in all levels of academic learning, i.e. from primary schools to large universities (http://moodle.org).

References

ABET (2017a) Accreditation Board for Engineering and Technology (ABET) (2017a) *Online Programs*. [Online] Available from: www.abet.org/accreditation/new-to-accreditation/online-programs/ [Accessed 31 August 2017].

ABET (2017b) Accreditation Board for Engineering and Technology (ABET) (2017b) *Online Programs*. [Online] Available from: www.aabrt.org/accredation/new-to-accredation/online-ptograms/#online [Accessed 3 September 2017b].

Allen, I.E. & Jeff, S. (2008) *Staying the Course by Online Education in the United States, 2008, ISBN:976-1-934505-07-6*. Sloan-C™, USA.

Bacon, L. & MacKinnon, L. (2016) The challenges of creating successful collaborative working and learning activities in online engineering courses. In: San, J. & Costa, R. (Eds) *14th LACCEI International Multi-Conference for Engineering, Education, and Technology Engineering Innovations for Global Sustainability*, 20–22 Jul 2016. Latin American and Caribbean Consortium of Engineering Institutions, San Jose, CA. pp. 1–9. https://doi.org/10.18687/LACCEI2016.1.1.151

Batanero, C., Fernández-Sanz, L., Piironen, A.K., Holvikivi, J., Hilera, J.R., Otón, S. & Alonso, J. (2017) Accessible platforms for e-learning: A case study. *Computer Applications in Engineering Education*, 1–20. doi:10.1002/cae.21852

Berdanier, C.G.P., Tate, R.H., Iwinski, T. & Kulkarni, A. (2017) Investigation of engineering student engagement and behaviour in an online second-year thermal science course. *Journal of Engineering Education Transformations*, 30, 3, 143–149.

Bourne, J., Harris, D. & Mayadas, F. (2005) Online engineering education: Learning anywhere, anytime. *Journal of Engineering Education*, 94, 131–146.

CDIO (2017) Conceive, Design, Implement, Operate (CDIO). Available from: www.cdio.org [Accessed 3 September 2017].

Castro, M., Tawfik, M. & Tovar, E. (2015) Digital and global view of engineering education using remote practical competences. In: *Tecnologias del Aprendizaje, IEEE Revista Iberoamericana de,10,3*, Aug 2015. IEEE. pp. 126–133 DOI: 10.1109/RITA.2015.2452651.

Duarte, A. & Craven, A.E. (2017) Blended learning: Institutional framewors for adoption and implementation. In: Pixel (ed) *Proceedings in The Future of Education*, LibreriaUniversitaria, Florence, Italy, 8–9 June 2017, pp 83–87.

EECom (2017) *Educating Engineers.com*. [Online] Available from: http://educatingengineers.com/online [Accessed 30 August 2017].

El-ZEin, A., Langrish, T. & Balaam, N.I.G.E.L. (2009) Blended teaching and learning of computer programming skills in engineering curricula. *Advances in Engineering Education*, 1(3), n3.

Everett, J., Morgan, J., Mallouk, K. & Stanzione, J. (2014) A hybrid flipped first year engineering course. In: *Proceedings of 6th First Year Engineering Experience (FYEE) Conference*, College Station, TX, Aug 7–8, 2014. pp. F2C-1–6.

Gonzalez, A.-B., Rodriguez, M., Olmos, S., Borham, M. & Garcia, F. (2013) Experimental evaluation of the impact of b-learning methodologies on engineering students in Spain. *Computers in Human*

Behavior, 29, 2, March, 2013. ISSN 0747–5632 doi 10.1016/j.chb.2012.02.003 url: http://dx.doi.org/10.1016/j.chb.2012.02.003. Elsevier Science Publishers B V, Amsterdam, The Netherlands. pp. 370–377.

Graham, C.R. (2006) Blended learning systems: Definition, current trends, and future trends. In: Bonk, C.J. & Graham, C.L. (eds) *The Handbook of Blended Learning: Global Perspectives, Local Designs*. Pfeiffer Publishing, San Francisco, CA, USA. pp. 3–21.

GOS (Guide to Online Schools) (2017) *Complete Guide to Top online Engineering Degrees*, [Online] Available from: www.guidetoonlineschools.com/degrees/engineering [Accessed 28 August 2017].

Jesiek, B.K., Borrengo, M. & Kacey, B. (2010) Advancing global capacity for engineering education research: Relating research to practice, policy and industry. *European Journal of Engineering Education*, 35, 2, 117–134.

Keller, C., Wass, S., Zetterlind, M., Ghassemali, E. & Seifeddine, S. (2017) Teacher roles in a blended learning materials engineering master program: "It's not a new role, it's a new way!". In: *Proceedings of 26th EDEN Annual Conference 2017: Diversity Matters*, Jönköping, 13–16 June, 2017. Projects Gjutmagistern 3.0 ORCID iD: 0000-0002-7527-719X. http://www.diva-portal.org/smash/record.jsf?pid=diva2%3A1111964&dswid=-2597

Kerr, B. (2015) The flipped classroom in engineering education: A survey of the research. In: *Proceedings of 2015 International Conference on Interactive Collaborative Learning (ICL)*, IEEE, Florence, Italy, 20–24 Sep 2015.

KAS (Keystone Academic Solutions) (2017a) [Online] Available from: www.onlinestudies.com/Courses/Engineering-Studies/ [Accessed 28 August 2017].

KAS (Keystone Academic Solutions) (2017b) [Online] Available from: www.onlinestudies.com/Bachelors/Engineering-Studies/ [Accessed 28 August 2017].

Kinney, L.S. (2015) *Faculty Perceptions of Online Learning in Engineering Education*. Doctoral dissertation. https://repositories.lib.utexas.edu/handle/2152/30923, [Accessed 12 September 2018].

Kootsookos, A., Alam, F., Chowdhury, H. & Jollands, M. (2017) Offshore engineering education: Assuring quality through dual accreditation. *1st International Conference on Energy and Power*, ICEP 2016, 14–16 Dec, RMIT University, Melbourne, Australia., Energy Proceedia 110, pp. 537–542.

Lucke, T., Brodie, L., Brodie, I. & Rouvrais, S. (2016) Is it possible to adapt CDIO for distance and online education? In: *CDIO 2016: 12th International Conference: Enhancing Innovation Competencies through Advances in Engineering Education*, 45-Research Reports. June 2016, Turku University of Applied Sciences, Turku, Finland. pp.10.

May, D., Terkowsky, C., Haertel, T. & Pleul, C. (2017) Using E-portfolios to support experiential learning and open the use of tele-operated laboratories for mobile devices. *Engineering Education*, 4, 47–65.

McCaslin, S. & Brown, F. (2014) Case study: Challenges and issues in teaching fully online mechanical engineering courses. In: *New Trends in Networking, Computing, E-learning, Systems Sciences, and Engineering Volume 312 of the series Lecture Notes in Electrical Engineering*. Springer International Publishing, Switzerland. pp. 575–579.

OSA (2017a) *Best Engineering Management Masters, Online Study Australia (OSA)*. [Online] Available from: https://onlinestudyaustralia.com/courses/degrees/engineering-management-masters/ [Accessed 21 August 2017].

OSA (2017b) *Engineering and Built Environment Degrees*. [Online] Available from: https://onlinestudyaustralia.com/courses/degrees/engineering-built-environment/ [Accessed 15 August 2017].

OSA (2017c) *Engineering Degrees*. [Online] Available from: https://onlinestudyaustralia.com/courses/degrees/engineering/ [Accessed 15 August 2017].

OSA (2017d) *Online Universities in Australia*. [Online] Available from: https://onlinestudyaustralia.com/best-accredited-universities-list/ [Accessed 21 August 2017].

OSA (2017e) *University of New England*. [Online] Available from: https://onlinestudyaustralia.com/courses/profiles/university-new-england/ [Accessed 21 August 2017].

Pinnacle Career Institute (2017) [Online] Available from: www.guidetoonlineschools.com/school-list?pid=7000&lvl=32 [Accessed 28 August 2017].

RHEA (2017) *RHEA*. [Online] Available from: www.utwente.nl/en/ce/bachelor-programme/bachelor_thesis/available_assignments/concurrent-collaborative-design-and-engineering.pdf [Accessed 22 September 2017].

Rutledge, A. & Sugumaran, V. (2017) Application of flipped classroom techniques in online education: Experience with an introductory MIS course. *MWAIS 2017 Proceedings*, 39. https://aisel.aisnet.org/mwais2017/39

Santiago, J.M.J., Guo, J., Kasley, K. & Phillips, P. (2017) Introduction to engineering using google docs and interactive video in support of an online classroom approach. In: *2017 Pacific Southwest Section, American Society for Engineering Education*, Tempe, AZ, USA, 20 Apr 2017. pp. 15.

Sowells, E.R., Ofori-Boadu, A.N., Douglas, J.R., Tsay, L. & Brown, D.R. (2017) Works in progress: Analyzing educational methodologies for electronic technology students, PaperID#18159. *American Society for Engineering Education (ASEE)*, 8.

Spracklin-Reid, D., Koenig, C., Hurley, S. & Phillips, P. (2014) Teaching with technology in engineering education. *Proceedings of 2014 Canadian Engineering Education Association (CEEA14) Conference*, June 8–11, 2014. Canmore, Alberta, Canada. p. 4.

SREG (2017) SR Education Group (SREG) [Online] Available from: www.guidetoonlineschools.co./degrees/engineering [Accessed 15 August 2017].

Sun, L., Chen, H. & Pang, S. (2017) Students' perception of flipped classroom design in engineering courses. *American Society for Engineering Education*, 6.

Teachonline (Contact North/Nord) (2016) *A 2016 Look at the Future of Online Learning – Part 2*. [Online] Available from: www.teachonline.ca/tools-trends/exploring-future-education/2016-look-future-online-learning-*part-2* [Accessed 21 September 2017].

Terkowsky, C., Haertel, T., Bielski, E. & May, D. (2017) Bringing the inquiring mind back into the labs – A conceptual framework to foster the creative attitude in higher engineering education. *Engineering Education*, 4, 937–947.

Top Online Colleges – Guide to Online Schools (2017) *2017 Top Online Colleges, Best Value – Engineering, SR Education Group*. [Online] Available from: www.guidetoonl;ineschools.com/degrees/engineering?pid=7000&page=2#best-value [Accessed 28 August 2017].

Wait, I.W. & Nichols, A.P. (2015) Virtualization of engineering laboratory experiments: Opportunities and limitations. *QScience Proceedings* (Engineering Leaders Conference 2014). p. 34. http://dx.doi.org/10.5339/qproc.2015.elc2014.34.

Chapter 8

Use of innovative technologies in enhancing students' learning outcomes

S. Shrestha

School of Computing, Engineering and Mathematics, Western Sydney University, New South Wales, Australia

ABSTRACT: Traditional methods of learning and teaching (L&T) in higher education institutions using lectures and tutorials are proving to be increasingly less effective in engaging today's students who have ready access to learning technologies. The challenges faced by these institutions, principally due to rapidly evolving technologies resulting in changing educational environment and student expectations, have been highlighted by many recent studies. One of the strategies to improve student engagement resulting in enhanced student learning is to replace the traditional lecture and tutorial spaces with collaborative learning spaces (CLS), making use of this improvement in educational technology. Students can use these spaces to collaborate and work on assigned tasks. Student responses to such spaces have resulted in positive learning outcomes. This chapter presents the use of the CLS in a core civil engineering subject at the Western Sydney University (WSU), Australia. L&T materials were developed to take advantage of the technology-driven learning space. The experience of the facilitator in the development of the L&T material and use of the technology-enhanced space are outlined. Student experiences are also discussed. The opportunities to improve student engagement and the challenges faced by both the facilitator and the students are highlighted.

8.1 Introduction

Traditional methods of learning and teaching (L&T) in universities make use of lectures complemented by tutorial and practical sessions. In this approach, lectures are usually delivered in large theaters principally to disseminate information. Students in these sessions usually take on the role of a "sponge," trying to absorb and digest the excessive information imparted upon them. Tutorial sessions are then held in small groups where interaction with students, in the form of student-student and student-tutor, occur. Laboratory sessions are subsequently used to impart firsthand knowledge to students of the theory presented in lecture sessions and reinforced in tutorial sessions. Student engagement in this approach is increasingly being questioned and is evidenced by continual decline in their attendance patterns (Rahman and Al-Amin, 2014).

Tertiary education institutions have been trying to address this issue of declining student interest in lectures and tutorials by changing the education technology, delivery method and physical environment where students are better engaged in learning (Alam, 2014; Rahman, 2017; Wood *et al.*, 2012). Termed as collaborative learning space (CLS), higher educational institutions have devoted substantial resources to their research and development. Compared to what learners can achieve in traditional teacher-centered classrooms, these newly developed and technology-enhanced learning spaces focus on learning activities that are student-centered with learning through collaboration among the peers (Beichner, 2014).

Western Sydney University (WSU) in Australia designed and constructed a prototype of collaborative learning space supported by the state-of-the-art technologies. Students can attend sessions in this space to collaborate and work on the assigned tasks. Students also have the opportunity to "zoom in" and actively participate in the sessions held in this CLS remotely. The zoom facility also enables specialist "expert" guest lectures to be delivered from any corner of the world. Availability of this room presented a unique opportunity to trial a new L&T approach in a core civil engineering subject (unit) in Autumn 2016 semester at WSU.

L&T materials were developed to take advantage of the technology-enhanced learning space. The challenges faced by educators in making such a transition have been discussed by Best and MacGregor (2017). The WSU experience of the facilitator in the development of the L&T material and use of the space for this unit are presented. Student experiences are also highlighted. The challenges faced by both the facilitator and the students are outlined. Potential opportunities to improve student engagement using the CLS are presented.

8.2 Learning space design philosophy

A learning space is typically a lecture theater, tutorial room or a lab area where students can learn with the input from an instructor. A prototype learning space (room) has been designed at WSU to facilitate teamwork in the class (Fig. 8.1). The room consists of 10 movable desks

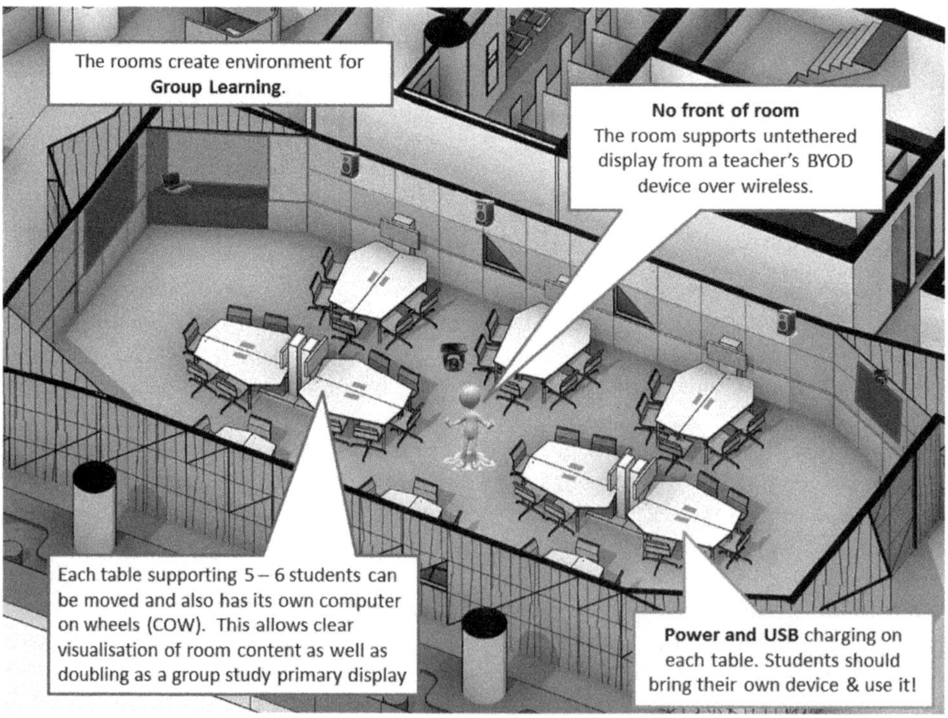

Figure 8.1 An example of collaborative learning space

with a monitor at the end of each desk. Each desk can accommodate up to six students. The room is supported by a series of whiteboards spread throughout the room and a whiteboard camera. It also has a document camera tucked away on the lectern with a PC and a monitor. The software to display and share information (peer-to-peer and class sharing) and the zoom facility are also provided.

This CLS was designed to cater for various forms of learning interactions. Some of the possible learning interactions include (Shrestha *et al.*, 2016): (i) contribute and compare; (ii) group work; (iii) present and discuss; (iv) highlight and share peer work and (v) remote collaboration.

The utilization of learning space as a "socio-material" process has been discussed by Johri and Olds (2011) who examined the underlying learning theories in the context of engineering education. They specifically highlight the importance of teamwork and collaboration in engineering education. Reiterating, the CLS provides the perfect opportunity to develop and implement the L&T material to test this theory. The details are presented elsewhere in this chapter. Technical features of the learning space with a short description of each features is presented in Table 8.1.

Table 8.1 Features provided in a typical collaborative learning space

Feature	*Description*
Lectern	A "presentation station" is tucked away in one corner of the room. This station swings out, when needed, to enable students and academics to operate a plug-in device or PC without turning away from the class. From the presentation station, a presenter can prepare content for sharing onto the primary screens from the following resources: one's own mobile devices, lectern PC, document camera, whiteboard camera, and/or wired inputs (VGA/HDMI/USB).
Computers on wheels (CoWs)	At each group table, there is a computer on wheels (CoW) for one student from each group to login using a wireless keypad (a combination of a keyboard and touch-pad). Except for the big screen to optimize viewing, the CoW works like a normal PC and it enables all students sitting around the table to collaborate on various learning activities. Students can also choose to use third party applications, and all students in the group can contribute simultaneously from their own devices, while the main screen can be used to display the outcomes.
Audio system in the room	Discrete microphones are located on the ceiling around the room to pick up audio across the learning studio. A lapel microphone is available to help maximize volume. It is especially useful for video conferencing sessions. Headsets are also available for hearing impaired students.
Whiteboard and whiteboard camera	Magnetic whiteboards are located around the walls in the prototype room. Students and facilitators can use them to share and discuss ideas by drawing/writing on the whiteboards or by simply sticking pieces of papers. The whiteboard camera installed on the ceiling in the middle of the room can be used to capture the content on the whiteboards and share with the rest of the class. There are buttons on the walls to switch whiteboard camera views. Students sitting at the tables can view the content on their CoW screens.

(*Continued*)

Table 8.1 (Continued)

Feature	*Description*
Document camera	A document camera is available on the lectern for sharing hard copy contents. The document camera can also take snapshots or videos of the shared content. Students sitting around the tables can view the content on their CoW screens.
Solstice	With the Solstice-enabled display in the room, any number of users can instantly connect, share and control the display, fostering collaboration and decision-making. Students in the CLS prototype room can use Solstice to share content from a wide range of sources with the rest of the class. Students sitting at different tables can view the content on their own CoW screens.
Zoom	Utilizing Zoom in the prototype room not only provides a backup sharing solution in addition to Solstice, it also helps connect other remote experts or learners to the prototype room. A Zoom session can be set up from the "Lectern PC." The camera installed at the end of the room can also be used to capture the room view and share with remote users.

8.3 Learning and teaching material design and delivery

In order to take advantage of the prototype CLS, the L&T materials were redesigned for a third-year core civil engineering subject, Surface Water Hydrology. While the learning outcomes did not change (to comply with the Engineers Australia accreditation norms), the traditional L&T materials were redesigned to fit the problem-based learning (PBL) pedagogy using WSU's blended learning strategy. The delivery method was also changed from weekly lectures and tutorial sessions to evenly spaced full-day workshop sessions. A total of five full-day (each 8 hours long) workshop sessions replaced the weekly 2-hour lecture supplemented by weekly 2-hour tutorial sessions; minimizing the variation in the total face-to-face contact hours. Following the student feedback, this has since been changed to 2-hour lecture sessions followed by 4-hour workshop sessions in alternate weeks. The response from the students thus far has been positive to this change. This chapter is based on the longer duration workshop sessions implemented in Autumn 2016.

The learning design made use of several types of interactions supported by technical features in the prototype room. Learning materials were developed to fit "flipped classroom (FC) approach" in a PBL subject.

8.3.1 Out-of-class preparation activities

Following the FC pedagogy, students were required to complete preparatory work before attending each workshop session, except for the introductory workshop session. Pre-recorded and sourced videos (durations ranging from 3.5 minutes to 66 minutes) were posted on vUWS, the learning management system (LMS) used at WSU. PowerPoint files were also posted regularly on vUWS to disseminate information on workshops. This allowed for the workshop sessions to be devoted to meaningful discussions and question-and-answer sessions.

Students were required to complete a series of preparatory tasks, including individual and group journals before attending subsequent workshop sessions. Entries made on individual journals, along with peer interactions during workshop sessions, were used to gauge individual student's engagement in the unit material – a nominal mark for "participation" was awarded for this component. This approach (awarding a nominal mark) allowed for implementation of compulsion for completion of out-of-class activities, implicitly forcing students to engage with the L&T material in a timely manner.

8.3.2 *In-class interactive sessions*

Students were required to attend a total of five pre-scheduled full-day (8-hour-long) workshop sessions during the semester. These sessions were used for in-class research, peer (student-student) interaction and student-facilitator interaction. The time in workshop sessions was used to clarify any confusion and doubts the students may have in completing the assigned real-world engineering project. In addition, these sessions were used to gauge student progress and their understanding of the L&T material and their ability to communicate with their peers. These sessions were effectively used to provide "just-in-time" supplementary information as identified during each session and as reflected in student journals.

The teaching team applied various L&T activities repeatedly and selectively throughout the semester. Some of the strategies used included (i) a "jigsaw activity" described by Aronson (2010); (ii) the "role playing" described by Bartz and Deaton (1996) and Ponsa *et al.* (2010) and (iii) the "one minute paper" strategy of Stead (2005).

8.4 Collaborative learning space

The layout of desks (shown in Fig. 8.1), known as CoWs (computers on wheels) facilitated peer-to-peer interaction during workshop sessions (see Table 8.1). The technology in the room allowed for the students to undertake live research and share their findings with their peers during the workshop sessions. This provided opportunities to share and learn from each other as identified by Gómez Puente *et al.* (2013). This also encouraged the students to take ownership of their own learning (Bell, 2010) and achieve better learning outcomes (Shrestha, 2016a). It may be noted that other researchers have argued that such an approach results in improved student engagement and learning (Kim *et al.*, 2014; Mason *et al.*, 2013).

8.5 Implementation and experience

The redesigned unit, Surface Water Hydrology, was delivered in the prototype CLS at the Penrith campus of WSU during Autumn 2016. The unit is a third-year core civil engineering unit that focuses not only on technical learning outcomes but also on a number of soft skills required for Stage 1 Competency Standards for a professional engineer (Engineers-Australia, 2013). The soft skills this unit focused on were effective teamwork skills, oral communication skills and written communication skills. Additional learning outcomes included technical skills in the specified area (engineering hydrology), problem-solving skills and independent research skills.

A total of 95 students were enrolled in this unit in Autumn 2016. The students were divided into 25 teams; the number of team members ranged between three and five. The students themselves were responsible for picking team members, following the presentation of the Belbin model (Fisher *et al.*, 1998) during the introductory workshop session and after completion of the self-assessment exercise during the session.

8.5.1 Student experience

Both the CLS and the hybrid L&T approach (PBL supplemented by FC) were new to the student cohort and the facilitating team. To ensure that the students were progressing well during the semester, they were encouraged to email regular feedback to the teaching team. They were also encouraged to post their personal reflections on their personal journals on a regular basis. These were in addition to the end-of-semester student feedback on the unit (design, delivery and space) as well as teaching effectiveness. One of the emails a student sent at the beginning of the semester stated,

> The opportunity to be in the first class to use the Prototype Room is one that seems interesting, along with the project to be completed in a "real world" type environment is one that I am very much looking forward to.

As per the room design and the technology available in the room, another student wrote,

> I liked how the room was designed to satisfy the purpose of team working as the tables were set up nicely and the computer on the tables were really fun as it had the new tools where you can see what the people share with you.

The discussions that were taking place between peers and the questions being asked of the facilitators suggested that the L&T approach adopted in this unit was working well. The assessment format that required team members to make a presentation before their peers was insightful and was found to be very helpful in building confidence. This also required the students to work effectively in teams and put additional efforts to solve the problem on hand, as one student put eloquently, "no one likes to look stupid before your classmates"; the fear that their colleagues will judge them harshly encouraged them to work harder to achieve a respectable outcome for their projects.

As with any new technology, there were regular hiccups in use of the technology in the CLS. The main problem was related to Wi-Fi connectivity in the room; it kept dropping out, causing frustrations among students and teachers throughout the semester. In addition, there were difficulties with the Solstice software used to display and share contents. As with Wi-Fi connectivity or lack thereof, problems with the display software was a major cause of discomfort and frustration to students and teachers and this lasted throughout the semester.

8.5.2 Facilitator experience

PowerPoint slides and video materials were developed, sourced and collated by the unit coordinator (who was also the lead facilitator) and posted on vUWS (WSU's learning management system). The workshop sessions were facilitated by two facilitators, and the journals were assessed by these facilitators. Both the lead facilitator and the associated facilitators had practical engineering experience, in addition to being experienced tertiary educators. However, neither of them had experience of the hybrid model nor did they have experience in teaching in the technology-enhanced CLS. In other words, both the facilitators and the learners were new to this system of L&T.

The facilitators met after each workshop session to discuss their experiences and feelings on student engagement. These discussions along with the reflections described in student journals identified the improvements needed as the semester progressed. Both the facilitators felt that the students were engaged and involved in deep learning activities. They also observed that the students were able to find innovative solutions to the real-world engineering problems they were trying to solve. They also felt that the CLS was contributing to better learning outcomes for the students, mainly from successful development of soft professional skills, e.g. teamwork and communication skills.

8.6 Discussion

Learning outcomes were gauged from three different assessment tasks – student journals, in-class (in-semester and end-of-semester) oral presentations and a technical report. The student journals and oral presentations were individually marked, whereas the technical report was assigned a group mark, considering peer assessment to gauge individual student contributions. The results are presented in Jones and Shrestha (2016). It was observed that the CLS contributed to student engagement, and the results indicated improved student learning outcomes (e.g. Jones and Shrestha, 2016; Shrestha, 2016b). The students enjoyed the opportunity the CLS provided to collaborate and share their findings with their peers. The room was also found to be helpful in providing additional avenues to find answers to the questions the students had difficulty during their studies.

Workshop attendance and participation as well as frequency of journal entries were used as indicators of student engagement in the unit. Three of the 95 students enrolled in the subject did not attend 69% of the classes. Attendance for the remaining 92 students was 100% throughout the semester (highly unusual for a senior-level engineering subject). The facilitators observed the students engaged and remained busy during workshop sessions – the constant discussions during the sessions were clear indication of student engagement. While student engagement was influenced by mandatory attendance requirement and the design of the activities as well as the CLS, the results of the detailed analysis of end-of-semester feedback (48% response rate) suggested that the new learning environment made a significant contribution (further details can be seen in Jones and Shrestha, 2016).

Performance in the subject by previous students was used to assess performance of the current cohort. The two student cohorts had the same learning outcomes and standards-based grading. The grade distribution of the current cohort indicated that the peer support available within and outside the workshop sessions helped the academically weaker students. This was also evidenced through a number of comments in the end-of-semester feedback. One student wrote, "The ability to work as a part of a team to work through and solve problems relating to the project" was one of the best aspects of the approach adopted in the subject. The grade distribution also shows a higher proportion of students receiving distinction (D) and high distinction (HD) grades (Shrestha, 2016a).

One of the most interesting and encouraging results was the performance of the cohort – every student who remained enrolled in the subject achieved the positive learning outcomes in the subject, in sharp contrast to previous offerings of the subject; as stated earlier, this was a challenging subject. Part of this success must be attributed to the design and effective use of the CLS.

In addition, the students appreciated the opportunities the CLS provided in terms of collaboration and the ability to engage with their peers and the teaching team. However, the constant deleterious issues with Wi-Fi connectivity and the display software Solstice were major impediments. Both the students and the teaching team felt that the L&T style provided real-world engineering experiences and the learning space assisted greatly in ensuring that the students were able to achieve the unit learning outcomes.

8.7 Conclusion

This chapter presents the use of the collaborative learning spaces (CLS) in a core civil engineering unit at the Western Sydney University (WSU), Australia. The L&T materials were developed to take advantage of the technology-driven learning space. The experience of the facilitators in the development of the L&T materials and use of the technology-enhanced space are presented. End-of-semester student feedback and performance of students gauged through grades achieved by the students indicate that the CLS achieved improved student learning outcomes. It is also recognized that the L&T material needs to be developed to suit the CLS and that the workshop facilitators need to be well versed in use of the technology. The teaching team needs to ensure that their delivery methods are pedagogically driven, taking advantage of the technology and student needs. Students need to be open-minded and prepared to embrace the new method of L&T and new space. Also, the L&T material needs to be carefully designed to cater for efficient delivery using meaningful assessment strategies. Students, who are used to the traditional delivery model, will need to be informed of the new learning modality and reassured throughout the semester of its utility. This implementation has shown how a state-of-the-art technology-enhanced CLS, despite some technical glitches, can contribute to an authentic and effective socio-material learning process.

8.8 Acknowledgements

The author acknowledges the feedback by the students enrolled in the Surface Water Hydrology subject in the Autumn 2016semester at Western Sydney University, Australia. The author also acknowledges the contributions made by the associate facilitators during the workshop sessions and the comments made by the reviewers, which have helped to improve the quality of this book chapter.

References

Alam, F. (2014) *Using Technology Tools to Innovate Assessment, Reporting, and Teaching Practices in Engineering Education*. IGI Global, Hershey, PA, USA.

Aronson, E. (2010) *Cooperation in the Classroom: The Jigsaw Method*. Pinter & Martin Limited, London, UK.

Bartz, M. & Deaton, R.J. (1996) Role playing in engineering education. *Age*, 1, 1.

Beichner, R.J. (2014) History and evolution of active learning spaces. *New Directions for Teaching and Learning*, 137, 9–16.

Bell, S. (2010) Project-based learning for the 21st Century: Skills for the future. *The Clearing House: A Journal of Educational Strategies, Issues and Ideas*, 83(2), 39–43.

Best, M. & MacGregor, D. (2017) Transitioning design and technology education from physical classrooms to virtual spaces: Implications for pre-service teacher education. *International Journal of Technology & Design Education*, 27(2), 201–213.

Engineers Australia (2013) *Stage 1 Competency Standard for Professional Engineers*. Engineers Australia, Canberra, Australia.

Fisher, S., Hunter, T. & Macrosson, W. (1998) The structure of Belbin's team roles. *Journal of Occupational and Organizational Psychology*, 71(3), 283–288.

Gómez Puente, S.M., van Eijck, M. & Jochems, W. (2013) A sampled literature review of design-based learning approaches: A search for key characteristics. *International Journal of Technology and Design Education*, 23(3), 717–732.

Johri, A. & Olds, B.M. (2011) Situated engineering learning: Bridging engineering education research and the learning sciences. *Journal of Engineering Education*, 100(1), 151–185.

Jones, L. & Shrestha, S. (2016) Use of flipped classroom approach in project based learning: Student perspective. *Proceedings International Conference on Engineering Education & Research*, 21–24 Nov 2016, Sydney, Australia.

Kim, M.K., Kim, S.M., Khera, O. & Getman, J. (2014) The experience of three flipped classrooms in an urban university: An exploration of design principles. *The Internet and Higher Education*, 22(0), 37–50.

Mason, G.S., Shuman, T.R. & Cook, K.E. (2013) Comparing the effectiveness of an inverted classroom to a traditional classroom in an upper-division engineering course. *Education, IEEE Transactions*, 56(4), 430–435.

Ponsa, P., Vilanova, R. & Amante, B. (2010) The use of role playing in engineering curricula: A case study in human-automation systems. *Education Engineering (EDUCON)*, 2010. IEEE, New York, USA. pp. 1335–1341.

Rahman, A. (2017) A blended learning approach to teach fluid mechanics in engineering. *European Journal of Engineering Education*, 42, 3, 252–259.

Rahman, A. & Al-Amin, M. (2014) Teaching of fluid mechanics in engineering course: A student-centered blended learning approach. In: Alam (ed.) *Using Technology Tools to Innovate Assessment, Reporting, and Teaching Practices in Engineering Education*. IGI Global, Hershey, PA, USA. pp. 12–20.

Shrestha, S. (2016a) Flipped classroom and project based learning. *Proceedings of the 27th Annual Conference of the Australasian Association for Engineering Education*, 4–7 Dec 2016, Coffs Harbour, Australia.

Shrestha, S. (2016b) Collaborative learning spaces: Opportunities and challenges. *Proceedings International Conference on Engineering Education & Research*, 21–24 Nov 2016, Western Sydney University, Australia.

Shrestha, S., Wang, H. & Russell, C. (2016) Student response to a technology enhanced learning space. *Proceedings of the 27th Annual Conference of the Australasian Association for Engineering Education*, 4–7 Dec 2016, Coffs Harbour, Australia.

Stead, D.R. (2005) A review of the one-minute paper. *Active Learning in Higher Education*, 6(2), 118–131.

Wood, P., Warwick, P. & Cox, D. (2012) Developing learning spaces in higher education: An evaluation of experimental spaces at the University of Leicester. *Learning and Teaching*, 5(2), 49–72.

Chapter 9

Quality assessment of industrial design curriculum

Discovering employability attributes from job advertisements

A.A. Mahmud[1], B. Kuys[1] and O. Mubin[2]

[1]Centre for Design Innovation, School of Design, Swinburne University of Technology, Melbourne, Australia

[2]School of Computing, Engineering and Mathematics, Western Sydney University, Australia

ABSTRACT: This paper reports the key knowledge and skills sought for industrial design jobs in Australia. The aim is to find out the required levels of skills that are wanted by prospective employers for industrial design jobs. Job advertisements were collected from the Australian popular job site seek.com.au. A total of 109 job advertisements were identified, which were later used for content analysis. The most frequent job titles were "industrial designer" ($n = 25$), "industrial design engineer" ($n = 17$), "product design engineer" ($n = 8$) and "product designer" ($n = 5$). Several key skills for the design jobs were identified and observed that in addition to domain-specific skills, majority of the jobs required very high proficiency in software tools, for example SolidWorks or CAD. Finally, key implications of the findings for design students and educators are described.

9.1 Introduction

Industrial design is a trans-disciplinary profession, which covers a wide range of subjects such as engineering, business, ergonomics and aesthetics. Furthermore, it includes social, environmental and cultural issues (Erkarslan, 2007; Giard, 2000; Yang *et al.*, 2005). According to the World Design Organization (WDO), industrial design is defined as "a strategic problem-solving process that drives innovation, builds business success, and leads to a better quality of life through innovative products, systems, services, and experiences" (WDO, 2018). Industrial designers focus on people's needs and on how to create safe and usable products that can be adjusted to the varying lifestyles of people. Thus, product designers or industrial designers emphasize the way in which people live and deal with concept, appearance, performance and human factors that play a key role in influencing design. On the other hand, engineers focus on details, functionality, performance and production and tend to concentrate more on the problems of making a product function better for the tasks it is designed to do while optimizing its design for production (Owen, 2004). The field of industrial design optimizes function, value and appearance of products and systems for the mutual benefit of users and manufacturers (IDSA, 2018). Therefore, the industrial design profession not only covers the knowledge and skills of engineering, commerce and aesthetics but also involves the social and cultural issues of design (Giard, 2000).

From a general perspective, there seems to be a gap between what is taught in design schools and what design industries are looking for as reported in research (Yang *et al.*, 2005). There are several notions about design education regarding its approach and focus, which also depends on the design school. As pointed out by researchers, design education is very much tied to basic aesthetics and shapes, and therefore the balance between theory and practice has become a vital issue for the development of the design curriculum (Kolko, 2005). Incorporating practice into design education is important for connecting with industries and producing industry-ready graduates, and integration of industry with education has been extensively discussed in the literature (Boyarski, 1998). However, the issue is still dominant in design education due to the dynamic nature of the needs of the design industries. A primary concern of design educators and professionals is the issue of the competencies of industrial designers, and the quality of the graduates is often considered by prospective employers to be below their expected level (Kaufmann, 1998). Consequently, there exists a gap between what students learn at design schools and what they are required to do in the professional fields after graduation (Ball, 2002; Yang *et al.*, 2005; Yeh, 2003). However, it is believed that over the years, the curricula of major industrial design schools have changed to meet the ongoing demands of industries, and most of the design schools have updated their curricula to connect their graduates with industries, resulting in industry-ready learners. Therefore, the study reported in this paper investigates the issue in Australian design jobs to compare and inform design schools and graduates about the possible skills sought by employers.

Methods to educate or train designers differ across different countries and design schools. Some design schools provide generalist-oriented programs to educate designers, and some other programs train people from multiple disciplines in a specialized manner. From our experience, we would like to stress that the method of education would depend on the policy of the institution. However, the most important requirements are the quality of designers as produced by the quality design education and the competencies of the designers for industries. Furthermore, a multidisciplinary approach to design education and close collaboration with industries and research institutions will produce the future competent designers who will lead the industries. Above all, after students finish their tertiary education, it is their industrial design skills that are very important for finding a suitable job.

The notion of design and the employment opportunities of designers have become more important as new concepts continue to be introduced every day. During the product development process, the key tasks are planning, designing, prototyping and engineering. The successful completion of each of these tasks depends on certain skills such as planning, understanding users, ideation, creativity, sketching and drawing, as well as model-making abilities for making complex prototypes of appearance models, operating models and mechanism models (Frascara, 2017). These attributes are definitely the key competencies of industrial design jobs. These competencies also vary depending on the job responsibilities and different levels of proficiency, which are required for the different jobs.

There are some prior research studies in this domain, for example, the work of Wang (2008), Yang *et al.* (2005) and Ramirez (2012). The work of Wang (2008) focused on Chinese job markets, whereas the work of Yang *et al.* (2005) focused on Taiwanese job markets. The study conducted by Ramirez (2012) focused on employability factors for industrial design graduates and focused on US job markets. Those studies figured out the key competencies required for industrial design jobs; however, none of those studies focused on the required levels of proficiency for industrial design jobs. It is very important to achieve a certain level of proficiency in certain areas. Furthermore, it is not known whether the findings of the above studies are relevant in the Australian context. The study reported here investigates the level of proficiency required for industrial designers as sought in job advertisements. This study aims to understand the various job titles currently adopted in the industrial design profession in Australia and help design educators guiding industrial design students to plan their careers.

9.2 Method

A descriptive research design was used in this study. Job advertisements available from the Australian job site seek.com.au were retrieved over 3 months. Several keywords were used, such as industrial design, product design, industrial design engineering and product design engineering to extract relevant jobs. Content analysis (Hsieh and Shannon, 2005) was used to analyze the required levels of skills for each job. If the title of the job was ambiguous or more than one title was given for a job advertisement, those titles were checked against the contents of the advertisements to determine the accuracy of the titles. The advertisements were coded for job titles, job locations, job types, required competencies, other skills and portfolios. Frequency analysis was conducted for each of the coded items.

9.3 Results

9.3.1 Overview of the job titles

In total, 109 job titles were collected that were later coded into 45 categories. The most frequent job titles are "industrial designer" ($n = 25$), "industrial design engineer" ($n = 17$), "product design engineer" ($n = 8$) and "product designer" ($n = 5$). Several job titles were used synonymously for industrial designers. Out of 109 job titles, 18 were for the mechanical or industrial designers. Furthermore, some jobs were for entry-level positions, such as graduate design engineer/design internship, and some were for more advanced levels, such as design manager/senior design manager/project manager. Furthermore, the terms designers and engineers were used interchangeably in those advertisements. Some job titles were misleading; for example, the job title "design consultant" requires several years of experience, which is not clearly understandable from the collected job titles. Four job descriptions were identified as entry-level jobs:

- Graduate industrial design engineer/product design engineer
- Graphic design intern
- Junior industrial designer
- Industrial designer internship

Other jobs ($n = 12$) were also suitable for new industrial design graduates; however, those were not identifiable from the job titles, as those titles did not include the words intern or graduate. Thirteen jobs were identified as senior-level positions:

- Design manager
- Brand manager: industrial designer/engineer
- Senior design manager
- Senior design architect
- Regional design manager
- Experienced industrial designer

Out of 109 jobs, 44 were full-time and ongoing positions. More than 50% of the jobs were located in Melbourne and Sydney, and almost half of the job titles mentioned a salary range for the advertised jobs.

9.3.2 Required competencies mentioned in the job advertisements

Generally, the job advertisements were classified into several parts, such as tertiary qualification, work experience (i.e. prior work experience, currently working or not), toolset experience. Other skills such as report writing skills and communication skills were sought in each of the job advertisements. Furthermore, there were domain/job-specific requirements. Finally, a requirement written on the most job advertisements was the portfolios. Below, the commonalities and differences observed in those job advertisements are discussed.

9.3.2.1 Qualification

Most of the advertised jobs asked for a tertiary qualification or equivalent industrial design or product design/engineering degree obtained from local universities or abroad. Several keywords, such as recognized degree and minimum degree-level qualification, have been used to determine the tertiary qualification. For the graduate industrial designer, the end result from tertiary qualification was also given preference, for example distinction or honors from a reputable university. Most of the other jobs ($n = 80$) just mentioned tertiary qualification or degree-level qualification and did not ask for any other attributes to quantify it. In those jobs, tertiary qualification was considered a priority in conjunction with other key skills and experience. For the position of a design manager job, strong design management background was sought.

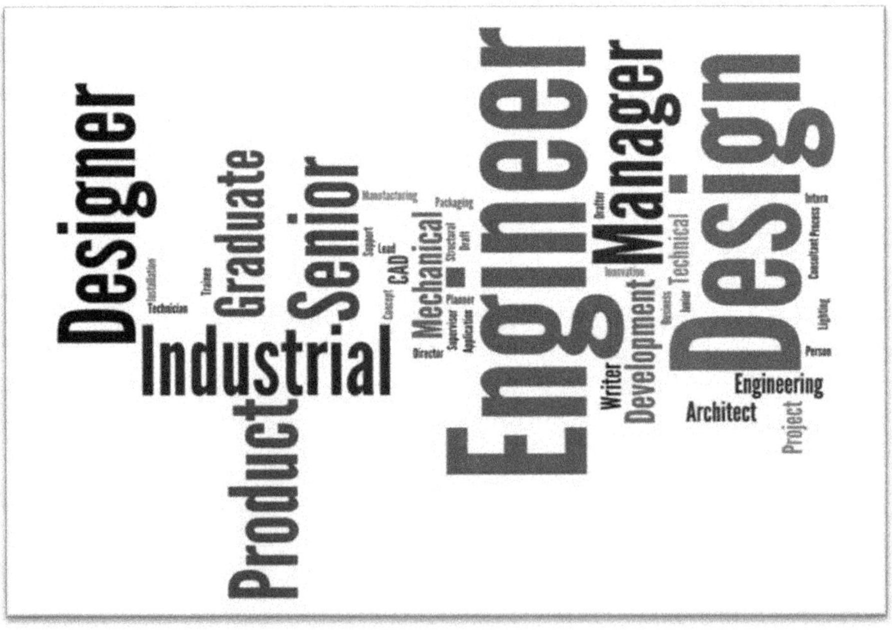

Figure 9.1 Word cloud for the collected job titles

9.3.2.2 *Work experience*

The required work experience varied across job titles. All jobs asked for recent work experience except for the graduate jobs or internship positions. For example, for a graduate product designer or industrial designer position, employers asked for candidates who graduated in the last 2 years. Other jobs asked for 1 to 8 years of job experience; the minimum number of years of experience was articulated by writing the word "minimum" or "at least," which clearly indicated the strength given to candidates having the right amount of work experience. Consultant jobs or jobs which are considered for experienced people or managerial positions require at least 5 years of experience in the similar role. For industrial design jobs on average, 2 years of work experience was sought.

It has also been observed that some jobs asked for continuous experience by mentioning "preferably continuous and current/relatively recent experience." In order to strengthen the job requirements, the words "must" or "preferably" have been used. For example, a job advertisement for industrial designers mentioned "demonstrable design experience and must either be currently working or have been working in a similar position within the past six months."

9.3.2.3 *Experience with software tools*

The levels of competencies were defined in various ways. Several adjectives/keywords were used to define different levels of competencies such as good proficiency, high proficiency etc. SolidWorks, CAD and Adobe suites were the top three skills advertised for the jobs. Some jobs looked for a software tool experience, which was mandatory, while other job advertisements were looking for toolset experience, for example, a high or good proficiency in SolidWorks or CAD. It has been found out that there are different levels of expertise required for different job profiles.

9.3.2.4 *Other skills*

In total, 14 other skills were identified, such as communication ability, taking initiative, ability to work independently and in a group, report writing, detailing, problem solving, time management, work ethics, attitude, analytical ability, multitasking and organizational skills, managing capabilities (i.e. conduct and coordinate) and networking capability. Those skills were considered mandatory in addition to tertiary qualification and toolset experience. Table 9.2 shows the summary of key skills extracted from the job advertisements.

Table 9.1 Experience and required levels of proficiency related to software tools

Software tools	*Level of proficiency*	*Required*
SolidWorks, CAD, CorelDRAW, SketchUp, Autodesk Inventor, Adobe Photoshop, Illustrator, InDesign, MS Excel, MS Word, 3D rendering package (V-Ray, 3DS Max), Photoshop rendering	High proficiency, experience using a particular software, skilled, excellent knowledge, substantial experience, competent, advanced knowledge, excellent proficiency, intermediate skills, competent, capable, strong, extensive	Mandatory, must have, prerequisite, should have, highly desirable, essential, beneficial, desirable, required

Table 9.2 Skills required for the collected jobs

Other skills	*Examples*
Communication ability: clearly, accurately, concisely, outstanding, excellent, fluent *"Must have"*	Must be able to communicate technical issues and ideas accurately, clearly and concisely Ability to accurately, clearly and concisely communicate technical issues and ideas both verbally and in written for
Self-initiation: natural ability, inbuilt drive *"Must possess"*	Must possess a natural ability to use initiative to drive developments to their conclusion along with an inbuilt drive to overcome obstacles and gain results.
Independence/work independently: ability *"Essential"*	Ability to take developments across all stages of the design cycle Ability to work unsupervised and/or part of a team
Report writing and presentation: experience, good, excellent *"Required"*	Experience in the generation of production and technical documentation Good documentation skills – read, interpret and develop technical drawings
Detailing: strong, high, great, good, ability, exemplary *"Required"*	A good eye for detail Exemplary attention to detail
Problem solving *"Have"*	Creative problem-solving skills Proven problem-solving skills
Time management: ability, highly, good *"Should have"*	Highly organized, disciplined Ability to work tight deadlines
Work ethics: excellent *"Should have"*	Excellent work ethic
Attitude: positive, open, pro-active	Positive attitude, open minded, pro-active approach to work challenges
Analytical: highly	Highly analytical candidate
Multitasking and organizational skills: ability, strong	Strong multitasking and organizational skills Ability to manage multiple projects
Conduct and coordinate	Conduct and coordinate onsite meetings
Decision-making: ability	Ability to take decisions
Networking: ability	Able to interface with clients

9.3.2.5 Job-specific requirements

Some job advertisements were more specific in terms of the experience of the candidates, though the basic qualifications of those jobs (n = 17) were industrial designer/mechanical engineer/product designer. Almost half of the job advertisements mentioned additional job-specific skills, which should be demonstrable by the job applicants. Job-specific requirements mentioned included sound knowledge, deep understanding, knowledge of, consideration for etc. For example, a job for graduate industrial design was looking for skills such as sound knowledge of injection molding, die casting and sheet metal forming processes. Another entry-level industrial design job asked for knowledge on sheet metal fabrication, which would be an advantage. These requirements clearly show that domain-specific knowledge is required and advantageous for the prospective candidates. However, from the job advertisements, it is clear that employers also want to have demonstrable experience on these components.

9.3.2.6 Portfolio

A portfolio was mandatory for half of the jobs, as those advertisements mentioned "must provide." Generally, a portfolio should demonstrate products and components to show design and technical ability and explanation of project involvement. Some job advertisements asked to provide a hard copy of the portfolio, whereas other advertisements asked to send a web link to it. Some of the job advertisements just asked to provide the best work, such as "a sample of your best work," "copies of previous work" or "a strong portfolio of previous work," while other job advertisements asked specific things. For example, a portfolio should include a sample of work in Adobe suites or a sample of engineering drawings and freehand sketching. For some industrial design jobs ($n = 20$), showing a physical portfolio during the interview was mandatory.

9.4 Discussion

In this section, the implications of our results for students, design educators and design schools are described. Some of the job advertisements, which we extracted from the seek.com.au website, were very elaborate and some were very brief. Clearly, there were some similarities and differences in the way the job advertisements were written and presented on the seek.com.au website. However, the extracted data and the analyses show the clear trend of the competencies required for industrial design jobs.

The first and foremost requirement was the tertiary qualification and degree, which was regarded as the basic requirement and the primary qualification of the intended job. The second key requirement was relevant job experience. For entry-level jobs, one year of experience was required, which could be compensated by doing internships in the relevant industries. Generally, design schools provide opportunities to do internships during a university degree. One of the remarkable requirements was experience with software tools and other relevant skills, which are sought for industrial design jobs.

It has also been witnessed that technical skills such as 2D and 3D software skills are very important for industrial designers. Other researchers who conducted studies in China (Wang, 2008) and Taiwan (Yang *et al.*, 2005) raised a similar issue. However, we also found out that the levels of proficiency required for those jobs are very important to consider. As evident from the results of our study, candidates need to have a high proficiency, which was regarded either as mandatory or essential for certain jobs.

The software tools (Table 9.1) identified from the job advertisements are taught in design schools, and students are required to deliver their assignments using those tools. However, our personal experience while teaching design students and evaluating their assignments show that students underestimate those tools and therefore do not realize the level of skills required for professional practice. Partly the reason could be that the use of software tools is required to accomplish one portion of their design assignment. That means that while students do design assignments, they spend a substantial amount of time in brainstorming, ideation, concept selection and detailed design. They eventually end up with the last part of the task where they need to produce technical designs using SolidWorks and render using Adobe products, but have little time at hand to produce quality outputs using those tools. On the other hand, if students have not achieved a high level of proficiency in using the software tools from a particular course dedicated to master those tools, it becomes difficult for them to show proficiency in their design assignments due to a lack of time and expertise. Therefore, it is advised that students should spend additional time mastering those software tools.

The 14 skills identified from the job advertisements were also considered essential for industrial design jobs. As evident from Table 9.2, most of them were required on a very high level, as advertisements used keywords such as "required, essential, should have, have, must possess etc." Most of the other skills are practiced during a design education, especially by the use of design project-based learning environment. However, it might not be possible for industrial design graduates to practice all those skills in one project. Again, it is imperative to realize for the prospective students that those skills should be achieved with high standards, as most of the advertised jobs require skills with varying levels of proficiency range from basic to exemplary skills. In addition, it would be a good idea for students to have an awareness of the attributes (Table 9.2) to act proactively in design assignments or industry-engaged design assignments. These skills are keys to the competency-centered learning approach (Hummels *et al.*, 2011; Voorhees, 2001). In addition to core subjects, design schools should offer specific courses for professional development for design students. It could be a separate module taught by professionals who have experience in providing personal development training. Furthermore, core design courses could be offered where design educators and personal development trainers can be engaged to help students achieve their professional skills.

Generally, job-specific skills are mostly taught during a tertiary education. However, job-specific skills are related to current or prior job experience. Having a thorough understanding of the industrial design profession through a tertiary education would help students to fulfill the job-specific requirements. Doing internships or gaining similar industry experience will help them achieve the remaining skills. For industrial design students, it would be useful to gain a deep understanding of the domain-specific knowledge and be able to demonstrate those in their design assignments as a discernible experience of this component.

The portfolio is regarded as the proof of competency of the designers. Design students generally produce portfolios for each design course they take in order to deliver their design assignments. As observed from our results, potential job seekers should be able to create portfolios and present their best works as requested by the employers. However, it is important to realize that selecting one's own best designs and organizing them requires knowledge, as most often the job advertisements collected do not indicate what things are needed to be included in the portfolio. Based on the job advertisements, our suggestion would be to demonstrate products and components to show design and technical ability and explain project involvement. Since sometimes employers ask for an online link to the portfolio, it would be wise to consider working on web portfolio. Our experience shows that some design schools ask students to create online portfolios for their design assignments, which are assessed at the end of the semester. While it is a requirement for every design student to create a design portfolio for each course, it would be beneficial to investigate how to make a portfolio for a potential employer, where only the very best design outcomes can be shown.

9.5 Conclusion

In this study, the assumption was that the contents of the job titles were correct representations of required skill demands of the employers. It was observed that some jobs were more elaborately described than others were. The reason could be the lack of experience in writing job advertisements. This study focused on Australian design job advertisements, which are related to industrial/product design. Though care has been taken to extract the relevant jobs from the seek.com.au website, it could be that some relevant jobs might have been missed.

This study has revealed that most industrial design job advertisements looked for a very high-level competencies (such as qualification, work experience, and software skills) and the competencies were sought by using phrases such as "must have" or "must possess." Several attributes were identified for industrial design jobs such as competency areas and levels of proficiency. The required competencies differ between various design tasks as well as between different types of design organizations. However, it has been found that a very high level of toolset experience is required for most of the industrial design jobs. It is expected that these findings will be useful for students and design educators to update their curriculum. In the future, it would be interesting to compare industrial design curricula of Australian design schools to assess industry readiness of industrial design graduates.

References

Ball, L. (2002) Preparing graduates in art and design to meet the challenges of working in the creative industries: A new model for work. *Art, Design and Communication in Higher Education*, 1(1), 10–24.

Boyarski, D. (1998) Education: Designing design education. *Special Interest Group on Computer and Human Interaction Bulletin*, 30(3), 2.

Erkarslan, O. (2007) Inter-disciplinary characteristics of design profession: Bridging the gap between design education and industry. *Proceedings Designtrain Congress Trailer I*. pp. 213–218, 10–12 May 2007, Amsterdam, the Netherlands.

Frascara, J. (2017) Design, and design education: How can they get together? *Art, Design & Communication in Higher Education*, 16(1), 125–131.

Giard, J. (2000) Industrial design values: Focus the toast, not the toaster. *The 2000 IDSA National Education Conference (CD ROM)*. Industrial Designers Society of America, Herndon, VA, USA.

Hummels, C., Vinke, A.A., Frens, J.W. & Hu, J. (2011) Competency-centered education for designing interactive and intelligent products. *Creation and Design*, 13(2), 4–17.

Hsieh, H.F. & Shannon, S.E. (2005) Three approaches to qualitative content analysis. *Qualitative Health Research*, 15(9), 1277–1288.

IDSA (2018) Industrial Design Society of America (IDSA) Available from: www.idsa.org/news/dblog/what-id [Accessed 17 February 2018].

Kaufmann, J. (1998) Why design education? Infrastructure issues affecting the future of industrial design education. *The 1998 IDSA National Education Conference* (CD ROM), Industrial Designers Society of America, Herndon, VA, USA.

Kolko, J. (2005) New techniques in industrial design education. *Proceedings of 6th International Conference of the European Academy of Design*. Available from: www.jonkolko.com/projectFiles/writing/ead06_id115_2.pdf [Accessed February 17 2018].

Owen, C.L. (2004) *What Is Design? Some Questions and Answers*. Illinois Institute of Design white paper. Available from: www.id.iit.edu/artifacts/what-is-design-some-questions-and-answers/ [Accessed 20 October 2016].

Ramirez, M. (2012) Employability attributes of industrial design graduates. *Proceedings of the 5th International Conference of Education, Research and Innovation (ICERI2012)*, Madrid, Spain. pp. 2462–2471.

Voorhees, R.A. (2001) Competency-based learning models: A necessary future. *New Directions for Institutional Research*, 110, 5–13.

Wang, K. (2008) Research of the recruiting qualifications of the ID profession and their implication in ID education. *Computer-Aided Industrial Design and Conceptual Design, CAID/CD 2008*, Kunming, China. pp. 851–854.

Yang, M.Y., You, M. & Chen, F.C. (2005) Competencies and qualifications for industrial design jobs. Implications for design practice, education, and student career guidance. *Design Studies*, 26(2), 155–189.

Yeh, W.D. (2003) The demand and the evaluation of industrial design profession from the industries. *The 6th Asian Design Conference (CD ROM)*. Institute of Art and Design, University of Tsukuba, Japan

WDO (World Design Organisation). Available from: http://wdo.org/about/definition/ [Accessed 17 February 2018].

Chapter 10

Blended learning in engineering education

Students' and lecturers' perceptions and achieving learning outcomes

D.S. Liyanapathirana and O. Mirza

School of Computing, Engineering and Mathematics, Western Sydney University, Kingswood, NSW, Australia

ABSTRACT: In this chapter, blended learning initiatives put forward by the Western Sydney University for engineering units are discussed emphasizing students' and lecturers' perceptions. One aim of this initiative is to provide a flexible student-centered learning environment for those who may not be able to attend lectures regularly due to work commitments, utilizing the technology in terms of recorded lectures and tutorials. Based on the feedback from students and lecturers, this research investigates how far we can move away from traditional face-to-face teacher-centered education, while meeting professional competencies set out by Engineers Australia and achieving unit learning outcomes.

10.1 Introduction

Internet and wireless technologies are advancing rapidly in recent years, and it is becoming convenient to access them anywhere in the world whenever necessary. This trend has enabled the younger generation of school leavers to be extremely versatile with computer use and internet technologies. As a result, universities around the world are striving to make courses more interesting, attractive and flexible for this "tech savvy" new generation by introducing computer-based e-learning technologies. As a result, the term "blended learning" has emerged in the higher education realm.

The most common definition of blended learning is the effective integration of computer technologies to enhance the teaching and learning experience of both lecturers and students. However, Oliver and Trigwell (2005) suggested that the term blended learning is still ill defined, and the application of a number of pedagogic approaches for teaching, regardless of technology usage, can also be considered as blended learning.

Literature about blended learning applications in higher education shows the benefits of e-learning integration in tertiary courses (e.g. Abraham, 2007; Ekwunife-Orakwue and Teng, 2014; El-Mowafy *et al.*, 2013; El-Zein *et al.*, 2009; Gonzalez *et al.*, 2013; Wardenski *et al.*, 2012), but some researchers pointed out a few weaknesses in the blended learning approach (e.g. Ellis and Calvo, 2004; Sethy, 2008). Engineering education needs both psychomotor and cognitive skills to be developed, but computer-aided learning can enhance only students' learning experience and not their psychomotor skills (Jones and Chew, 2015).

Although there are benefits of these new educational strategies to the teaching and learning experience of students, they should be affirmed cautiously, and implementation should be carried out after broader discussions of the context of applications and the expectations of students (Wardenski *et al.*, 2012). Although utilization of web-assisted content offers cost efficiency in terms of reduced face-to-face time for academic staff as well as time and travel expenditure for students, it is an open question whether blended learning can enhance achieving desired learning outcomes in engineering courses.

According to Debnath *et al.* (2014), the higher education community regards online courses with some uncertainty. This may be a result of a lack of institutional culture and knowledge (Wardenski *et al.*, 2012) as well as mixed results found in the literature from case studies. Another reason is that the lecturers are not familiar with the practice of creating and delivering the same courses they learned as students and have delivered for years in face-to-face settings through virtual learning environments (Porumb *et al.*, 2013). Especially for engineering education, lecturers still do not consider the e-learning methodologies stable enough or powerful for delivering courses integrated with practice-oriented activities such as laboratory experiments and group design projects. Nevertheless, integration of computer-based technology in tertiary education has been considered as a promising approach to combine with face-to-face teaching, is growing rapidly in Australia and is predicted to become the new traditional model in university education (Graham, 2004), though it is still at its preliminary stages of exploration.

According to Twigg (2003), there are five blended learning models: (1) supplemental model; (2) replacement model; (3) emporium model; (4) fully online model and (5) buffet model. In the supplemental model, the traditional face-to-face course structure and class meeting times are retained and out-of-class activities are introduced to enhance student learning. In the replacement model, class meeting time is reduced and the face-to-face time is replaced by online activities for which students can participate from anywhere at any time. As discussed by Twigg (2003), the emporium model is different to other models because it replaces all class meetings with a learning resource center with online materials and on-demand personalized assistance. For mathematics teaching, the emporium model has been successfully adopted by Virginia Tech University and Alabama University in the USA. These learning centers are equipped with computers and other specialized equipment and study spaces. Development of these learning centers is exhaustive and needs instructional software, interactive tutorials, computational exercises, practice questions, online quizzes and solutions to frequently asked questions. These centers are staffed by graduate students and senior undergraduate students, and student involvement is tracked to measure the peak times. Based on these data, teaching assistants are allocated to the learning centers. The Virginia Tech math learning center is open 24 hours a day and 7 days a week. The fully online courses are delivered online and all the assessment submissions, class tests and exams are run online without any class meeting times. In the buffet model, the teaching approach used is different to other methods. The other four models are cost effective and the students are considered the same; in other words, all students complete the courses with the same assessments and learning resources. In the buffet model, learning is customized based on the needs of the students. More options are available for students to choose, such as lectures and individual discovery laboratories (in-class and online), group discovery laboratories, individual and group review (in-class and online), group study sessions (online and in-class), videos, oral and written presentations, large-group or individual problem solving, and individual and group projects (Twigg, 2003). Depending on the choice and strength of individual students, they will enter a learning contract at the beginning, selecting their preferred options. They receive an initial orientation, and during the term progress is tracked.

Each blended learning model discussed above has its own merits and demerits, when considered within the delivery of engineering education. The fully online model may not be suitable for engineering education due to requirements of laboratories and computer software tutorials. Laboratories are difficult to offer online because of the traditional desire for the direct operation of instruments. Also, software licenses can generally be used only on university computers, and they may require computing power and graphics that are not readily available off campus (Bourne *et al.*, 2005). Hence, class meetings cannot be completely eliminated, and the other four methods can be considered as suitable blended learning models for engineering education. However, the blended learning models currently in place for engineering education can be categorized under supplemental and replacement models.

A blended learning model was introduced at the Western Sydney University (WSU) in 2013 within the undergraduate (BEng) and postgraduate (MEng) engineering courses via the e-learning system. This chapter investigates the students' and lecturers' perceptions of the blended learning model adopted at WSU and how this model affected the student learning and achievement of professional competencies set out by Engineers Australia.

10.2 Blended learning model used at WSU

The blended learning model introduced at WSU engineering units is more aligned with the supplemental model; however, the university aims to move towards a replacement model due to the multi-campus offerings planned for engineering (both undergraduate and postgraduate levels). Currently in the majority of undergraduate units, lectures are recorded and posted in the e-learning system known as vUWS a few hours after the lecture. The system, which records lectures automatically in many lecture theaters and posts in vUWS, is called "lectures online." In some units, face-to-face delivery was partially replaced by recorded lectures and tutorials. Tutorial recordings were not made during the normal class meeting times. They were recorded manually outside the scheduled tutorial class meeting time, e.g. in Fluid Mechanics as reported in Rahman (2017). In some units, online quizzes, online assessment submissions, online feedback on assessments, recorded laboratory experiments and discussion boards were also facilitated through the e-learning system.

After the introduction of blended learning, a significant decline in class attendance of students was observed. For example, if about 80% of students enrolled in a unit attended lectures before the university introduced lectures online, it was reduced to about 25% of attendance for many units after the introduction of blended learning. This decline in student participation and concerns regarding students' achieving learning outcomes and meeting professional competencies set out by Engineers Australia, such as team-building skills and collaborative problem-solving abilities, are the motivation for this research. Research was carried out using survey data collected from both students and lecturers as outlined in the following sections.

10.3 Survey for students

The survey issued for students consisted of only eight questions, because a lengthy survey of a large number of questions may not be suitable to get a genuine response from students. If the survey is long, then students tend to write something without much thought just to complete it quickly and some students may not complete the survey at all. The length of questions were also kept short and simple. Questions required either "Yes" or "No" answers, and descriptive questions were included requiring them to demonstrate the reason for selecting either "Yes" or "No" as their answer. In this study, the survey delivered to students aimed to get feedback on student perceptions about three areas: (i) face-to-face teaching; (ii) lectures online and (ii) flexible learning as shown in Table 10.1.

The survey forms were distributed to students, and they were given the freedom to complete and return the forms anonymously. Out of all the distributed forms, only 92 surveys were returned. From the first and second years, only 49 students responded, and from the third and fourth years, 43 students responded. In this section, results are presented from the composite group of first to fourth years, but results were analyzed for the first to second years and third to fourth years in two separate groups. The idea was to investigate whether the junior- and senior-level undergraduate students have the same understanding on face-to-face teaching, lectures online and flexible learning. When discussing these results, it is important to note that the surveys were issued to students during the lectures. Since the lecture attendance of students drastically declined after the introduction of lectures online, the sample size of students who participated in this survey is small. Another factor that contributed to the small sample size is when it was issued to students, which was towards the end of the semester in week 10 when student attendance for lectures is typically low.

According to the data collected in this survey, as shown in Figure 10.1(a), 94% of the student cohort agreed that the face-to-face teaching is the most useful mode of delivery. The common reason given by those who answered "Yes" is the opportunity students get in asking questions in real time, when they have issues in understanding the concepts. Also, students appreciated the opportunity to engage with other students studying the same unit during the lectures. Students liked the study environment within the lecture theater and accepted it as a more conducive environment in which to learn, rather than studying in isolation in front of

Table 10.1 Survey distributed to students

Face-to-Face teaching

1. Do you think face-to-face teaching is valuable in comparison to other modes of teaching?
   Yes No
2. Please explain the reasons for selecting the above answer.

Lectures Online

3. Do you access lectures online in vUWS weekly?
   Yes No
4. Do you see any value of lectures online?
   Yes No
5. Please explain the reasons for selecting the above answer.
6. Because of lectures online, do you think that your attendance to face-to-face lectures reduced?
   Yes No
7. If you selected "Yes" to the above question, how do you think this affected your learning?

Flexible Learning

8. What is your perception of flexible learning?

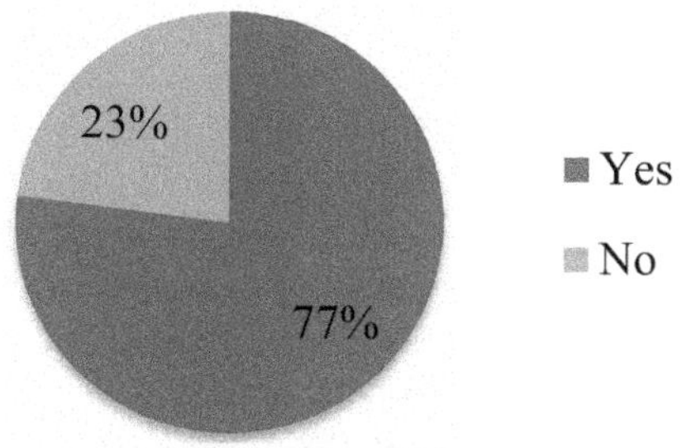

(a) Face-to-face teaching is useful in comparison to other modes (Q1)

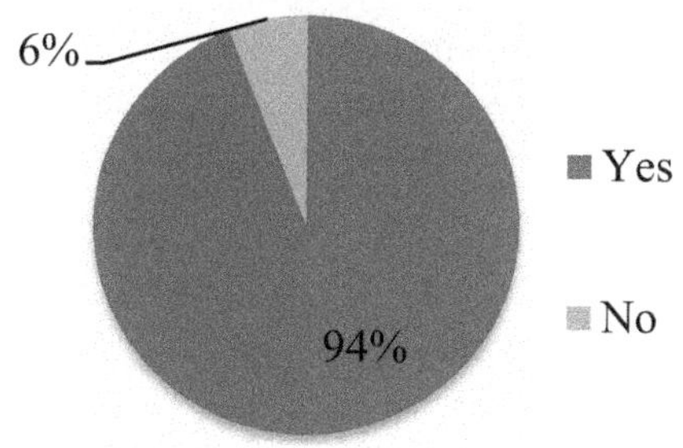

(b) Access lectures online weekly in vUWS (Q3)

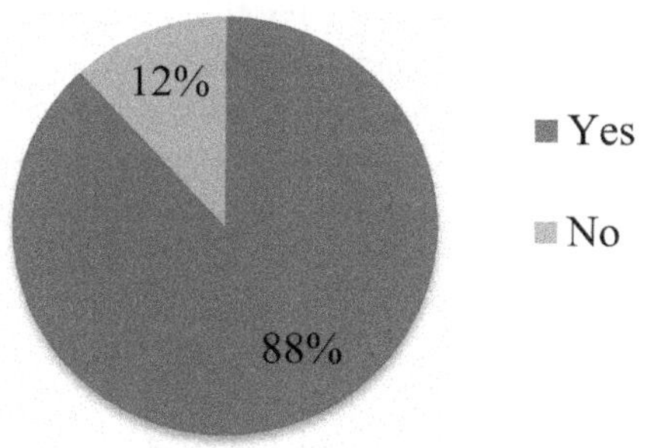

(c) Lectures online is useful (Q4)

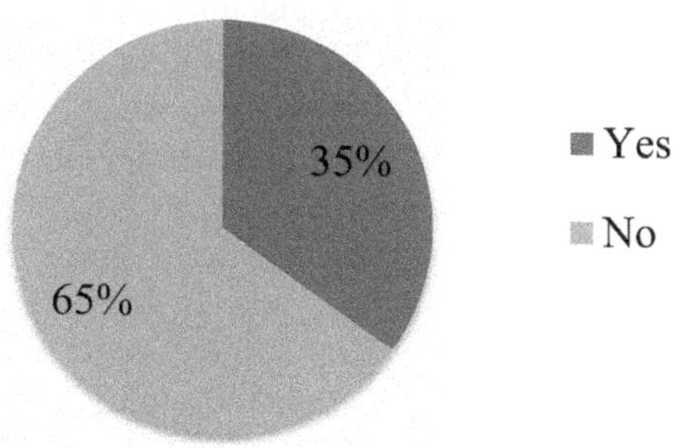

(d) Lectures online reduced attendance to face-to-face teaching (Q6)

Figure 10.1 Summary of quantitative questions in the survey given in Table 10.1

a computer. The main reason for the "No" answer was specifically about the lecturer. They gave the reason as communication problems, such as the language proficiency or the English pronunciation issues of the lecturer. However, the number of students who do not agree with the fact that "face-to-face learning is better than other teaching modes" is a very small percentage of 6%. Based on these results, it is a question whether a blended learning model, which reduces face-to-face teaching, is the best solution, even though it is obviously the most economical approach due to the ability in repeatability of lectures in a multi-campus setting.

The next section of the questionnaire was about the students' perceptions related to lectures online introduced by WSU as part of the blended learning initiative. As shown in Figure 10.1(b), only 77% of the students said they watched lectures online weekly. However, as shown in Figure 10.1(c), 88% of students agreed that they were a useful resource for their learning, which is a larger proportion compared to the 77% of students who watched lectures online in a weekly basis. The reasons they gave for answering "Yes" to this question are (i) by watching the lectures online, they could revise what their lecturer taught in class and (ii) if they missed the lecture, it was a good source to revise the material. Those who mentioned that the lectures online was not a useful resource mentioned that they selected that answer because (i) they preferred face-to-face teaching to e-learning; (ii) they liked the learning environment and (iii) it allowed them to interact with both peers and lecturers.

Figure 10.1(d) shows the response to the question, whether lectures online has reduced their attendance to weekly face-to-face lectures. For this question, only 35% answered "Yes." From this group, 65% said lectures online has not reduced their attendance. Students who mentioned that their attendance was reduced gave the main reason as the traveling distance from their homes to the university. Also, they mentioned that lectures online is a much better option for students who fall behind because they can repeatedly watch the online lecture if they cannot understand any concepts. A few students enjoyed the fact that they did not need to attend the lectures because all the necessary information is on lectures online. Those who said that their attendance was not reduced due to lectures online mentioned that they liked the interaction with other students within the lecture room. In addition, the attendance to lectures gave them an opportunity to see their peers studying and their level of understanding of the subject matter, which in turn motivated them to study.

The last question of the survey was about the students' understanding of the flexible learning. This is also important within the context of blended learning, because one of the ultimate goals is to provide a flexible environment for the students. Students who appreciated flexible learning said it was good when they had other commitments because the class attendance was not rigid as in a traditional face-to-face setting. They also said that flexible learning made them maintain a good life balance between study commitments and free time. Some students said flexible learning was not good because it reduced class attendance and made them less interested in learning, lowering their exam performance.

The main conclusion from this part of the study is that although a large proportion of students see the benefits of lectures online, they do not like to reduce the face-to-face class meeting times. Therefore, based on the blended learning models discussed earlier, a supplemental model is the one preferred by WSU undergraduate students. Martinez-Caro and Campuzano-Bolarin (2011) also concluded that a mix of classroom and online technologies could achieve the best results in terms of student satisfaction. However, for cost reduction and providing a flexible learning environment, a replacement model is preferable. In that case, the right proportion of e-learning and traditional face-to-face learning should be used. This proportion may not be uniform across disciplines and may depend on the academic abilities of the student cohort as well as the degree of difficulty of the unit. A blended learning model with lectures online and reduced class meeting times can be a weakness for students who do not have the right motivation to engage actively in learning. Two separate groups (junior and senior undergraduate students) analyzed here showed similar results, i.e. there are benefits of lectures online; however, reduction of the face-to-face class meeting time is undesirable.

10.4 Staff survey

This section of the paper examines the survey conducted for selected engineering academic staff at WSU. The questionnaire shown in Table 10.2 was completed by 17 lecturers. This was also an anonymous survey similar to the student survey, as mentioned before. The questions were mainly categorized under four main areas: (i) face-to-face teaching; (ii) lectures online; (iii) flexible learning and (iv) blended learning.

Figure 10.2 shows the summary of results for the quantitative questions in the questionnaire. For question 1, all participants selected "Yes" to indicate that the face-to-face teaching

Table 10.2 Survey distributed to lecturers in WSU

Face-to-Face Lecturing

1. Do you think face-to-face learning is valuable in comparison to other modes of learning?
   Yes No
2. Please explain the reason for selecting the above answer

Lectures Online

3. Do you provide lectures online in vWUS?
   Yes No
4. Do you think that your students take the full benefit of lectures online?
   Yes No
5. Please explain the reasons for selecting the above answer
6. Because of lectures online, do you think that student attendance to face-to-face lectures reduced?
   Yes No
7. If you answered "yes" to the above question, how do you think this affects the learning of your students?

Flexible Learning

8. What is your opinion on the "flexible learning environment"?

Blended Learning

9. What is your perception on blended learning in terms of student performance in your unit and interaction between lecturers-students as well as students-students?
10. Apart from recorded lectures, what other blended learning approaches have been implemented in your units?
11. Do you think new blended learning approaches introduced at UWS will affect achieving the professional competency levels set out by Engineers Australia (e.g., team membership and leadership skills, communication, team design projects)?
12. In what ways has the blended learning influenced the average assessment marks in your units?

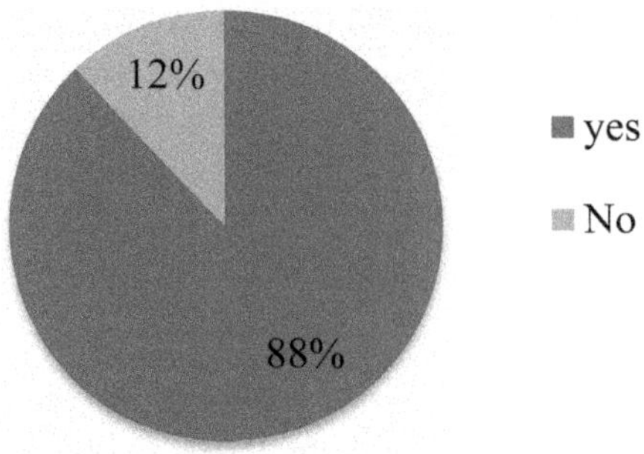

(a) Lectures were recorded and posted on blackboard (Q3)

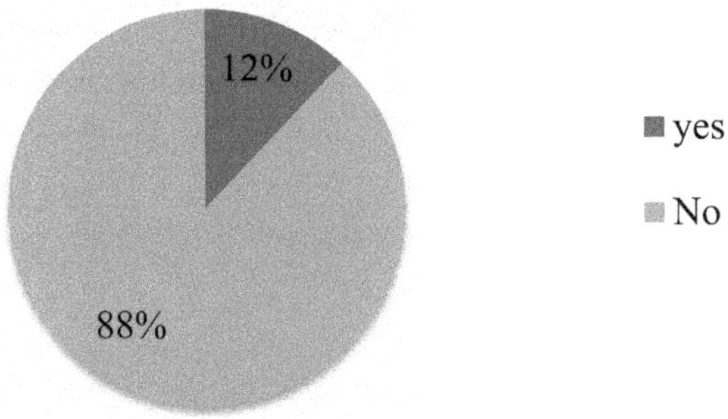

(b) Students take advantage of lectures online (Q4)

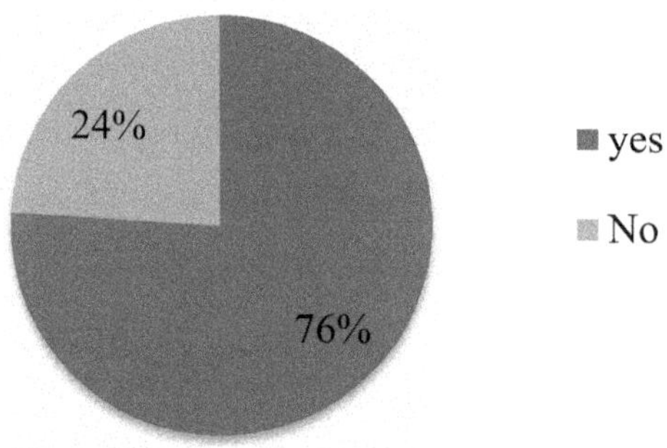

(c) Attendance to face-to-face lectures reduced (Q6)

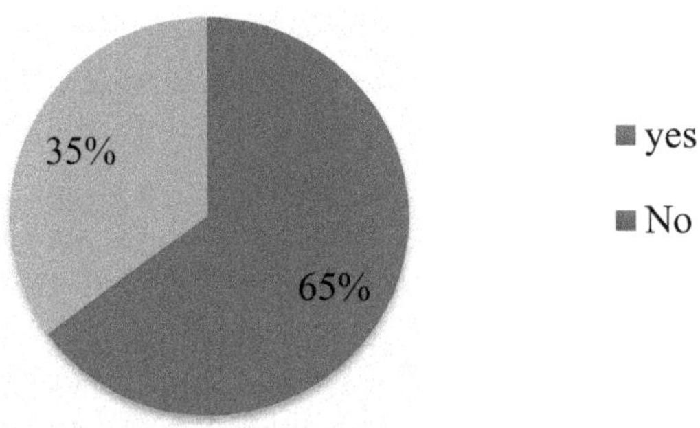

(d) Lectures online reduced assessment marks (Q12)

Figure 10.2 Summary of quantitative questions in academic staff survey in Table 10.2

is valuable in comparison to other modes of teaching. According to the staff, face-to-face teaching provides significant mentoring to students during lecturing. It is an interactive mode of teaching, and students get instant feedback for their questions and hence problems do not accumulate until closer to the exams. Listening to a lecture in a classroom setting is very different to listening it at one's home in front of a computer. In a classroom, students are able to focus on the lecture without external distractions. It also provides an environment for social interactions and discussions in which students develop ideas and create knowledge beyond the content in the lecture slides, because lectures online capture only the content in the slides. If the lecturer explains any extra details on the whiteboard, lectures online does not include that content. When students are among their peers, their desire to learn and to meet deadlines is enhanced. Also, lecturers are able to understand student progress while teaching a certain topic. Some staff members mentioned that the active learning should be promoted through more face-to-face tutorial classes and using a flip mode of delivery if the face-to-face lectures are partially replaced by the recorded ones. In the flip mode, students learn the subject matter outside the class by listening to a pre-recorded lecture and then come for a supervised tutorial class to work on problems.

About 88% of lecturers recorded all their lectures, as shown in Figure 10.2(a), and they were made available to students in the e-learning system (vUWS) after the lecture hour. However, according to Figure 10.2(b), 88% of lecturers considered that lectures online was not beneficial to their students. The main reason for selecting this answer for question 5 was that they thought lectures online discouraged attendance to lectures. Many students who do not attend lectures wait until the end of lecture weeks to learn the unit content via lectures online to pass the final exam.

According to the staff survey, a number of lecturers monitored student participation in lectures online using the statistical tracking activated in vUWS. They observed that the number of students accessing lectures online normally increased during the last two to three weeks of the semester. During the rest of the semester, less than 1% of students viewed lectures online for some units. Lectures were recorded by the university IT services, and lecturers did not have any control over the quality of recordings. If the quality of the recordings were bad, student evaluations on teaching (SFT) and unit evaluations for unit (SFU) scores suffered because students directly put the blame on the lecturer. Overall, all the staff said that many students did not attend lectures and did not interact much with others and hence had a shallow view of the unit content.

As illustrated in Figure 10.2(c), 76% of lecturers identified that the attendance to lectures drastically reduced after the introduction of lectures online. When students do not attend lectures and attempt to learn by watching the recorded lectures, they cannot participate in stimulating discussions, which occur in a traditional class setting, and they become a silent audience. As a result, the critical thinking ability of students may degrade. Students wait until the last few weeks of the semester to study for the final exams. Again, if they do not watch the videos, they have to rely on a set of slides with minimum information. On the other hand, overall student consultation times increase because in the classroom, all students can hear the questions and answers but when students study individually, many students come to the lecturer with the same question, increasing the overall student consultation times.

Similar to question 8 in Table 10.2, a question about the flexible teaching environment was asked to the lecturers. The majority of the lecturers said flexible learning was good only for motivated students. It may give benefits to the students who are engaged in outside work, but some students may suffer from procrastination which may hinder their learning. Some lecturers suggested introducing a minimum requirement of participation and setting thresholds to pass the main assessment tasks, including the final exam. Then students must complete the main assessments to meet the threshold criteria, instead of selecting only few assessment tasks to do well, which carry a higher weight towards the final mark.

The last section of the staff survey was devoted to blended learning. It is important for both lecturers and students to have a clear idea of the blended learning and how this should be implemented in teaching without downgrading the course standard and student learning outcomes. In this question, lecturers' perception on blended learning was asked in terms of lecturer-student, student-student interaction and student performance in units. The majority of the lecturers noted that the students did not have the correct view of blended learning–related e-learning initiatives, and as a result they think that they do not need to attend the lectures if lectures are recorded. For mature and motivated students, e-learning does not affect their performance, as they use e-learning as a supplement to face-to-face teaching rather than a replacement. In addition, e-learning reduces the opportunities for group assignments, which is a mode of learning involving students in a group motivating each other. Hence, some lecturers suggested that the exam performance of students degrades in blended learning.

Currently, engineering lecturers in WSU use a variety of e-learning resources in the e-learning system vUWS. These consist of links for resources available on the internet, online quizzes, pre-recorded videos for labs, discussion forums, collaborative sessions, recorded tutorials, and video solutions for tutorials, in addition to the recorded lectures. No units have all these resources, but all units have recorded lectures if the classroom has recording facilities. Despite the availability of these resources, it was evident that students did not use them efficiently to enhance their learning.

In question 11, it was asked whether lecturers think that the new blended learning approach introduced at WSU may affect achieving the professional competency levels set out by Engineers Australia (e.g. team membership and leadership skills, communication skills and team design projects). All the lecturers stated that achieving the above-mentioned professional competencies could be an issue, in particular when students do not interact with peers. Students do not master the skills necessary for teamwork or building relationships with peers or communication skills, which is very much needed in engineering profession. Also, in units with group design projects, students constantly approached unit coordinators claiming that they cannot form groups because they do not know others enrolled in the unit.

The response given for the last question is summarized in Figure 10.2(d). It is clear that a significant proportion of the lecturers' view is that blended learning has affected the student performance in their units. According to the responses provided by the lecturers, for some units with high mathematics-based content, failure rate had doubled. According to this response, e-learning may not be a successful mode in an engineering unit with high computational volume. However, for descriptive units related to humanities or management, which involves less critical thinking and problem-solving skills, e-learning methods may be more useful.

10.5 Conclusion

In this chapter, the perceptions of WSU's engineering students and lecturers on blended learning are discussed. The majority of the students surveyed accepted that the blended learning model adopted at WSU is useful; however, they identified that the face-to-face component of a unit should not be removed, because this allows students to interact with their lecturers and peers in a classroom and to get answers to their queries during the lectures, and because the classroom environment is conducive for learning. However, it should be noted here that the survey was carried out among the students who were present at the lecture theater on the day of the survey. Therefore, the student perceptions presented here are from a group of students who appreciated face-to-face teaching. From the point of view of academic staff, a significant number them stated that the lectures online are not beneficial to students because statistical tracking identified that only a small percentage of students watched lectures online throughout the semester – the numbers only increased during the last few weeks of the semester and during the study break. Also, student consultation times increased as more students requested to meet lecturers to ask similar questions. In some units, reduced class attendance was reflected in the poor results of the final examination, often with a higher failure rate. When team design projects were introduced, it became difficult for students to form groups, as many students did not know each other well. Hence, incorporating group assignments in units became difficult, making it harder to meet specified professional competency levels set out by Engineers Australia.

References

Abraham, A. (2007) Student centred teaching of accounting to engineering students: Comparing blended learning with traditional approaches. *Proceedings Ascilite*, Singapore, 1–9.

Bourne, J., Harris, D. & Mayadas, F. (2005) Online engineering education: Learning anywhere, anytime. *Journal of Asynchronous Learning Networks*, 9(1), 15–41.

Debnath, B.C., Rahman, M.M. & Hossain, M.J. (2014) Blended learning approach for engineering education – An improvement phase of traditional learning. *International Journal of Computer Science and Network Security*, 14(11), 85–90.

Ekwunife-Orakwue, K.C.V. & Teng, T.L. (2014) The impact of transactional distance dialogic interactions on student learning outcomes in online and blended environments. *Computers and Education*, 78, 414–427.

Ellis, R.A. & Calvo, R.A. (2004) Learning through discussions in blended environments. *Educational Media International*, 40(1), 263–274.

El-Mowafy, A., Kuhn, M. & Snow, T. (2013) A blended learning approach in higher education: A case study from surveying education. In: *Design, Develop, Evaluate: The Core of the Learning Environment. Proceedings of the 22nd Annual Teaching Learning Forum*, 7–8 Feb 2013, Perth, Australia.

El-Zein, A., Langrish, T. & Balaam, N. (2009) Blended teaching and learning of computer programming skills in engineering curricula. *Advances in Engineering Education*, 1(3), 1–18.

Gonzalez, A.B., Rodriguez, M.J., Olmos, S., Borham, M. & Garcia, F. (2013) Experimental evaluation of the impact of b-learning methodologies on engineering students in Spain. *Computers in Human Behaviour*, 29, 370–377.

Graham, C.R. (2004) Blended learning systems: Definition, current trends and future directions. In: Bonk, C.J. & Graham, C.R. (eds) *Handbook of Blended Learning: Global Perspectives, Local Designs*. Pfeiffer Publishing, San Francisco, CA, USA.

Jones, L.J.N. & Chew, E. (2015) Blended learning in engineering education: Curriculum redesign and development. In: Tang, S.F. & Loganathan, L. (eds) *Proceedings of the Taylor's 7th Teaching and Learning Conference 2014*, Springer Science+Business Media, Singapore.

Martinez-Caro, E. & Campuzano-Bolarin, F. (2011) Factors affecting students' satisfaction in engineering disciplines: Traditional vs. blended approaches. *European Journal of Engineering Education*, 36(5), 473–483.

Oliver, M. & Trigwell, K. (2005) Can 'Blended Learning' be redeemed? *E-Learning*, 2(1), 17–26.

Porumb, C., Porumb, S., Orza, B. & Vlaicu, A. (2013) Blended learning concept and its applications to engineering education. *Advanced Engineering Forum*, 8–9, 55–64.

Rahman, A. (2017) A blended learning approach to teach fluid mechanics in engineering. *European Journal of Engineering Education*, 42, 3, 252–259.

Sethy, S.S. (2008) Distance education in the age of globalisation: An overwhelming desire towards blended learning. *Online Journal of Distance Education*, 9(3), 29–44.

Twigg, C.A. (2003) Improving learning and reducing costs: New models for online learning. *EDUCAUSE review*, 38(5), 28–38.

Wardenski, R.F., Espindola, M.B., Struchinert, M. & Giannella, T.R. (2012) Blended learning in biochemistry education. *Biochemistry and Molecular Biology Education*, 40(4), 22–228.

Chapter 11

Evolution of Master of Engineering coursework degrees at Western Sydney University

Past, present and future

R. Liyanapathirana, G. Fang and C. Leo

School of Computing, Engineering and Mathematics, Western Sydney University, New South Wales, Australia

ABSTRACT: Engineering curricula are subject to changes, with academics and university administrators taking up prevailing ideas and responding to technological changes as well as operational challenges. There is a need to meet the interests of potential international and domestic students of varying backgrounds to pursue high-quality postgraduate studies. This chapter presents the genesis of the postgraduate coursework-based Master of Engineering program at Western Sydney University from its inception to the near future.

11.1 Introduction

Engineering curricula must respond to change as technological advancements and operational changes occur. Introduction of blended learning is a good example of a contemporary situation in Australian universities as they face funding shortages and the concomitant need to cater to increasing student cohorts while maintaining student-to-staff ratios at current levels. Technological changes occurring in the tertiary education and vocational training sector have a flow-on effect to the delivery of courses and how student learning targets can be met.

The formalization of such changes takes into account the mission, aims and objectives as well as strategic plans of the university. In the Western Sydney University (WSU) case, these are summarized in the strategic plan (WSU, 2015). Engineering courses are subject to accreditation and other requirements of their profession. Hence, a substantial change in the culture of the organizational unit responsible for the delivery of the curriculum is called for (Heywood, 2005). Thus, changes are often influenced by external factors.

According to Heywood (2005), the role of the curriculum designer is to determine (a) the aims and objectives (course learning outcomes [CLO] or learning targets) of the course to be offered, (b) ideal methods of achieving those aims and objectives, (c) the unit sequence, and assessing if, as a result of (b) and (c), they have been achieved. Traditionally the last process has been called evaluation. Evaluation embraces the assessment of student learning and determining if learning targets have been met and at what level. It detects mismatches between the formal learning environment and the experiences of students in that environment achieving desired outcomes. It also includes the evaluation of teaching performance, the continuing appraisal of goals in response to sociotechnical change and the attention to the core values of the course (program) (Heywood, 2005).

New learning models including problem-based learning (PBL), blended learning (BL) (Martinez-Caro and Campuzano-Bolarin (2011)) and service learning are introduced as engineering schools evolve towards a more student-centered approach. This is also brought about through the increasingly competitive higher educational environment and the need for optimal productivity in delivering courses to an ever-increasing number of students. Various factors affecting student satisfaction in traditional as well as blended approaches are taken into consideration in course design.

According to the 2015–2020 Strategic Plan (WSU, 2015), Western Sydney University is embracing the above challenges with a renewed focus on a number of areas that will be key to its success and future sustainability. Traditional approaches will cede to more creative solutions to advance world-class teaching and curricula, delivered in technology-rich high-tech classrooms and collaborative learning spaces. Western Sydney University will extend its international reach and global standing through its students and graduates, its research, and its expanding partnerships with industry.

This chapter presents evolution of Master of Engineering coursework degrees in Western Sydney University. At the beginning, the objectives of Master of Engineering coursework program at Western Sydney University are discussed. This is followed by an overview of the Master of Engineering coursework programs at Western Sydney University. The pathways, entry, academic credits and exit points are then presented along with proposed future changes.

11.2 Program objectives

The Master of Engineering coursework program at Western Sydney University has evolved in three distinct stages: Stage 1, Stage 2 and Stage 3. Stage 1 began in the early 2000s, which involves the creation of a 1-year generic Master of Engineering program where students are at liberty to choose specialized units offered in civil, environmental, mechatronic, electrical, computer and telecommunications engineering. In Stage 2, during the mid to late 2000s, units were rationalized and consolidated together with the creation of distinctive key areas in civil, environmental, mechatronic, electrical, computer and telecommunications engineering. The program remained as a 1-year program, and the student numbers were modest across the range of disciplines.

In Stage 3, during the mid-2010s, the 1-year program was restructured to a 2-year program (not including a 1-year preparatory study) with the addition of new core and specialized units. Moreover, provisional Engineers Australia (EA) accreditation was sought and approved. The specialization areas were amended to civil, environmental, mechanical, mechatronic, electrical and telecommunications engineering. Further changes in the future are being planned to create an increased number of flexible pathways for potential domestic and international students interested in pursuing coursework masters at Western Sydney and in preparation for seeking full EA accreditation for the program. Thus, by focusing on "student-centered" learning, a more flexible and demand-driven unit structure may be introduced with minimal cost increases.

The restructuring undertaken in 2013–14 was aimed to:

1 Comply with Australian Qualifications Framework (AQF) requirements.
2 Achieve EA professional accreditation of Stage 1 competencies, which means that some foundation/developing engineering and capstone units are required to be taught.

3 Renew the Master of Engineering program to provide for a rounded as well as a more specialized learning experience (with a number of sub-specialties) for both domestic and international engineering students, two groups who are not necessarily attracted to the program for the same reasons and aspirations.
4 Establish a presence of postgraduate engineering at a major urban center – Parramatta city campus of WSU.

> The AQF is the national policy for regulated qualifications in Australian education and training. It incorporates the qualifications from each education and training sector into a single comprehensive national qualifications framework. The AQF was first introduced in 1995 to underpin the national system of qualifications in Australia encompassing higher education, vocational education and training and schools.
>
> (AQF Second Edition, 2014)

The master's program was designed to be compliant with AQF Level 9 to meet accreditation requirements of the agency and, concurrently, to meet EA accreditation requirements as detailed in EA (2015).

Accreditation was a critical aim of the restructure since it provides an internationally benchmarked standard for judgment of postgraduate engineering education programs. The accreditation process assures the competence of graduates and provides a guarantee of standing that is independent of the education provider. There were only a few programs accredited at the postgraduate level in the Australian tertiary education sector, and there was a need to develop new course offerings in response to international market demand.

11.3 Course overview

The suite of postgraduate coursework programs in engineering consists of 160 credit points (CP) Master of Engineering, 120 CP Graduate Diploma in Engineering (exit program only) and 80 CP Graduate Certificate in Engineering. The programs provide opportunities to develop specialized technical and research method skills for application in professional engineering practice. It is structured with facilitated learning and directed research aimed at technical- and research-based skills development that encourages critical review, analysis, consolidation and synthesis of knowledge, and critical thinking. Students analyze realistic situations and adapt proposed outcomes based on their understanding of the theory and the related body of knowledge. In addition, the course provides opportunities to apply and adapt knowledge and skills in diverse contexts and environments, thus fostering increased awareness of sustainability, collaboration, communication and uptake of responsibility and accountability in line with expected professional practice.

Table 11.1 highlights the breakdown of the key learning components covered in each coursework program. These components are designed to achieve specified learning outcomes for students to develop the required competencies within their profession. Sustainable engineering units and units with soft skills development remained the same across all the programs, changing slightly from one discipline to another, contingent upon the delivery of project units. The major headings in Table 11.1 are explained below.

Specialization: This component is designed for teaching of specialist skills and knowledge in the chosen specialization.

Table 11.1 Key learning components in postgraduate coursework programs in WSU

Course	*Key Learning Components (Credit Points)*					
	Specialization	*Sustainability*	*Communications and Professional Practice*	*Project*	*Research*	*Electives*
Grad Cert in Eng	≥30 CP	≥10 CP	10 CP	0	10 CP	10 CP
Grad Dip in Eng (exit program only)	≥40 CP	≥10 CP	10 CP	≥0	10 CP	20 CP
Master of Eng	≥60 CP	≥10 CP	10 CP	20 CP	30 CP	20 CP

Sustainability: Understanding the principles of sustainable design and development to meet human needs in the present while preserving the environment for the needs of future generations.

Communications and Professional Practice: Understanding the ethical and moral responsibilities of the professional engineer, and the need for effective oral and written communication skills. There has been a tendency to increase learning targets and objectives associated with soft skills development in all disciplines.

Project: Application of established engineering methods in a holistic way to solve complex engineering problems within context of a capstone project.

Research: A systematic inquiry to identify and develop alternative concepts, solutions and methodologies. This area along with design and systems engineering aspects has received a lot of attention among course developers owing to market forces as well as demands of accreditation bodies such as Engineers Australia.

11.3.1 Master of Engineering

The proposed Master of Engineering (specialization) course is a 160 CP 2-year equivalent full-time program structured as follows:

1 8 × 10 CP Master of Engineering core units (including Master Project 1, Master Project 2, Advanced Design Project 1, Advanced Design Project 2 from chosen area of specialization)
2 6 × 10 CP specialized units (alternates) from chosen area of specialization
3 2 × 10 CP electives

This course is designed for both domestic and international engineering graduates within the same specialization. It is meant to be both AQF- and EA-compliant, the latter requirement being for accreditation of EA Stage 1 competencies. However, the more stringent requirements are mainly the result of the need for EA compliance. The competency elements and units are the main guideline in learning design, and they play a major role in course learning outcome and unit learning target development.

Three-year Bachelor of Science graduates and engineering graduates from another discipline are required to complete a preparatory study before articulation into the 2-year Master of Engineering program. The preparatory study is an 80 CP 1-year full time or 2-year part time non-award program. Alternatively, students in this category may undertake and complete an EA-accredited Bachelor of Engineering course to articulate into the Master of Engineering program. Advanced standing is granted for relevant prior studies in accordance with Western Sydney University policies.

As with the current version of Master of Engineering, sub-specialization streams are to be created as practicable as possible to mirror those offered in the undergraduate program. For example, telecommunication engineering, while not a specialization at the undergraduate level, had its own cohort at the postgraduate level to meet the needs of incoming students – mainly full-fee paying international students.

11.3.2 Graduate Certificate in Engineering

The Graduate Certificate in Engineering is an 80 CP 1-year equivalent full-time non-accredited engineering program completely embedded in the Master of Engineering. It consists of 4 × 10 CP core units and 3 × 10 CP specialized alternates from a chosen area of specialization and 1 × 10 CP elective. It also provided an exit point to those students who had to discontinue the full engineering course owing to financial, full-time work or other legitimate reasons.

11.3.3 Graduate Diploma in Engineering

The Graduate Diploma in Engineering is a non-accredited exit program only, awarded for successful completion of 120 CP units consisting of 6 × 10 CP core units, 4 × 10 CP specialized alternates from a chosen area of specialization and 2 × 10 CP electives.

11.3.4 Master of Engineering specializations

The specializations offered at the inception are civil, environmental, mechatronic, mechanical, electrical and telecommunications.

11.3.5 Course learning outcomes

The course learning outcomes and EA Stage 1 competencies, WSU graduate attributes and AQF specifications are all met following a judicious selection of each in core units and within the various specializations. Each unit will have the competency elements and units either introduced, developed or assured. Often, the first-year units introduced the relevant competency elements while the latter units developed and assured the achievements.

11.3.6 Preparatory study

Preparatory study is a non-award program for Bachelor of Science graduates and 4-year Bachelor of Engineering graduates from another specialization seeking articulation pathway into the Master of Engineering program. Advanced standing may be granted for units in preparatory study on a case-by-case basis, but it must be justified by support and verifiable evidence. An alternative pathway for these students is possible by undertaking and completing an EA-accredited undergraduate Bachelor of Engineering course. Preparatory study does not include specialties such as telecommunication engineering and environmental engineering.

These are embedded as part of the mainstream specialties of electrical engineering and civil engineering. Table 11.2 summarizes preparatory study and the following points are relevant to this.

1 All alternates are to be chosen from student's area of specialization.
2 Students with a recognized 4-year engineering degree or equivalent qualifications in the same discipline will enter Year 1.
3 Students with an EA-accredited 4-year Bachelor of Engineering degree may be granted advanced standing.
4 Students with a recognized Bachelor of Science or 4-year Bachelor of Engineering in another specialization or equivalent qualifications must complete the preparatory study. Alternatively, they may undertake and complete the 4-year Bachelor of Engineering course before moving to the Master of Engineering program.
5 Students with non-EA-accredited 3-year Bachelor of Science in the same specialization or equivalent qualifications must undertake and complete the 4-year Bachelor of Engineering course before articulation into the Master of Engineering program.
6 Students with an EA-accredited 3-year Bachelor of Engineering Science in the same specialization or equivalent qualifications may be given provisional admission to the Master of Engineering, but must successfully complete the Graduate Certificate of Engineering before transferring to the Master of Engineering.
7 Twelve weeks of industrial training required for Master of Engineering program.

Table 11.2 Preparatory studies for each discipline

Preparatory Study Semester 1 (40 CP)	*Preparatory Study Semester 2 (40 CP)*
Civil • Intro to Engineering Practice • Mechanics of Materials • Fluid Mechanics • Soil Mechanics	*Civil* • Environmental Engineering • Introduction to Structural Engineering • Pavement Materials and Design • Hydraulics
Mechatronics • Intro to Engineering Practice • Mechanics of Materials • Digital Systems 1 • Kinematics and Kinetics of Machines	*Mechatronics* • Automated Manufacturing • Dynamics of Mechanical Systems • Power and Machines • Advanced Mechanics of Materials
Mechanical • Intro to Engineering Practice • Mechanics of Materials • Industrial Graphics 2: Transition • Fluid Mechanics	*Mechanical* • Automated Manufacturing • Dynamics of Mechanical Systems • Thermodynamics and Heat Transfer • Advanced Mechanics of Materials
Electrical • Intro to Engineering Practice • Circuit Theory • Digital Systems 1 • Signals and Systems	*Electrical* • Control Systems • Microprocessor Systems • Power and Machines • Engineering Electromagnetics

11.4 Pathways, entry, academic credit and exit points performance analysis

11.4.1 Pathways

Students from a range of backgrounds apply for admission to postgraduate engineering, which necessitates the inclusion of suitable pathways into the Master of Engineering course. In order to maintain a high standard and a quality-controlled program, careful selection criteria must be in place. This is challenging when students come from non-accredited international educational systems. The course designers (directors of the academic program and the academic course advisors) must take into consideration prior experience with students from a diverse background in selecting pathways for the proposed program.

11.4.2 Entry, academic credit and exit points

The course offers multiple entry and exit points as shown in Figure 11.1. The applicants are categorized to five bands, A to E, based on their undergraduate qualifications. These are shown on the left in the figure as entry points. Bands B and D who do not possess a degree from the same discipline are offered an 80-CP preparatory year. This is a non-award program which prepares students to enter Year 1 of the MEng program. Students have the option of exiting with a Graduate Certificate qualification on completion of 80 CP, or preferably proceed to Year 2 of the MEng program. Those who complete a minimum of 120 CP are eligible to exit with a Graduate Diploma in Engineering. Table 11.2 summarizes entry, academic credit and exit points. The following points are relevant to Table 11.2.

1 Applicants will be required to select specialization and preferred sub-specialization (if any) in the application form.
2 Block academic credit of up to 40 CP may be granted for graduates of EA-accredited 4-year Bachelor of Engineering of the same discipline or equivalent qualifications. Equivalent qualifications may extend to a student with non-EA-accredited 4-year Bachelor of Engineering degree who has achieved at least EA graduate membership and 2 years of full-time professional employment. This may also apply to degrees complying with the Washington Accord.

Admission requirements for Master of Engineering

Students must have a minimum of a recognized 4-year undergraduate degree in engineering or equivalent qualifications in the same engineering specialization. Advanced standing may be granted for relevant prior studies in EA-accredited Bachelor of Engineering courses or equivalent.

Those with a recognized 3-year Bachelor of Science or 4-year Bachelor of Engineering in another specialization or equivalent qualifications must complete the non-award preparatory study before articulation into the Master of Engineering program. Alternatively, they may undertake and complete the 4-year EA-accredited Bachelor of Engineering course in the same engineering specialization as an articulation pathway to the Master of Engineering program.

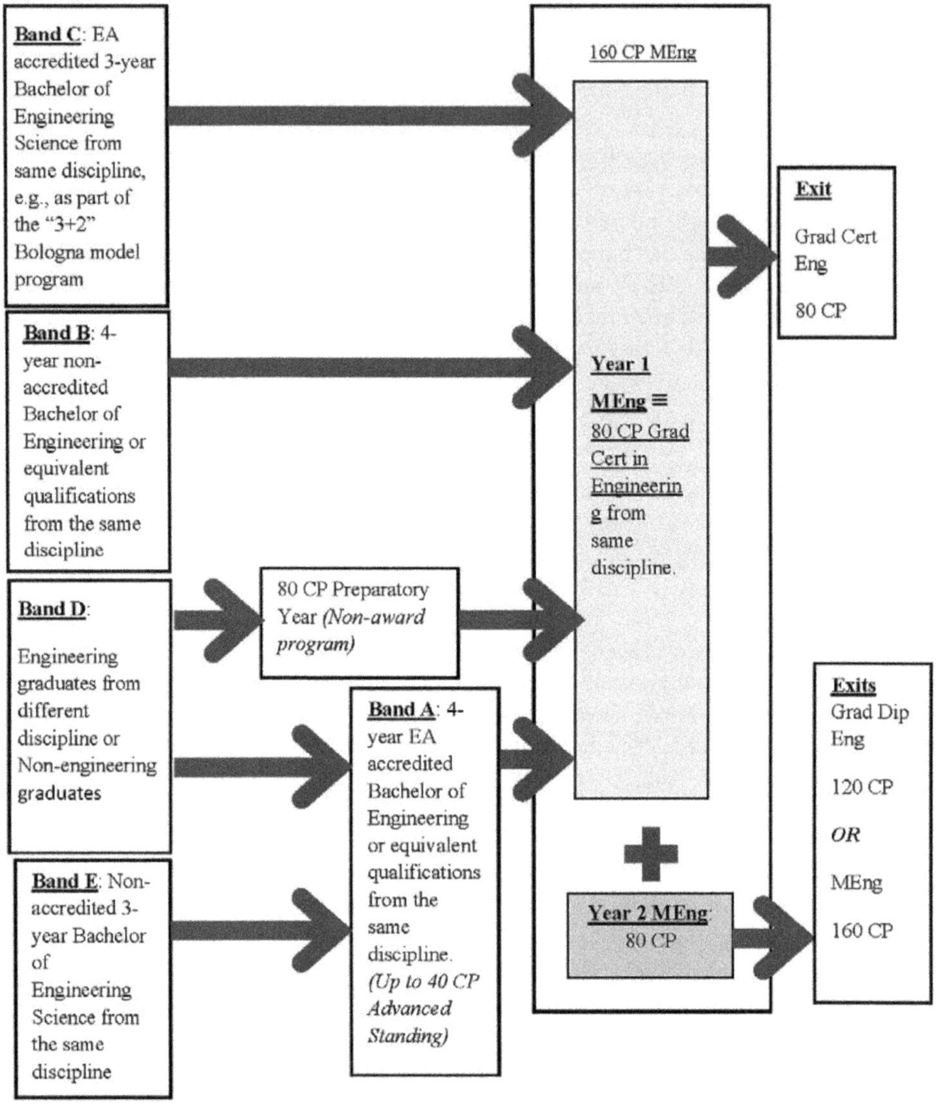

Figure 11.1 Pathways for admission to Master of Engineering programs

Students with an EA-accredited 3-year Bachelor of Engineering Science or equivalent qualifications in the same specialization may be given provisional admission to the Master of Engineering program, but must first enroll in and successfully complete the Graduate Certificate of Engineering before transferring to the Master of Engineering program.

Students with a non-EA-accredited 3-year Bachelor of Engineering Science or equivalent qualifications must undertake and complete an EA-accredited 4-year Bachelor of Engineering course in the same specialization as an articulation pathway to the Master of Engineering program.

Table 11.4 Course entry, academic credit and exit points

	Year 1 Semester 1 (40 CP)	*Year 1 Semester 2 (40 CP)*	*Year 2 Semester 1 (40 CP)*	*Year 2 Semester 2 (40 CP)*
Entry point	Non-accredited BEng (Specialization), AQF Level 7 or equivalent	Accredited 4-year BEng (Specialization), AQF Level 7 or equivalent Accredited 4-year BEng (Specialization) (Honors), AQF Level 8 or equivalent		
Block Academic Credit (for Grad Cert in Engineering, Master of Engineering)	Block Academic Credit for Semester 1 EA-accredited 4-year BEng (Specialization), AQF Level 7 of same discipline or equivalent EA-accredited 4-year BEng (Specialization) (Honors), AQF Level 8 of same discipline or equivalent			
Exit points	*At completion of Semester 1 and Semester 2 (80 CP)* Grad Cert (Specialization)		*At completion of Semesters 1, 2 and 3 (120 CP)* Grad Dip (Specialization)	*At completion of 4 Semesters (160 CP)* MEng (Specialization)

Admission requirements for graduate certificate in engineering

Students must have a minimum of an EA-accredited 3-year undergraduate degree in Engineering Science or a recognized 4-year undergraduate degree in engineering or equivalent qualifications in the same specialization.

Those with a recognized 3-year Bachelor of Science or 4-year Bachelor of Engineering degree in another specialization or equivalent qualifications must complete the non-award preparatory year study before enrollment in the Graduate Certificate in Engineering. Alternatively, they may undertake and complete the 4-year EA-accredited Bachelor of Engineering course in the same specialization as an articulation pathway to the Graduate Certificate in Engineering.

Students with a non-EA-accredited 3-year Bachelor of Engineering Science or equivalent qualifications must undertake and complete an EA-accredited 4-year Bachelor of Engineering course in the same specialization as a pathway to the Graduate Certificate in Engineering.

11.4.3 Future changes

A 2-year full-time 160 CP master's-level program accredited by Engineers Australia is clearly seen as an attraction for international students, and it has been marketed as such. The attraction of the program to domestic students depends on the de-linking of the postgraduate unit from the associated undergraduate unit and the potential of high-caliber industry professionals contributing to the teaching. The former is a "chicken-and-egg" problem which depends on class size, and the latter will need to rely on the industry connections of the academics. With the recent introduction of the postgraduate courses to a campus located in a major urban center (Parramatta), sustainability of the units has also improved, leading to a more stable course offering. Future changes include the introduction of more student-centered activities within units and consolidation of blended learning initiatives.

11.5 Conclusion

Western Sydney University's Master of Engineering postgraduate coursework program's evolution since its inception over a decade ago has allowed for a 1-year program to be revamped into a 2-year program with the addition of new core and specialized units, as well as other features, to ensure a solid learning platform. The creation of distinctive key programs encompassing the areas of civil, environmental, mechatronic, electrical and telecommunication engineering has led to the rationalization and consolidation of several units. Moreover, provisional Engineers Australia accreditation being sought and approved has confirmed that the course meets Stage 1 competency elements and units. The program is currently under review for full accreditation.

References

AQF (2014) *Australian Quality Framework*, 2nd ed. Available from: www.aqf.edu.au/ [Accessed 20 September 2016].

EA (2015) *Engineers Australia Program Accreditation*. Available from: www.engineersaustralia.org.au/about-us/program-accreditation [Accessed 20 September 2016].

Heywood, J. (2005) *Curriculum Design, Implementation and Evaluation*. Wiley-IEEE Press, New York, USA.

Martinez-Caro, E. & Campuzano-Bolarin, F. (2011) Factors affecting students' satisfaction in engineering disciplines: Traditional vs. blended approaches. *European Journal of Engineering Education*, 36(5), 473–483.

WSU (2015) *Western Sydney University 2015–2020 Strategic Plan*. Available from: www.uws.edu.au/__data/assets/pdf_file/0004/844672/OVP5222_Securing_Success_Strategic_Plan_Rebrand_v3.pdf [Accessed 20 September 2016].

Chapter 12

Mastering mathematics in engineering by critically reading engineering texts

G.A. Tularam[1] *and O.M. Hassan*[2]

[1]*Mathematics and Statistics Discipline, Griffith University, Brisbane, Australia*
[2]*Environment Futures Research Institute, Griffith University, Brisbane, Australia*

ABSTRACT: The skill of reading mathematics language and its related contents, including abstracts and higher mathematical forms in textbooks, may soon become a forgotten art for some students, as most educators and teachers continue to focus on concrete-type "doing" (kinesthetic) and/or visual-type "seeing" activities. Action or visualization is often espoused by educationalists rather than placing an emphasis on abstractions, so it is not surprising that students do not get appropriate levels of practice in abstract thinking. The process of and focus on reading to make sense (understanding) and learning from oral presentations (or lectures) (auditory) all tend to be demonized, but this is only at the expense of a group of students who favor the act of learning by critically reading books and/or gaining their knowledge via the auditory or "lecture mode" of learning. Clearly, teaching should include and accommodate student learning styles, but what is not appropriate is to place an inappropriate or low level of emphasis on the development of students' weaker attributes, such as their less-preferred learning styles. Rather, to increase the overall ability of students to learn mathematics, we must develop and strengthen their weaker styles to an appropriate level, as shown by the results of this study. Both the qualitative and quantitative aspects of this study greatly increase our understanding of the nature of mathematics learning. It is noted that gently pushing students to critically read and interact with mathematical materials helps in building self-confidence as well as in the development of higher-order skills; this in turn increases their overall performance. Increasing the expectations of students in teaching moves their learning forward. The educationally sound goal of presenting materials favoring students' preferred learning styles may, in fact, be doing them a disservice. Thus, to meet our goal of providing a well-rounded education in mathematics, it is imperative that even from an early age, teaching should focus on developing and strengthening attitude, perseverance, independence and self-motivation. The focus on students' weaker attributes or styles of learning, such as the ability to abstract from the critical reading of texts, even when another style is preferred by a learner, is also an educationally sound goal.

12.1 Introduction

Reading continues to be an enjoyable activity for many in the real world (Anderson *et al.*, 1984; Barton and Heidema, 2009; Grimm, 2008; Hermida, 2009). Even in the computer-focused visual effects age, there are many readers, including children, who continue to buy novels and historical autobiographies not only for leisure but for learning as well. The act of reading is nothing new for most of us, and we start learning to read at an early age (McNamara, 2009). As such, we can make use of this prior knowledge to help students learning mathematics as well – for example, mathematical content, procedures and strategies, including aspects of "mathematical know-how," are often presented in textbooks (Goddard *et al*., 2000; Grimm, 2008; Gutstein, 2006; Tularam, 2011, 2013, 2014, 2015).

There is now a strong push for the learning of mathematics and related sciences; while not new, it is now led by Australian Prime Minister Turnbull (*The Sydney Morning Herald*, 2016). During the past 30 years, there have been a number of calls for an improvement in the level of mathematics of our students in primary and secondary schools and in colleges and universities. There is much evidence showing that in Australia and in many other parts of the world, first-year university students are underperforming in mathematics and related sciences (Reyna and Brainerd, 2007; Tularam and Hulsman, 2015). Also, many students in high schools are now choosing not to study mathematics or sciences. There is much concern about the level of our primary students in mathematics when compared to other OECD countries such as Canada (Australian Broadcasting Corporation, 2016).

As a response to the lower levels of attainments in mathematics in Australia, in 2016 Prime Minister Turnbull requested that math be compulsory in secondary schools (Australian Broadcasting Corporation, 2016). A similar call was also made by the then minister of education, Christopher Pine, and others such as Professor Prince, the director of the Australian Mathematical Sciences Institute (Tularam, 2016; Tularam and Hulsman, 2015). Importantly, the talk around and about improvement in science and technology was particularly prominent before and during the recent 2016 elections in Australia – in fact, the election was fought on the theme of innovation, growth and jobs. More generally, the low key indicators in the sciences in schools have led many politicians to focus clearly on mathematics and its applications as well as on the sciences, which often depend on facility in mathematical thinking. It is important that the prime minister of Australia and his minister of education, among others, have started to address the issue of low levels in mathematics. In fact, the call for "making learning of mathematics compulsory in high school" has been rather an important one in recent times, given the universality of applicability of mathematics. A certain level of mathematical competency seems to be a critical factor in the future growth of Australia, and this means we need to attract many more students into mathematics (and related sciences), particularly in schools, and then, importantly, to higher mathematical and science-related studies in university degree programs.

In recent times, educational research has focused particular attention on the learning styles of students, and this has tended to bind teaching design to the students' strengths (Rohrer and Pashler, 2010). This is of course important, but there are also many other weaknesses of students that need to be focused upon while designing courses in mathematics. The research into learning styles may need reassessment, in that learning and teaching in the classrooms of today should be more balanced than simply providing opportunities to reinforce what students may already be proficient in (Riener and Willingham, 2010; Rohrer and Pashler, 2010). The emphasis should be on the development of an adaptable and flexible learner who is able to cope with multiple styles of delivery – as in the learning of mathematics, which will also be the case when working in real-life contexts after graduating (Tularam, 2016). It is, therefore, just as important to focus on the weaker aspects of students' styles of learning, and thus to design teaching methods that strengthen both the preferred and non-preferred modes of delivery, rather than solely relying on students' favored learning styles

(Rohrer and Pashler, 2010). Simply preparing material that suits the favored learning styles of learners in the teaching of mathematics may indeed be doing a disservice to students in terms of their mathematics and sciences education.

There is evidence that some students prefer certain learning styles but are not good in other styles (Riener and Willingham, 2010; Tularam and Hulsman, 2015). This has not been adequately addressed by educators and researchers. Researchers have not focused on the issue of the more "able" and "gifted students" in tertiary studies; these students enjoy and can cope with many different learning styles and are usually bored easily when a single general preferred style is imposed upon them. If only the individual-preferred styles are focused upon, then it seems that some learners can be disadvantaged in that they have not been given the opportunity to improve their weaker learning ways; that is, the students fail to learn/practice their weaker learning styles (Dweck, 2006; Riener and Willingham, 2010). It is well accepted that the goal of educational institutions is to provide all students with the opportunities to become competent persons ready to deal with real-life problems (Tularam, 1990, 1992, 1994a/1994b, 1996). Therefore, it is crucially important not only to enhance the students' favored learning styles, but equally to help improve the less favored and weaker styles of learning to develop flexible and adaptive learners and problem solvers for our society (Roediger & Karpicke, 2006; Tularam, 2013, 2014).

The main aim of this chapter is to study the importance of critical reading of mathematics texts, because this aspect of learning style has been less favored in modern times, perhaps due to technological advances. The first study interviews experts in the field of mathematics and reports on experts' reflections (in focus groups) upon the definition of critical reading and then its importance in mathematics learning through reading. The second study is a mixed qualitative and quantitative study that examines the effects of critical reading in first-year mathematics on performance – student learning from a new text (mathematics written in prose/verbose form) and its relation to performance. The text was specially written for this level, but in a rather wordy form, and as such differs from most available material. In this text, step-by-step explanations of concepts and solutions are provided in prose form, more like a lecturer talking to the student from within the book. Also, when compared to other texts, this has a different approach in that students may write in the book itself while solving problems; asking other students or questioning themselves is encouraged when seeking logical flow in the work. The book facilitates students to ask for help from close friends to solve problems to move forward when striking barriers in thinking. This style of learning motivates the reader to read over before getting help from a tutor or lecturer to solve problems. That is, a student learns problem-solving skills, and thereby is better prepared to learn higher-level mathematics, for example. The chapter addresses the essential content and problem-solving requirements of the first-year non-mathematics majoring students, covering contents up to simple separable differential equations.

12.2 Background

One way to improve a student's mathematical learning is to focus teaching not only on solving routine/non-routine problems daily, but to also learn from usefully "reading of mathematical material" presented in written or prose form (Fite, 2002; Friere, 1983; Kelson and Tularam, 1998a/1998b, 1998, 2003; McNamara, 2009; Metsisto, 2005; Riccomini *et al.*, 2015; Riener and Willingham, 2010; Silver *et al.*, 1997; Tularam, 1997a/1997b/1997c, 1998; Tularam and Amri, 2011). In fact, much of the work in mathematics is presented in written form, for example in textbooks or e-books. Students can learn by reading material presented to them if they are guided to do so, given that much of the work in higher mathematics has been explicated in written form as in research articles, textbooks and the like (Billmeyer and Barton, 1998; Hermida, 2009). It is clear then that we need to make certain that our students can actively read written mathematical work and make sense of the content presented (Friere, 1983; Paschler *et al.*, 2009).

Many of the studies of the past have shown that students can learn from reading, but the nature of the reading style is less clearly defined (Barton *et al.*, 2002; Billmeyer and Barton, 1998; Friere, 1983). It is argued that students can use a "critical reading" method to learn to successfully negotiate mathematical material supplemented by lectures in first-year courses (Barton and Heidema, 2009; Barton *et al.*, 2002). Being prepared to critically read, with the sole purpose of trying to make sense of the work, requires much persistence and motivation on the part of the learner, particularly in modern times when reading books is not the most common mode of learning (Friere, 1983). However, mathematics learning requires critical reading of the work, with a focus on trying to make sense of others' work or understanding the author's intention, as well as questioning/delving into the reasons why the work has been presented in the manner that it has; for example, that chosen by the author, or indeed whether the work presented is correct and follows logical reasoning.

Critical reading involves the learner being prepared to revise and relearn or literally learn by working backwards – that is, to go back to revise or learn any prerequisite work that may be necessary before learning the work presented in texts or articles (Friere, 1983; Metsisto, 2005). Critical reading also means that the student should then know when he or she has successfully learned the material to an appropriate competency level (Chi *et al.*, 1989, 1994). This can be easily checked by the student's self-examination when returning to the text and questioning their understanding; that is, whether the student can understand the new work presented by others, or whether he or she can solve the related problems within the section. In the end, there may be a need for much more work for the learner when critically reading, particularly if the student is to gain an appropriate understanding of the written work or crucial knowledge that has been presented in other mathematical texts (Cooper, 2004; Davey, 1983). In line with the above, Friere (1983) said, "Reading involves critical perception, interpretation, and the conscious transformation of what is read into practical action" (p. 1).

Clearly, the internet has changed the way in which many students think they can learn. There are well-organized presentations in auditory, visual and concrete formats that can be used for learning mathematics, and many of them are freely available on the Web. While these are useful for learning, students will also need to be rather critical as they go about learning from such websites, particularly in relation to the "correctness of working" presented in them. However, what is important is that a student not only sees the video but that there is also some written material presented in the video; thus, he or she needs to read any presented material to try to understand the work. Therefore, equal importance must be placed on both critical reading and videos/visual presentations of mathematical work as in classrooms, if students are to creatively learn by reading the written material presented, or by watching online videos, that is, if they are to learn to become active/creative problem solvers.

The process of critically reading mathematical work is particularly important for those students who learn from the mathematical material that is presented online. This is another reason for preparing all students for such learning. The process of critical reading must be highlighted in classrooms if students are to gain appropriate knowledge of the intended content, skills and procedures (Hermida, 2009). Since reading is something that students are already familiar with, they can easily learn to refine the process involved in reading, for example to utilize it for the learning of mathematics. In the end, students will have to learn to value the process of critical reading as a form of learning tool in mathematics. The skills involved then need to be included in the teaching of mathematics generally, so the students can acquire, learn and practice critical reading.

In more recent times, there are not many researchers who have said that learning mathematics by reading is as important as learning by doing (concrete activities) or seeing (visual formats) although, clearly, they often go hand in hand (Grimm, 2008). In classrooms, not only those teaching mathematics, most teachers at some stage during a lesson, class or lecture will also use worksheets that are presented in a written form to explain the work that the students need to do during the lesson or class. It is then self-evident that students must be able to read to understand the work that has been presented to them; this is also important when reading from texts. Students need to learn to appreciate the importance of critical reading and realize that it can indeed help them to better understand the problems posed in texts (electronic or otherwise); and by critical reading of solved examples students may be able to transfer their knowledge and, thus, solve different problems. In the process, there is a need to practice the skills of identifying errors or mistakes in working presented in texts; that is, students must practice to improve their ability to locate and identify errors in presented material. This may be done by presenting material that could include erroneous material or otherwise, such as illogical arguments etc.

All mathematics textbooks (as in other fields) are presented in written forms (e-books) and the learners have to learn how to make sense of the work presented in them. In the majority of mathematics texts, the steps are not presented in much detail, particularly in prose form; or if the step-by-step explanations are given, the work is often not presented in prose form but in technical language which does not encourage the students to read the work critically (Dewitz *et al.*, 1987; Fite, 2002). Mathematics is also a technical language and, as such, students will have to learn the language that is used in the texts. Indeed, mathematical language can be learned from critically reading the work presented in the mathematical texts (presented in any type of written form) (Dweck, 2006). For example, the nature of a variable can be known; such as u, v, x, y, z are noted as variables, and that each variable may belong to the set of integers. Also, the reading of expressions such as x and y or $xy + z$ is required in mathematics, among other mathematical expressions that can be learned by critically reading mathematical texts (University of Oregon, 2005). There are mathematical expressions such as those above, and there are mathematical sentences such as $x + 2 = 10$ etc. Students will need to learn to read equations or sentences in mathematics such as $x + 7 = y$ and $uv - 1 = 0$, for example. The mathematical sentences or equations can only be fully understood if the language used in mathematics is practiced more through critical reading of the texts generally, and reading algebraic work in particular, all of which need to be well known or mastered if students are to successfully solve real-life tasks in the sciences (McNamara, 2009). It is then important for students to learn critical reading so that they can understand mathematical language and the related material normally presented in texts (Lapkin *et al.*, 2003).

Simple surface reading does not help in mathematics learning, and this is probably because usually there is much less story form of presentations in mathematical texts (Friere, 1983; Marzano, 2005; Winstead, 2004). This is not to say that stories should be encouraged, as the mathematical forest may be lost for the trees, but rather, the mathematical language is important, and it must be taught in addition to other aspects (Riccomini *et al.*, 2015). Instead, teaching is usually done using technical examples with a little, if any, explanation. And even if detailed explanations are presented, students are not encouraged to critically read, mainly because they have had insufficient experience and practice (NCEE, 1983). In most cases, most of the work presented in books is in terms of solving equations or word problems, and this is its sole focus.

Although visual representations in geometry and graphical work are used to help in the learning of important concepts, in most cases, it is only possible to understand the work if the written work in textbooks is critically read afterwards as well – the esteemed authors of most texts have spent much time writing the work in a manner that facilitates student learning; but in most cases they tend to use fewer words and explanations for the steps presented. In particular, the texts are usually not presented in prose form. In any case, learning the knowledge from texts requires focused, persistent and motivated critical reading of the mathematical work presented to understand the work completely; this is particularly true when students have to learn the higher order as well as the cognitively demanding abstract nature of many of the mathematical concepts that follow.

There is an historical side to the learning of mathematics that concerns the contributions of various cultures throughout the world which have helped develop our mathematical knowledge to date, such as the gift of zero as a decimal number (Smithsonian Magazine, 2013). The contributions to mathematical learning over the centuries need to be appreciated in order to put a human face on mathematical learning. Our students need to read about these mathematicians and to learn more about the general aspects regarding the traditions and culture of mathematics itself in addition to its utility. This is mainly acquired through the reading of popular material on mathematics, physics and sciences, often presented in books, video and recorded forms – that is, the works developed over centuries through various cultures need to be learned if a student is to fully appreciate the history of the growth and importance of mathematics to our civilization. In particular, the utility of mathematics in the arts and sciences is often clearly outlined in these cultures. Appreciating the importance of mathematics in a historical manner may help students develop an intrinsic and possibly long-term motivation for the learning of mathematical contents. There is also a future benefit for when their children will be encouraged to enjoy the difficulty of learning mathematics, instead of being forced away from the sciences simply because they dislike them. It is when their sons/daughters reach university that the parents' mostly negative beliefs about mathematics tend to put students off mathematics and the sciences more generally. But if parents of our children master the historical significance of mathematics and science, their personal beliefs can change, and this may positively encourage their children to more diligently pursue science and mathematical studies.

Popular series on the utility of mathematical knowledge can be found in many bookstores online. Books such as *God and the New Physics* written by Paul Davies (1984) and the more recent *Our Mathematical Universe*, written by Max Tegmark (2014) which questions the nature of reality together provide great reading material (mathematical) for the general public. They present interesting insights into the nature, structure and utility of mathematics and how it has been, and can be, applied to the understanding of our universe – a topic of interest to all of us. Such general and easy-reading texts on mathematics, including other books on the history of mathematics, can help to improve the level of interest and trust the public can place on the importance of mathematics generally. Clearly, the important goal of lecturers or teachers is to instill in their students a lifelong yearning to master mathematical content that is suitable for their level of study. At the first instance, this means a focus upon the basic content, procedures, skills and strategies that students must master in order to develop an appreciation and understanding of some simple applications, so that they may indeed experience the solving of some of our real-life problems as well, using mathematics.

12.3 Method

This study involved a two-stage procedure, where the first was a focus group study of experts in the area of mathematics, while the second was a mixed-method procedure – both qualitative and quantitative data were gathered using first-year university students. The first was a short study of a group of 10 expert mathematics and applied mathematics majors (PhDs). This group had completed a large amount of mathematics during their undergraduate and master's courses and were considered experts. They were selected and interviewed as a group. Their responses were noted with regard to the acquisition of knowledge over time, from high school to university.

The second study analyzed the performance of some randomly selected novice students with a focus upon their scores on weekly tests and mid-semester and final exams. A large sample comparison was also done from two yearly groups. These groups were selected from a large number of non-mathematics major first-year students studying science (100). Ten groups of 10 students were also observed over the semester, and from these only 20 students were involved in focus group discussions. These 20 students were also involved in in-depth focus group interviews/discussions, where the students were questioned in some depth in the group. In tutorials, the observations and responses of students were noted as in the focus groups during the semester. All information regarding the teaching, including other more general observations, was gathered by the researcher and by three tutors who were trained to be facilitators in tutorials, rather than teachers. In terms of quantitative analyses, about 100 student performances and scores were examined from the 2nd year groups and their scores compared using statistical tests.

Specifically, a "physical" mathematics textbook was used to learn from throughout the semester in the first group (Method A), and the performance of students in this group was compared with student results of the past years' groups who did the same course but mostly in a computer-based learning environment, using videos and PowerPoint presentations (Method B). In Method A, no PowerPoints lessons or videos were provided in class or on the internet blackboard. Only recorded videos of the actual lectures of the course were available, but these have been available to all years. In Method A, the content presented in lectures was from the book prepared for the study. The assignments, tests and exams used in this study were very similar to those presented in other years and were similar to the problems presented in the book.

12.3.1 Expert study results – qualitative analysis

In mathematics, it is important to emphasize the importance of learning by reading from textbooks, although students may find the process of reading to be cognitively demanding or even a little boring. In the end, however, the experts say, "it is by reading material that we all learn in general anyway . . . more particularly, in mathematics" [Expert A]. When a number of mathematicians were asked how they acquired or learned the mathematical content material and skills during their careers, many of them said by reading others' or experts' works (10/10).

Indeed, it is noted that Newton also said in his book that he achieved his greatness by standing on others' shoulders, those who were the giants in their fields. The experts further detailed that they used "*repeated reading to make sense*" [Expert C] and as a way of getting to know the difficult and crucial aspects of their respective specializations. In particular, they read to critically "investigate the material presented in the form of words and symbols . . . that is, work presented in the usual mathematical language" [Expert F]. This is in line with many others who argue that the value of "actually" reading works of others has provided great insights, and in many cases such knowledge has led to new discoveries in mathematics (see for example, the works of Einstein and others).

It is the act of reading for understanding that has also led mathematicians and others to apply mathematical concepts to almost all areas now known to humankind. The experts interviewed said that "*by reading to make sense of the mathematical work presented in specialist articles allowed them to delve into the actual thinking processes*" [Expert C] of the authors, and "through this process I was able to learn higher aspects of their field of work" [Expert C]; that is, most of the past cutting-edge works in science are presented in written form in top mathematics and science journals.

Experts say that "*by reading carefully learners can critically question and reflect upon the work presented in the articles, lectures or texts and then try to make sense of the work written material*" [Expert G]. If the work is not well understood, learners can delve into related works that are pitched at a lower or easier level to cope with. Often such an action will lead to further investigation of related work, as well as further reading to relearn – referred to as the background knowledge required for the work. It is only "*upon returning to the source of problem that was encountered was the earlier work better understood*" [Expert H].

Such a process allows the learners to advance their work by using various strategies to improve their understanding of concepts in mathematics. It is this cyclic nature of learning and relearning that appears to be the key to gaining expertise in mathematics, as noted by the experts.

Evidently, it is through critical reading of other individuals' works worldwide that many of the world's prominent mathematicians have gained the knowledge and insights that in turn have led them to develop the complex and sophisticated fields in pure and applied mathematics (Newton). So, it should be the case with others such as novices, but the approach taken should be at their level of sophistication. In other words, all the written mathematics and explanations presented to novices must be just out of reach of their level of understanding. From the expert analysis, we can see that the act of critical reading of mathematical literature and, thus, the "nature of the reading process" itself is what is crucial to the learning and understanding of the mathematical contents. The process of reading for learning can then be seen as a developmental process that is evolving; the experts had continued this process of reading and learning and, thus, improved over time. This suggests that to reach an appropriate level of expertise as required in mathematics learning, we need to focus upon the critical reading of mathematical work. In fact, it ought to be taught in mathematics classrooms much earlier to help prepare students for their higher education studies, instead of spending a large proportion of teaching time focusing on "seeing and doing" types of mathematical activities only, although these aspects are also important.

12.3.2 Study on the effects of critical reading in mathematics achievement

The learning of work from lectures and by reading textbooks has now taken on new meaning as we move to the age of computer-based e-learning. It is the computer presentations and video-recorded lectures that provide the bulk of the learning material, at least for those at the tertiary level. However, with over 10 years of work by Dr Tularam in this area, it is becoming clearer that the results of learning and understanding from videos and computer presentations alone do not deliver the level of knowledge and understanding that is needed at the first-year level for most novice and middle to marginal students (Tularam, 2011, 2013, 2014). The recent work done by the author with such students shows little evidence of appropriate understanding of mathematical concepts, let alone a satisfactory recall of content procedures and strategies, except in the case of some highly competent students and/or those who are highly persistent and motivated (Tularam, 2013; Tularam and Hulsman, 2015).

A lot of money has been spent on computer education in many parts of the world, but it seems that any improvement in learning overall has not been satisfactorily demonstrated, particularly in mathematics and statistics (see Australian and New Zealand governments' expenditures in the past 10 years). More generally, a recent Australian report has shown little longer-term benefits in education over a 10-year period, even after spending more than billions of dollars; this analysis was based on the key indicators (The Organisation for Economic Co-operation and Development, 2016). This is an overall education spending figure, and perhaps the amount spent has much to do with the installation of the relevant infrastructure and computerization of learning and teaching in classes. But what is interesting is that the computerization of all aspects of teaching has been the focus of spending in education for a long while now, particularly in schools (during the past 20 years). This would suggest that the benefits of such computerization for students would show clearly in university studies. In the past 10 years of work, Dr Tularam has noted no real change in learning and understanding of mathematics and statistics by students entering tertiary education during this time, with the exception of the fact that the power of computers and the ability to conduct calculations has helped students to solve many problems that would have usually taken longer.

The tertiary sector has also spent much on the computerization of learning and teaching facilities, but in the field of mathematics there appears to be no clear evidence that shows student understanding is better now than when computers were not present, when "chalk and talk" was the main way of teaching mathematics. The spending on teaching facilities continues to grow and is now certainly at a higher level than previously, as it should be. But there is not much evidence of improvement overall; rather it seems that the standard of mathematical learning is at a less-than-satisfactory level in terms of performance. This is particularly true for novices (where it would seem that much help has been available), mid-level students and the majority of non-mathematics majoring students (even when much remedial help is available). The author has much evidence regarding the low level of application and transfer levels of the first-year undergraduate scores in mathematics and statistics over many years (Tularam, 2014; Tularam and Amri, 2013; Tularam and Hulsman, 2015).

12.3.3 Qualitative analysis and results of reading study

A recent change was made in the teaching of novices and non-mathematics majors, and this was the introduction of a new book (Essentials of Applied Mathematics for the Sciences, 2016) from which to learn, and the reading of the book as the focus of this study. The results of the study are encouraging, and it seems that with little cost, some progress is possible in this regard after all. The analyses and findings are presented in the following section. A new method of teaching mathematics to the first-year students was introduced, with a rather different focus in lectures and tutorials. The computer was used only as a tool for the provision of recorded lecture materials and for some calculations, but the learning focus was placed mainly on the reading and understanding of a new textbook (a physical book, not online) that was designed, written and developed to include all the work of the first-year mathematics course for non-mathematics majors. The book was written to appropriately accommodate the level of work of first-year students, including the mathematical content, procedural skills and related problem solving as per the requirements in the degree program in environmental sciences.

In addition to the usual lectures which were closely aligned to the text, all students were instructed to read their texts to learn the content; and if they were unsure of any of the concepts, skills and procedures, the students were told to use the book for help in their learning. The problem-solving work and course assignments and exams were based on the content of the book. The book was written in a manner that includes first-person prose, with many written explanations and solved examples, together with a simplified set of notes in prose form for easier comprehension, in the hope that this leads to a much better understanding of the many concepts, skills and procedures required in the first-year course.

12.4 Results

The analysis showed that students who applied the critical reading approach performed much better than those who did not. The previous year's method was e-learning based, and when the new approach was compared in terms of the overall results, a significant difference was noted at the 1% level (see later). The yearly groups compared were deemed to be of a similar level and background (based on entry scores). The cohorts had the same intake characteristics to the new group; essentially, the previous entry scores matched reasonably well, and the comparison groups were considered to be appropriate for this statistical analysis (see Fig. 12.1). The focus group interview method was also used in this part of the study, and based upon this analysis, the new method group showed a better level of understanding overall, and indeed, a better than satisfactory recall of content procedures and use of problem-solving strategies, as shown by in-depth analyses of the students' examination scripts. Also, when compared with other measures, overall the student performance was better, for example, on weekly problem-solving tests as well as based on mid-semester and final exam scores (Tularam and Hulsman, 2015).

The weekly tutorials during the semester were designed to encourage certain students to first read the text before asking questions (of tutors who were always prepared to help guide them in this process). The gentle push to require students to read their texts met with some initial resistance, but in general what it did to the attitude and work habits of students was interesting. The emphasis on reading their books for gaining understanding led many students to "move onto an inquiry-based learning approach." For example, as soon as the students realized their weaknesses or lack of knowledge, that is, after reading the work in the book, a number of actions were taken to search and read the related earlier works with the text in more detail, in a particularly learning-focused manner. Simply knowing they would need to search by themselves for parts of the work that would help in their understanding before asking led students to think more critically while reading the text. For example, in searching for answers, the students were actively seeking different approaches and strategies for learning the content by going back to earlier work and attempting to understand the examples and written explanations, even when they had done this in earlier learning. This preparation occurred especially if the tutor was not immediately available to give a helping hand, as

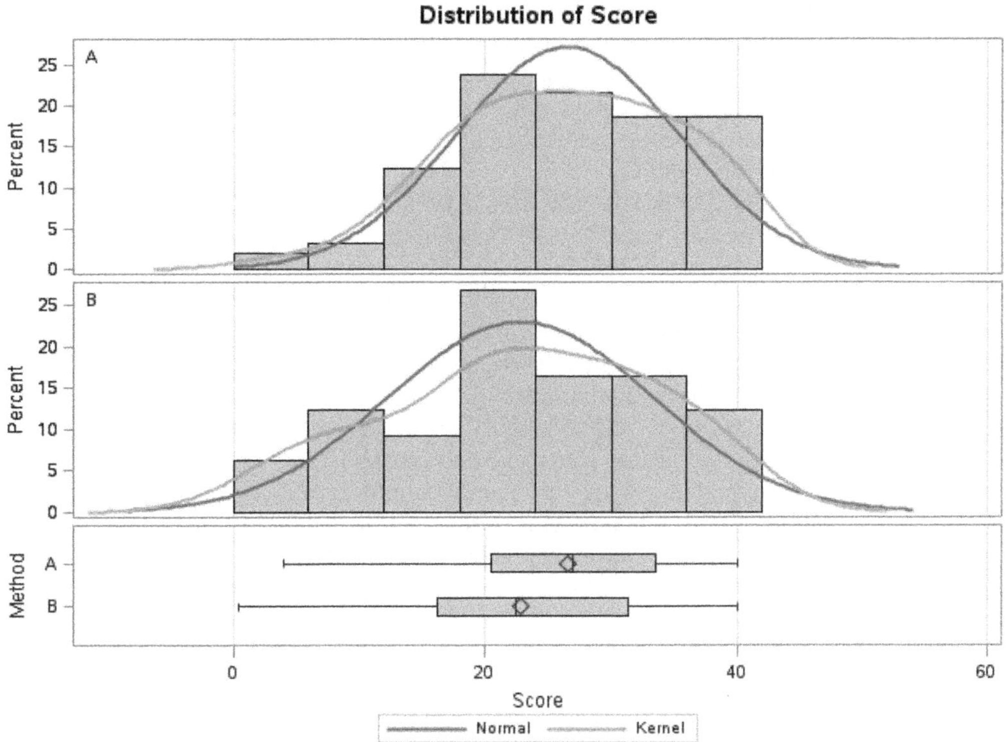

Figure 12.1 Variance and normality

was usual in other years when students would constantly ask for help, believing the solution would be provided for them. Instead, using in this new design, the tutor gently guided the students to see if they could solve the problem by seeking strategies for themselves, such as re-working or reading the appropriate sections of the text (Bournot-Trites *et al.*, 2001).

With regard to the new method and approach, some interesting comments were captured from student discussions. That is, upon reflection in focus groups, some students' learning experiences of using a textbook (Method A) were analyzed and the data were placed into groups. These are presented in the following:

Teaching method: Students believed that the teacher should always show them the simplest and easiest methods and that understanding at times was not desirable and takes longer anyway. A comment by a student summarized this feeling. The teacher should "show the easiest way to some math problems" (Student A). Around 10/90 comments were in relation to this aspect when tutorial data was collated.

It was particularly concerning that whenever mathematical explanations were given in lectures regarding the underlying theory and/or a sketch or an initial proof of a mathematical problem, a few students believed this sort of proof work was inappropriate for their learning – that is, how one can convince oneself that a rule or method being used in the text was in fact appropriate or indeed correct was believed to be "*a longer and a round-about way of teaching the rules or methods.*" These students simply wanted to be told the rules that would help them obtain the correct answers. The learning of rules was their main driver for learning the content, and it was a concern for these students that the teacher was teaching using a method that was different to what they had been used to earlier; thus, the conclusion was that it must be a longer way of teaching the same work. These students did not show any level of curiosity at all about whether or how the rules or methods in the text were constructed in the first place. This means many students would not experience the underlying theory of work with any level of deep explanations and proofs. This may indeed be a major hurdle to higher learning and research in mathematically related areas. These students felt that the underlying aspects were not really necessary for their learning and really believed that the teacher was making things more complicated when a rule was so much easier to learn and use. But when asked to explain the rule or method in any level of detail, the students showed very little knowledge, as noted in their explanations (or written work in solutions) when they tried to explain the meaning of the relations in the method or rule.

Learning from the textbook: The text was written differently from other texts with a focus on clear and written explanations to make understanding of work being studied better for students, as it was hypothesized that the "wordy" or prose nature of a new mathematical textbook may enhance the performance of first-year non-mathematics majors. The interactive prose-style book is a new and novel way of presenting texts to first-year students. It was hoped that the first-year students would learn mathematics by critically reading, given that they are experienced readers of novels and other popular literature (prior knowledge base). A new book with in-depth and detailed step-by-step explanations was written for the benefit of students to learn the "baby steps" of mathematical work, before learning the more difficult steps. In addition, the book included gently guided extensions of work, with many routine/non-routine and higher-order thinking questions posed to the learner. This was to promote students' investigation of important mathematical ideas (such as a slope of a line, rate of change, etc.) or to confirm the logical nature or the appropriateness of working presented, such as overseeing proofs etc.

Writing in the textbook: Working space was specifically provided in the text to prompt the students write in the book. This is a new idea, and while the students initially felt that this was not appropriate, for this was a textbook on mathematics and it should not be written in, their ideas soon changed when they saw other students writing in their books – that is, by following others they felt happier and at ease when writing appropriate details when learning in their actual texts. It was noted that the text was used by most of the students to learn their first-year mathematics. This was confirmed by the majority of the positive comments made by the students (80/97). Nonetheless, there were still a number of the students who did not wish to write in their textbooks. It is hoped that this will change in the future when more and more real-time books are presented in the market for students to interact with the texts and thus own their learning process. It is hoped that students will write in the texts and then feel that they also own the mathematics they have learned, which means they may keep their written texts for later mathematics learning, and they may be reminded of how to learn when the need arises. In this way, it may be possible for students to also realize that mathematics involves critical reading, writing and thinking – this is what mathematics is about – not just arithmetic.

With regard to the above, comments that highlight a number of important learning and teaching issues that arose during discussions in focus groups are presented below. The students' comments are numbered alphabetically for convenience only – the comments were given during and within discussion groups but not in this order. Upon analysis, the work has been presented in the following manner:

Student A said that "the textbook provided was more like an English textbook and was hard to follow. I quickly got lost in some of the explanations in the book."

Student B said, "the textbook was a bit more helpful, minus the errors. I feel like it had a little too much depth/writing for a text in mathematics . . . as needed to understand the concepts . . . worked examples are way more effective than words."

Learning style: Some students felt that reading books was not their style of learning. This is one of the major shortfalls of our teaching – it seems that we have focused so much on the learning styles of the learners. Rather than equally developing their weaker attributes and/or styles of learning, we have focused on their preferences, and this is now showing at universities where critical reading becomes a major tool for learning independently for deeper understanding. These students believe they have a style of learning and that is the only way they can learn (this is their attitude), but this attitude is not conducive to novel and creative activities that may require other styles such as reading. It is very important to focus on the positive aspects of students' learning styles, but equally we should focus on the less experienced aspects of any students' learning styles in order to develop them to an appropriate level, that is, if students are to do well learning generally, and learning mathematics in particular. The use of different teaching methods was espoused by student C, but it also shows that when reading is important to learning, the student did not wish to change her style. In mathematics learning, we need to constantly change our learning approach or attitude, or else the work given cannot be learned so easily. The following comments appear to confirm this.

Student C said there should be "*different ways of teaching, not telling students to just look at the book if they have trouble, as many of us didn't understand the book.*" The case of "many of us" was noted but was not forthcoming in the analysis of the number of comments, as only 7 out of 90 students commented that the book was difficult to understand. The majority said that the book was easy to learn from. It seems that it was not really the book but rather the mathematical work in the book that was difficult for the student. When asked further it was noted that the student was not focused or driven to critically read or understand, rather to ask for help first when in trouble, and that is when help should have been given. This is also noted next when the tutor did not help immediately.

Student D said, "*text he constantly refers to . . . is badly written and does not help.*" The student refers to the tutor who was gently pushing him to read the work before simply asking for help. The student believed the work was badly written, but clearly if the student understood the work, then this is a great comment by the student, as it shows creativity and the critical ability to decipher what was intended in the book and what was written. However, this was not the case, as the student did not understand the mathematics being taught, as noted by the tutor.

Student E said, there should be "*more examples . . . in real time and not just encouraging us to read the book . . . this is an issue for visual learners.*" The comment regarding more examples was made because in this text, fewer questions were provided for practice at the end of the section. This aspect is different to other texts and therefore is also a different style to other texts, which usually have lots of examples, some going from a few easy ones to hard ones very quickly. Instead, in this text only around 5–10 questions were at the end of each section. The students had to solve a few examples at the end of each section, and this was because we expected students to master work taught in the section by successfully solving only 5 to 10 examples of a similar nature.

Even when the lectures provided many of the different presentation approaches of the work that was in the text, which were available in recorded video, the student believed that in the tutorials they should also be given the same opportunity; so, the work should also be done on the board and students should not be encouraged to read the book for answers. In other words, the tutor should also provide the solutions on the board for the students; but this approach was not allowed, at least initially, to gently encourage students to independently seek answers by critically reading the text.

Student F said the tutor "*sometimes just referred you to the textbook if you didn't understand him . . . if you did not understand him, you often didn't understand the textbook either*." The author was also the lecturer, and this is a comment regarding the presentation of the work in lectures and that is then learned again in tutorials. Since the work was taught from the text, the content followed in the tutorials, so even when presented differently to the way it may have been presented in the book, the student found it difficult to understand the mathematical work. The similarity of the work presented in lectures and tutorials meant that the work presented had been repeated for the student, but the student believed the similarity meant that they still could not understand and gave up, rather than trying to critically read the text over again, as is required in the learning of mathematics.

Although the students may have felt that the learning style used was not appropriate, it was hoped that the reading level of students would have been at an appropriate level when students had done all the required work to gain a position at university. As such, prior knowledge of reading can be transferred to mathematical books, in that reading should be adapted easily to all content areas; at least the reading aspect. Given that some university students could not cope with mathematical reading as the comments noted above, we perhaps need to rethink the teaching of mathematics at earlier levels to include reading as a major tool for learning. In addition, we should be developing students' ability to cope with and/or adapt to various ways of learning, not simply their preferred styles, for any type of style may be imposed upon them in future; for example, in a workplace. Also, we should not be labeling students visual learners, concrete learners and auditory learners etc. since labeling can be dangerous. The adaptability of students to cope with changing world circumstances is what makes them better problem solvers. So it is in mathematics. At times, we are dealing with reading word problems in arithmetic reasoning problems, while at other times reading and translating problems based on abstractions of the same, for example in terms of algebraic rules and methods. This becomes abstract reasoning. Further still, we may translate drawn figures as in geometrizing the problem; that is, the thinking and reasoning of earlier problems can now become visual pictures to enhance students' earlier understanding. But these may not be done in any order, for students should engage with them as they see fit. Importantly, what this example also shows is that to be successful in mathematics, we need to be able to cope with all types of learning styles, particularly the ability to critically read mathematical material to not only apply but also abstract and visualize in order to learn higher mathematics.

Identification of procedural errors in prose, working and solved examples: This was an important part of the textbook teaching method. Often students were often asked to read over and check the solutions just to test their understanding. They were asked to critically examine the notes in the book together with its *solved solutions and proofs*, in order to identify errors and mistakes if there were any. Not surprisingly, some students (3/90) believed that it was inappropriate to have a book with errors in it. This is not usually seen in texts, and this made this textbook different to others. Clearly, most solutions did not include any mistakes but purposefully a few did, and this was mainly to develop discussions amongst students, which it surely did. Yet some students did not read critically enough or believed all that was written was fact, and so did not realize that there were any errors at all in the solutions or proofs. This was mainly because of their belief that the solutions should not include errors, as this would be inappropriate in mathematics books. It was only occasionally that students were asked to identify errors, but only after the correct solutions were presented of similar examples. However, it was noted that many students (10/90) believed that mathematics texts should always be presenting correct accurate work for students to learn from. Although this is an important aspect and a valid argument, we should nevertheless equally assess students' ability to be engaged in the mathematics they are learning by critically examining solved problems to identify errors. That is, students should learn to use their higher-order thinking skills to identify illogical and inappropriate arguments within mathematical explanations, solutions or arguments, as we often naturally do in daily conversation (prior knowledge). In mathematics, we should always learn to self-check and be aware of what we are doing, whether we are being illogical or erroneous in our arguments, and if so, we should be happy to reflect and learn from our mistakes. This aspect is another vitally important part of mathematics learning, showing students that we can commonly make mistakes in our working (we have made many mistakes over centuries of the development of science), but what is crucial is that we revisit the work to check its appropriateness. Mathematicians and scientists mostly reflect on their thinking and reasoning and only then identify some new ways or errors and mistakes in earlier work, and in this manner learn to identify the errors/mistakes and correct them. This is most commonly done by going over solved solutions and critically identifying the working steps that may have errors in thinking, reasoning or working, etc. That is, we as mathematicians do a lot of learning and relearning as we climb the mountain of mathematical knowledge.

Some examples of students' comments appropriate to this aspect of error identification in problem solving were noted in the following:

Student G said there should be "no mistakes in the textbook you should be providing a text book free of errors."

Student H said, "*the textbook was designed for the student to identify errors, this did not help me at all as I need the textbook to teach me when I don't know what I'm doing, and I just end up doing things wrong as a result. . . . I asked the tutor and he said just work through the text book and you'll be fine.*"

Student J said, "*I don't understand how I'm supposed to answer a question that mathematically isn't correct if I barely understand how to the questions in the first place? This really frustrated me.*"

Student K advised "*not putting mistakes in the textbook.*"

Upon conceding that many students were learning from the books or have been doing well in the course by using books, two students felt that it did not work for them. Two (out of 90) of the general comments about their experiences regarding the issue of being gently pushed to read critically to learn from a specially written textbook are as follows:

Student L said, "*while I'm sure this works for some, it's definitely did not work for me as I really didn't even know where to start, let alone be able to work through the book.*"

Student M said, "*it was good if you understood what was going on . . . if you didn't it just made you more confused . . . especially regarding the questions that were wrong.*"

As an author, this aspect of presenting errors can be easily fixed, and the few errors can be easily removed from the proofs and solutions. But what is more important is that there were only a few problems that were of that type, and secondly, similar problems were solved earlier anyway, and students were simply required to identify errors by comparison from examining a similar related solved problem later. It was noted that students in this group tended to not go back and relook at the worked solutions and mostly gave up or tended to ask a tutor for the answer or how to solve problem. Some asked for an answer before even starting on a solution by themselves.

Student N said, "*the textbook serves little help as it can be easily shown online.*" In this case, the reference was to the physical text (preferred online text) but this was made by a student who was rather a high-scoring student. The analysis showed that the majority, however, would rather have a text to work with and write on, even though they liked to work online or get help online. And this particular book allowed them to do just that, to read and to learn and to write in the book (80/97). While this was noted that there was only one student who made this comment, it was useful, and students should be encouraged to get help with their work online.

In the end, however, only a few students (10/97) thought that the work presented in the steps of the text were "far too easy and we could have done much more . . . by covering more content and challenging the higher group of students" (Student N). Some students were doing the mathematics for an additional 10 credit point course for their degree programs and had done much mathematics earlier, such as Math B and C in Queensland. For such students, this work may be rather easy and straightforward, while as noted earlier, most of the students were non-mathematics majors and had done general mathematics and Math A in Queensland. It was not surprising to note that some students in this study may have found the work somewhat easier than the majority of students who were non-mathematics majors.

Some students disliked being asked to read and search for answers themselves and instead felt that the answers should be given to them when asked. The majority, however, soon realized that the way of learning in the class had changed. Soon, all of the students followed the routine of what was required in classes, trying first to work independently and then with a tutor if needed. The qualitative data showed that ultimately such a process led to much better levels of understanding overall in the student cohort (tutor's analyses). There was indeed a significant improvement in test scores using the "critical reading to learn" as a tool rather than just simply learning from the videos or lecture presentations of the same material, or learning from internet, as had been done earlier.

In the past, the low scores of the novices occurred even when the presentations using advanced techniques in computer presentations were preferred. In fact, some students even regressed from simply learning from videos by focusing on the non-important and surface features of the video, because when examined they could not explain the work presented or what they had learned or even written based on what they had seen. Perhaps this is not surprising, as some students fail to engage critically when watching mathematical video presentations. An intense mental cognitive process appears to be a necessary aspect in learning for understanding, particularly in learning the various mathematical concepts.

It seems then that much intense mental work is similarly required if a student is to apply their learned skills and techniques. That is, to apply skills that are necessary in real-life problem solving in the application of knowledge as and when needed, such as when students are asked to solve non-routine mathematical problems. While the fancy visual presentations clearly maintain attention, students who were simply learning by watching videos did not appropriately demonstrate understanding of the content. The process of watching videos of solving problems did not help students in later applying their knowledge to successfully solving similar routine problems.

Clearly, the process of learning from videos is beneficial and useful, but it may not be the most important learning tool for students who are novices, particularly in mathematics. The proviso is that if the student is very highly motivated, they tend to apply or engage all types of strategies to fully understand the work that is presented to them for learning. What has been noted is that students who are learning through visualization methods may feel that they have understood, but what seems to happen is that the video learning process provides students with a "false feeling of understanding" – where students "feel" that they have learned because they saw it working in front of them, and thus believe that they have understood the work, but when tested or questioned afterwards, a satisfactory level of understanding is not demonstrated consistently.

This also applies to watching videos on the internet, for example, as many students use the internet to learn from instead of PowerPoint notes or lecturer-provided video lectures. While internet learning is to be encouraged, what is noted is that clearly student learning does indeed occur in some cases, but this only occurs when a student is usually highly motivated and remains persistent in relearning the work involved. In which case, he or she often redoes the work or critically rewatches the presentations or videos. This process may be repeated a number of times by a student before an appropriate level of understanding is acquired.

12.4.1 Results based on quantitative data analysis

The quantitative analyses involved two groups of students, and their performance was compared. The groups were studied using a two-sample independent t-test, as the two groups were selected so that they were not related in any way, apart from being two first-year groups of successive years. The null hypothesis was that the mean score of working from a new textbook method (A) was the same as the mean score of the e-learning (as used in this study) method (B). The alternative was that the mean score of the new textbook method was greater than the mean score of the e-learning method (as defined in this study). The e-learning was defined to be learning from recorded lectures and work in tutorials, using teacher-prepared PowerPoint notes and/or video presentations that were also available 24/7 on the university blackboard; whereas the new textbook was the only source of notes available for the students during the semester of study at the university website for Method A. It was used during lectures and tutorials by the lecturer, and recorded work only involved the notes and examples from the text. All tutors were trained to be facilitators rather than teachers of "chalk and talk," and they were required to gently push students toward studying and learning as well as solving problems presented in the book for the sake of learning mathematics. The students were to solve their difficulties on their own in the first instance, or perhaps ask friends for a better understanding by asking them to explain the work presented in the textbook.

Two random samples of 97 students were chosen from two classes of sizes of more than 180 students. Normality tests were conducted; testing and statistical tests were used to determine whether there was a difference in their final average scores (Fig. 12.1).

12.4.2 Individual sample (A and B): summary and normality test results

Table 12.1 shows that means in groups A and B present some variations, while the standard deviation appears to show less variability. It can be concluded that both groups are from the same population with some similar characteristics.

The variances are equal as the hypothesis of equal variances is not rejected.

Table 12.1 The t-test procedure

Variable: Score

Method	*N*	*Mean*	*Std Dev*	*Std Err*	*Minimum*	*Maximum*
A	97	26.5232	8.8082	0.8943	4.0000	40.0000
B	97	22.7990	10.4244	1.0584	0.5000	40.0000
Diff (1−2)		3.7242	9.6502	1.3857		

Method	*Method*	*Mean*	*95% CL Mean*		*Std Dev*	*95% CL Std Dev*	
A		26.5232	24.7479	28.2984	8.8082	7.7191	10.2579
B		22.7990	20.6980	24.9000	10.4244	9.1355	12.1402
Diff (1−2)	Pooled	3.7242	0.9911	6.4574	9.6502	8.7740	10.7224
Diff (1−2)	Satterthwaite	3.7242	0.9906	6.4578			

Method	*Variances*	*DF*	*t Value*	*Pr > |t|*
Pooled	Equal	192	2.69	0.0078
Satterthwaite	Unequal	186.8	2.69	0.0078

Equality of Variances

Method	*Num DF*	*Den DF*	*F Value*	*Pr > F*
Folded F	96	96	1.40	0.1005

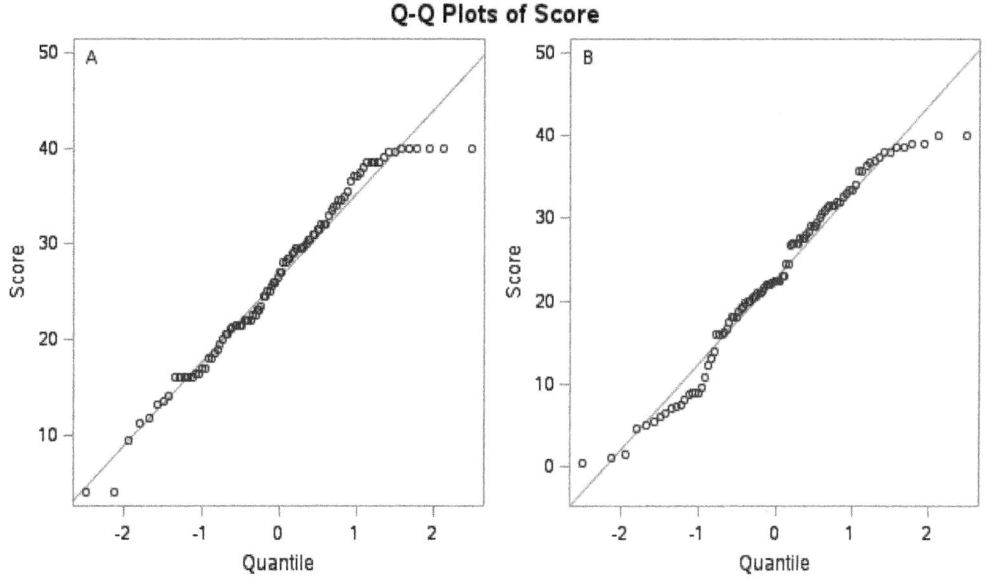

Figure 12.2 QQ plots

12.4.3 Manual t-test analysis

Method A: Used new textbook only in lectures and learning throughout the semester and only some notes copied from the textbook were available online.

N_1: 97

$df_1 = N-1 = 97-1 = 96$

M_1: 26.52

SS_1: 7448.14

$s^2_1 = SS_1/(N-1) = 7448.14/(97-1) = 77.58$

Method 2- Used prepared PowerPoint and video presentations available on university blackboard.

N_2: 97

$df_2 = N-1 = 97-1 = 96$

M_2: 22.8

SS_2 :10432.2

$s^2_2 = SS_2/(N-1) = 10432.2/(97-1) = 108.67$

T-value calculation

$$
\begin{aligned}
s^2_p &= \left(\left(df_1/\left(df_1+df_2\right)\right) * s^2_1\right) + \left(\left(df_2/\left(df_2+df_2\right)\right) * s^2_2\right) \\
&= \left((96/192) * 77.58\right) + \left((96/192) * 108.67\right) \\
&= 93.13
\end{aligned}
$$

$$s^2_{M1} = s^2_p/N_1 = 93.13/97 = 0.96$$

$$s^2_{M2} = s^2_p/N_2 = 93.13/97 = 0.96$$

The observed $t = (M_1 - M_2)/\sqrt{(s^2_{M1} + s^2_{M2})} = 3.72/\sqrt{1.92} = 2.69$ which has a p-value of 0.0039 associated with it in the not equal to the hypothesis case. The critical t value from t distribution table is also the same, confirming the test results. However, the p value associated is 0.0039, in this case due to the two-sided alternative used in the manual case.

At the 5% significance level and using a one-tailed, two independent samples *t*-test showed a significant difference between the methods with Method A (new text) being superior to B (e-learning). The null hypothesis is rejected in favor of the alternative at the 5% level. There is a possibility of type one error, and more intensive highly controlled studies should further improve this first initial level study of the importance of critical reading – that is, critically reading mathematical texts can indeed help improve student overall performance for most students, but this does require a gentle push by the teachers towards the case for reading to learn. In fact, the value is still in the critical region at the 0.01 level of significance since the p value is still less than this amount.

Further, a chi-squared test was also calculated to examine the strength of the association between "high" and "not high" scores with the type of method used. The chi-squared analysis showed that there was a significant association between high (scores greater than 28/40) and not high (medium/low – otherwise) with the type of method used in the teaching (Table 12.2).

The test of association was conducted by the grouping shown above; where A was related to the critical reading of textbook method and B represented the alternative. Table 12.3 shows significance at the 6% level. Further, it is tested for high scores and not high scores that are

Table 12.2 Method of learning and high/not high scores*

Method	*Score I*		
	HIGH	*LOW*	*Total*
A	47.00 24.23 48.45 58.02	50.00 25.77 51.55 44.25	97 50.00
B	34.00 17.53 35.05 41.98	63.00 32.47 64.95 55.75	97.00 50.00
Total	81.00 41.75	113.00 58.25	194 100.00

*Frequency row and column percentages

Table 12.3 Chi-squared test of association between performance and method of learning

Statistics for Table of Method by Score I				*Fisher's Exact Test*	
Statistic	*DF*	*Value*	*Prob*		
Chi-Square	1	3.5820	0.0584	Cell (1,1) Frequency (F)	47
Likelihood Ratio Chi-Square	1	3.5944	0.0580	Left-sided Pr <= F	0.9794
Continuity Adj. Chi-Square	1	3.0521	0.0806	Right-sided Pr >= F	0.0402
Mantel-Haenszel Chi-Square	1	3.5635	0.0591	Table Probability (P)	0.0196
Phi Coefficient		0.1359		Two-sided Pr <= P	0.0803
Contingency Coefficient		0.1346			
Cramer's V		0.1359			

equal to or less than 28/40. The significance appears to be noteworthy; it shows that the percentage of highs and not highs are significantly different when the methods are compared.

The high-level group based on scores above 28/40 provides a measure that relates to a much better understanding of the concepts, skills and procedures (examination of solutions by two tutors). Also, the high scores are related to better recall and solving of routine problems. In addition, the high group is also related to better application and transfer levels in terms of problem solving. The chi-squared testing confirmed this, as the test showed that the high group was significantly different (6%) to the not high (score < 28/40 – medium/low) group when the method of learning mathematics was taken into account. It is clear therefore that Method A was associated significantly with more numbers in the high labeled group when compared to Method B. Even when a simple number analysis is considered, there was a greater number in the not high (or low level) in the Method B group when compared to Method A (Table 12.2).

12.5 Reflection

It is to be noted that some of our students are often partially employed and working while studying and learning mathematics. Being part-time students means that they will be almost always either reading a text for learning or watching videos to catch up on lectures. Teachers and lecturers should prepare for this, where many of the higher education or university students may be mostly reading or viewing presented works in mathematics. Therefore, the students need to be taught how to read and write to learn from texts or e-books and PowerPoint presentations. An equal amount of time should involve reading texts and writing or communicating solutions, as well as engaging in activities that involve seeing and doing (solving) in classrooms (Sutton and Krueger, 2002). The use of computer technology can help in this regard in that when doing practical work, one has to start the work only after reading instructions, and then some writing ability is also necessary to summarize the mathematical concepts and related work that is completed (developing summary sheets of learning). The more the students experience the know-how involved in terms of reading critically, the better they can understand and comprehend mathematical concepts from a written set of notes or textbooks.

While inquiry-based learning and concretizing mathematical activities is in favor in modern times, what we really need in mathematical classes is to make certain students learn to think in abstract terms more often, or learn to play and practice mental and mind games (practice with abstract-based cognitive processes) constantly (Tularam, 2016). This particular experiential process appears to have been left out of the equation when teaching or preparing learning tools for tertiary students. Rather, it seems that we tend to "dumb" down the work to either make things easier for students to comprehend in a reductionist framework – where students can only learn to master baby steps to develop understanding – or by using tools that greatly simplify the process of abstract thinking itself, in which case the student does not have to undergo or engage in seriously abstract-thinking-related processes.

This means students often fail to experience higher-level cognitive or mental demands when learning or solving problems (Tularam, 2015). In the end, we do not tend to allow our students to work through the written paragraphs themselves much or solve problems posed on their own, all of which provides a very low level of "experience and practice" of abstract thinking. Indeed, students should constantly engage in cognitive play and thus experience mind games that cause them some mental stress. The evidence shows that continuous practice of such mind games will lead to sophisticated thinking processes, as is required in, for example, algorithmic development, code formation, network analysis, big data analysis, etc.

It is the higher-order thinking processes that we need to develop in our students of the future, and this can be done by students having to constantly use their minds to work with mathematical processes by engaging in critically reading about them (Tularam, 1990, 1997a/1997b/1997c, 2013, 2015). Thus, we need to not only focus on practical and concrete learning tools that help make the work somewhat easier, but rather to provide higher thinking opportunities to help students develop the ability to develop a mathematical thinking attitude in algebraic, geometric, calculus and other higher-order mathematics, which involves logical, deductive, creative and higher thinking generally (Anderson *et al.*, 1984; Barton *et al.*, 2002; Sutton and Krueger, 2002; Tularam, 2013). Such a learning process may be more confronting and mentally stressful to students than other methods, but it seems that the benefits will outweigh the difficulties and negativities the students may experience in the meantime (as indicated by the large number of positive comments by students, which are not presented here).

Although noble, it appears that our best intentions are those that may have let our students and us down over time; this may have had a significant effect on the learning and teaching of mathematics. It seems that we have allowed students to aim for a level of thinking that is lower than their potential in terms of mathematical thinking (Tularam, 2016). The teaching of "reading for understanding" concerning the critical processes described above may start during the early learning of mathematics but must continue with diligence over time to be effective. Some studies show that educators must aim high in STEM courses and expect high scores from our students, not the other way around, "dumbing down" the courses for university-level students who have already acquired a good 12 years of mathematics learning and general education (Tularam, 2013, 2014, 2015).

The evidence shows that the students who were less than satisfactory were not as engaged in the way that was necessary for deeper mathematics learning. The students have to undergo self-critical reading and then metacognitive mental/cognitive processing for learning to occur. Moreover, the process of understanding requires much relearning (practice) to take place regarding the mathematical topic or content being studied. The students need to search for where the problems are occurring in the learning process and then appropriately address the issues concerned. On the other hand, those students who were highly persistent and motivated were able to read the work a number of times and used their text examples in the learning more often than others. Also, students who had allowed extra time to spend studying did better than those who were working while studying university-level mathematics. In both the earlier year groups, including the Method A group, the highly persistent and motivated students also used the videos as learning tools; that is, as a way of realizing what they did not know, and they were able to read their texts often to learn what they did not know. Moreover, they also sought help from friends, tutors and lecturers. In essence, they were a group who were well prepared and in a specific mode of learning, those who critically sought to understand the text or videos, rather than just viewing them and doing little else about any lack of understanding. Of most concern is the fact that the majority of the students in the novice group did not seem show persistence or high motivation to work to find answers for themselves, rather they expected help almost immediately and asked for answers or help to solve their problems in mathematics immediately when asked. Instead, these students would have enjoyed the teacher doing their work on the board while they "sat and listened or quickly" copied down the answers. The problem with this "answer-providing method" (which was prevalent in many tutorials of the past) is that the method does little to help many of the less fortunate students (who are not coping or understanding) to become independent learners. The better students usually can cope with self-learning, and thus understanding of the concepts or solving of related problems (Tularam, 2014), while most of the others usually do not have enough practice with "independent" thinking and so find the development of a solution process difficult.

While the goal of making things easier and more user friendly for learners does help students "understand to a degree," what we have noted in our research is that often this type of facilitation process only makes student feel at "ease," particularly with the "harder" and more abstract work in mathematics, but it does not in fact help them to understand work for later use. It was noted that a "feeling of being at ease" with the course content is not the same as demonstrating a satisfactory level of understanding of the work being taught. Breaking down things into its parts to make them easier may seem a positive approach (at least to an expert), but seeing the parts of the whole often may provide the "feeling of understanding" of the parts, perhaps, but not the whole in most cases. This is what was noted when students were questioned on the topic studied or issues addressed in the manner suggested above. The level of student understanding was noted to be less than satisfactory in many of the mid- to lower-performing students, for many students just did not understand the linkages between the parts and the whole at all.

It seems that much more in-depth learning needs to take place before the whole emerges or the work is understood; that is, for a student to gain an in-depth understanding, such as the aha! moment. The data shows that there is no substitute for "getting into the work habit" of learning mathematics through critically:

1 Reading all work;
2 Self-questioning all work read;
3 Reading solved problems – both correctly or incorrectly solved examples (in search of understanding the nature of errors that may exist);
4 "Going back" and self-relearning and/or redoing the same or similar examples and
5 Summarizing the work read or presented or questions posed.

If such processes that are so important to mathematical learning processes are not "mentally undergone," then it seems much less is retained and/or gained, at least in the context of mathematics. The complex nature of learning means that students will need to become an apprentice to the teacher as well as to the course itself, and thus critically engage in the learning of mathematics through a process of engagement in the mathematical processes involved over a period of time (even from Grade 1 onwards). Clearly, this does not simply mean doing specific assignment questions or attaining high marks in routine assignment problems, but rather to being able to verbally explain solution processes and to demonstrate application of their knowledge with a level of understanding to friends, tutors and/or others, especially when asked to explain aloud the solving of non-routine problems.

So, in order to demonstrate mathematical expertise at a level required at undergraduate study level in all majors, the students should be able to:

1 Read any of the mathematical work presented at that level;
2 Use the mathematical methods to communicate to others in a satisfactory manner, thus demonstrate competency in the language of mathematics.
3 Solve routine problems and be able to learn how to solve them from solutions provided; for example, from solved problems in textbooks and
4 Apply their knowledge in a creative manner to solve novel or non-routine problems.

It seems that only if students can demonstrate the above can they transfer their knowledge to solve the actual or real-life problems that we face in the constantly changing world of today (Tularam and Hulsman, 2015). Clearly, the learning of higher-order skills is an essential part of mathematical thinking and problem solving in the processes involved in reflective thinking, logical and deductive thinking and creativity, but these skills can only be acquired through an intense and critical mental engagement in the process of learning mathematics itself, including the reading of mathematical material.

More generally, as good citizens we need to be able to examine information to evaluate various sources with a view to testing the validity of the information confronted. We observe that the more general skills that are needed to operate in real life are akin to those needed to learn mathematics. Indeed, the mathematical language reading skills are related to the general skills and processes that are involved in the processes of reading and gaining general literacy skills. The research literature in mathematics education has constantly suggested that the skills and processes of reading and writing activities are important in the development of communication, analyzing data and differentiating and interpreting mathematical concepts. Bloom's (1956) work has shown that communicating, comparing and contrasting can lead to the predicting or inferring of cause and effect that are higher-order skills and processes. Students can learn to develop routine as well as non-routine relationships or higher-order relationships once they realize that these skills are as important in reading as in the learning of mathematics. So the critical reading of mathematical material and texts then becomes an important aspect of learning mathematical language, concepts, skills and procedures which are often referred to as domain-specific discipline knowledge.

It is true that students find difficulty reading, but more of the students can read than are successful in mathematical knowledge, so it is important to allow students to read as much mathematical content as possible. That is, by allowing students to examine textbooks on mathematics that are written in a manner that allows students to provoke and engage their critical reading skills (prior knowledge), not only solving problems all the time; critically examining the mathematics used in other applications will help students to learn to use their mathematical knowledge in other contexts and thus become better problem solvers. The students will be able to learn from examples that show the applications of their acquired knowledge generally to other disciplines such as physics, chemistry, biology, ecology and so on. If we can assume that students are generally good readers of work, then when encouraged to read mathematics texts or written material, they may be more easily able to identify the structures that could make up the underlying mathematical knowledge that has been presented. It seems that this can only be achieved by critically examining the written work or by critically reading through the worked (and flawed) examples provided in texts (videos and/or verbal lectures).

This chapter shows just how important the critical reading ability of students is to the learning of mathematical content skills and processes. Most of the work of mathematics is indeed in the written form, and reading of the mathematical work can open many new pathways which would not only help in the promotion to the general population of mathematics itself, but also allow enhancement of the students' vocational lives. However, what we also need are writers and authors who may be able to present the work in a greater level of prose form that is mathematically sound and which students can read to understand. In this manner, students can realize that mathematics learning can also be a reading process and that the prose form of mathematics texts can be accepted as legitimate ways of writing mathematics textbooks. The authors can also think laterally and creatively to make textbooks more interactive, and one way of doing this is to allow space for students to write within textbooks as they are learning, thus providing opportunities for students to partially own the mathematics done and the textbook itself.

In this way, the mathematical texts (physical) will indeed get their "optimal use" – reading and writing students' own versions of the text information. The textbooks and the written material will become a major tool in mathematics learning in the tertiary sector. Commonly, students spend many hours in classrooms writing solutions to problems presented in texts, and in doing so it is hoped that students are practicing writing their work in a "logical and deductive manner." However, this study shows that equal or more time must be spent reading the explanatory work in textbooks. It is the duty of all mathematics educators to not only focus on writing and communication skills in primary, high school and tertiary mathematics courses, but to also focus equally on the critical reading of mathematical material.

12.6 Importance of blended learning

There are number of electronic learning issues to be considered to ensure achieving learning outcomes set by both instructor and learner. E-learning has number of definitions; Stockley's (2003) definition seems appropriate for mathematics teaching. "E-learning involves the delivery of a learning, training or education program by electronic means. E-learning involves the use of electronic device (e.g. computers and mobile phones) to provide training, educational or learning material" (Stockley, 2003). Further, online training or education, internet or intranet, CD and DVD are among the actual procedures used in e-learning environments. It is understood that one of the first areas that used e-learning in its delivery was distance education. Interestingly, the demand on the learning through e-learning has been increasing among the tertiary education students. E-learning overcomes many of the challenges that plague the modern tertiary students, such as attending lectures on time, avoiding parking at universities and other difficulties faced in traveling to places of learning etc. A combination of various methods, including traditionally based, online 24/7 lectures and tutorials can also be included in e-learning, but this type of combined effort is usually referred to as blended learning. Despite the increasing popularity of blended learning, researchers gave little attention on the effectiveness of combining traditional classroom learning environment and online learning, a practice that is generally referred as blended learning. Schechter *et al.* (2015) examined the potential benefits of a blended learning approach on the reading skills of the students from low socioeconomic backgrounds in the early grades of their university education.

According to Chaney (2017), the modest studies on blended learning to date tended to focus on higher education, leaving a significant gap in the research regarding K–12 education. Framed within Vygotsky's theory of social development, the objective of this causal comparative study was to assess if there were any significant differences when comparing charter school students who participated in a blended learning approach to reading and mathematics with students who studied the two subjects via fully online mode and with students who studied them in traditional classrooms with no online learning. The design was causal comparative with a nonrandomized control group. Chaney (2017) compared the archived 2014 State of Texas Assessments of Academic Readiness scores of 1797–2298 students in one charter management organization: students in a blended learning environment, students who received traditional classroom instruction and those students who used "fully" online learning. Analysis of variance (ANOVA) was used in combination with appropriate post-hoc comparisons to evaluate group means. No statistically significant relationships were noted between traditional, blended and fully online students and math scores, or between traditional and blended learning students and reading scores. However, fully online students and higher reading scores were statistically different. Therefore, any blended learning process should involve with the appropriate communication devices and encompass collaborative activities for a constructive critical thinking, reflection and discussion (Albano *et al.*, 2013).

12.7 Conclusion

The aim of this study was to compare two methods of learning at the tertiary level. But a brief examination of blended learning in reading and mathematics was also explored. The results show that students can be "gently pushed" to read and understand the work presented in the first year rather than spending time solving hundreds of problems. The reading for understanding method provided a much more comprehensive level of understanding of mathematical material, and the students seemed to enjoy the experience of reading a verbose mathematical text rather than one which was mostly solving problems at the end of a chapter. Many students said this was the first time they had read a mathematics text from the start to the finish. Hence, it seems that learning by critical reading can be used and focused upon in classrooms if textbooks are written in a way that a student can critically read and learn from (Tularam, 2016). In the process of reading and questioning the written work, the critical reader soon realizes what they do not know and then hopefully follows up the work by relearning the prerequisites. In this manner, the reading process allows more opportunities for mathematics students to ask questions about those aspects they cannot understand. Indeed, the students can learn by first asking their peers and friends who may or may not know. This provides an important opportunity for the learner to clarify their understanding by going further along the chain if necessary.

A student is then gently "forced to clarify their own understanding of mathematics," but this is only possible if the learner is persistent and motivated to progress their understanding of mathematics. The process of critically reading mathematical texts and example solutions will forcefully help students clarify, organize and solidify their own ideas and thoughts about the concepts presented in the textbooks to develop self-confidence. The students can then logically self-construct their knowledge in the process – this type of learning will be not only long-lasting but will also satisfy the needs of the constructivists – that is, allowing independence for the learner to construct their knowledge through providing experiences, because students will need to discuss and negotiate with others or ask questions during or after reading (if they do not understand). Therefore, the social discourse and cultural learning that will occur in the mathematical context by students having to converse with friends, neighbors, tutors, teachers and others in order to negotiate meanings of mathematical concepts will satisfy the social constructivists' ideas as well – where focus is more on social and cultural interactions as a means and way of learning. It is clear that the noble goal of all the mathematics educators is the satisfactory teaching and learning of mathematics material that lead to a well-structured domain-specific knowledge that is highly connected, coherent and accommodating, and is thus a transferable knowledge base. This can be further enhanced by the appropriate use of blended learning practice.

References

Albano, G., Coppola, C. & Pacelli, T. (2013) The use of e-learning in pre-service teachers' training. *Quaderni di Ricerca in Didattica (Mathematics)*, 23(1), 1–5.

Anderson, R.C., Hiebert, E.H., Scott, J.A. & Wilkinson, I.A.G. (1984) *Becoming a nation of readers. Becoming a Nation of Readers: The Report of the Commission on Reading*. The National Institute of Education, Washington, DC, USA. Available from: http://www-tc.pbs.org/teacherline/courses/rdla230/docs/session_1_sutton.pdf

Australian Broadcasting Corporation (2016) ABC: Election 2016: Malcolm Turnbull says maths, science should be prerequisite for university. [Online] Available from: www.abc.net.au/news/2016-06-19/maths-science-should-be-compulsary-in-year-12-turnbull-says/7524190

Barton, M.L., Heidema, C. & Jordon, D. (2002) Teaching reading in mathematics and science. Reading and writing in the content areas. *Education Leadership*, 60(3), 24–28

Barton, M.L. & Heidema, C. (2009) *Teaching Reading in Mathematics*, 2nd ed. Mid-continent Research for Education and Learning, Aurora, CO, USA. ISBN 978-1-893476-14-1

Best, R.M., Rowe, M.P., Ozuru, Y. & McNamara, D.S. (2005) Deep-level comprehension of science texts: The role of the reader and the text. *Topics in Language Disorders*, 25, 65–83.

Billmeyer, R. & Barton, M.L. (1998) *Teaching Reading in the Content Areas: If not Me, Then Who?*, 2nd ed. Mid-continent Research for Education and Learning, Aurora, CO, USA.

Bloom, B. (ed). (1956) *Taxonomy of Educational Objectives, the Classification of Educational Goals – Handbook I: Cognitive Domain*. McKay, New York, USA.

Bournot-Trites, M. & Reeder, K. (2001) Interdependence revisited: Mathematics achievement in an intensified French immersion program. *Canadian Modern Language Review*, 58(1), 27–43.

Chaney, T. (2017) *The Effect of Blended Learning on Math and Reading Achievement in a Charter School Context*. Doctoral Dissertations and Projects. 1392 Available from: http://digitalcommons.liberty.edu/doctoral/1392 [Accessed 12 October 2017].

Chi, M.T.H., Bassok, M., Lewis, M.W., Reimann, P. & Glaser, R. (1989) Self explanations: How students study and use examples in learning to solve problems. *Cognitive Science*, 13, 145–182.

Chi, M.T.H., de Leeuw, N., Chiu, M. & LaVancher, C. (1994) Eliciting self-explanations improves understanding. *Cognitive Science*, 18, 439–477.

Cooper, S.J. (2004) *Addressing Scientific Literacy through Content Area Reading and Processes of Scientific Inquiry: What Teachers Report*. Unpublished Thesis Presented at the University of Central Florida, Orlando Florida.

Davey, B. (1983) Think aloud: Modeling the cognitive processes of reading comprehension. *Journal of Reading*, 27, 44–47.

Davies, P. (1984) *God and New Physics*. Simon & Schuster, Sydney, Australia.

Dewitz, P., Carr, E. & Patberg, J. (1987) Effects of interference training on comprehension and comprehension monitoring. *Reading Research Quarterly*, 22, 99–121.

Dweck, C. (2006) *Mindset: The New Psychology of Success*. Random House, New York, USA.

Fite, G. (2002) Reading and math: What is the connection? *Kansas Science Teacher*, 14, 7–11. Available from: www.emporia.edu/dotAsset/9acbacde-104d-4b37-b13a-ffc1ec7885cb.pdf

Friere, P. (1983) The importance of act of reading. *Journal of Education*, 165(1), 5–11. Available from: http://serendip.brynmawr.edu/exchange/files/freire.pdf

Goddard, R.D., Sweetland, S.R. & Hoy, W.K. (2000) Academic emphasis of Urban elementary schools and student achievement in reading and mathematics: A multilevel analysis. *Educational Administration Quarterly*, 36(5), 683–702.

Grimm, K.J. (2008) Longitudinal associations between reading and mathematics achievement, *Developmental Neuropsychology*, 33(3), 410–426, doi: 10.1080/87565640801982486

Gutstein, E. (2006) *Reading and Writing the World with Mathematics*. Taylor and Francis, New York, USA.

Hedges, L.V., Pigott, T.D., Polanin, J.R., MarieRyan, A., Tocci, C. & Williams, R.T. (2016) The question of school resources and student achievement. *Review of Research in Education*, 40(1), 143–168.

Hermida, J. (2009) Importance of teaching academic reading skills in first year courses. *International Journal of Research and Review*, 3(3), 20–30.

Intercultural Research Development Association (2002) (IDRA): What is the importance of reading and writing in the mathematics curriculum? (2002) Available from: www.idra.org/resource-center/what-is-the-importance-of-reading-and-writing-in-the-math-curriculum/ [Accessed 12 December 2016].

Kelson, N. & Tularam, A. (1998a) Tutoring with higher mathematics and the use of technology. In: *Proceedings of the Effective Assessment at University Conference*, held 4–5 Nov, in UQ Brisbane. Teaching and Educational Development Institute (TEDI), UQ. Australia. Available form: www.maths.qut.edu.au/pub/1998centerpubs.html [Accessed 12 October 2017]

Kelson, N. & Tularam, A. (1998b). Implementation of an integrated, technology based, discovery mode assessment item involving an incubation period to enhance learning outcomes for engineering mathematics students. In: *Proceedings of the Effective Assessment at University Conference*, held 4–5 Nov, in UQ Brisbane. Teaching and Educational Development Institute (TEDI), UQ. Australia. Available from: www.maths.qut.edu.au/pub/1998centerpubs.html [Accessed 12 October 2017].

Kelson, N. & Tularam, A. (2003) Implementation of an integrated, technology-based, discovery mode assessment item involving an "incubation period" to enhance learning outcomes for engineering maths students. *Effective Assessment at University Conference*. Available from: www.tedi.uq.edu.au/conferences/A_conf/program.html [Accessed 12 October 2017].

Lapkin, S., Hart, D. & Turnbull, M. (2003) Grade 6 French immersion students' performance on large-scale reading, writing, and mathematics tests: Building explanations. *Alberta Journal of Educational Research*, 49(1), 6–23. Available from: http://search.proquest.com.libraryproxy.griffith.edu.au/docview/228620424?accountid=14543 [Accessed 12 October 2017].

Marzano, R.J. (2005) Preliminary Report on the 2004–05 Evaluation Study of the ASCD Program for Building Academic Vocabulary. Available from: www.ascd.org/ASCD/pdf/Building%20Academic%20Vocabulary%20Report.pdf. [Accessed 11 December 2016].

McNamara, D. (2009) The importance of teaching reading strategies. *Perspectives on Language and Literacy*, 35(2), 34–38.

Metsisto, D. (2005) Reading in mathematics classroom. In: Kenney, J.M., Hancewicz, E., Heuer, L., Metsisto, D. & Tuttle, C.L. (eds) *Literacy Strategies for Improving Mathematics Instruction. Association For Supervision And Curriculum Development*, Washington, USA. Available from: www.ascd.org/publications/books/105137/chapters/Reading-in-the-Mathematics-Classroom.aspx: [Accessed 11 December 2016].

National Commission of Excellence in Education (1983) *NCEE: A Nation at Risk: The Imperative for Educational Reform*. U.S. Government Printing Office, Washington, DC, USA.

Paschler, H., McDaniel, M., Rohrer, D. & Bjork, R. (2009) Learning styles concepts and evidence. *Psychological Science in the Public Interest*, 9(3), 105–119.

Reyna, V.F. & Brainerd, C.J. (2007) The importance of mathematics in health and human judgment: Numeracy, risk communication, and medical decision making. *Learning and Individual Differences*, 17(2), 147–159. doi:10.1016/j.lindif.2007.03.010

Riccomini, P.J., Smith, G.W., Hughes, E. & Fries, K.M. (2015) The language of mathematics: The importance of teaching and learning mathematical vocabulary. *Reading and Writing Quarterly*, 31(3), 235–252.

Riener, C. & Willingham, D. (2010) The myth of learning styles. *Change: Magazine of Higher Learning*. Available from: www.tandfonline.com/toc/vchn20/current [Accessed 12 October 2017].

Roediger, H.L. & Karpicke, J.D. (2006) The power of testing memory: Basic research and implications for educational practice. *Perspectives on Psychological Science*, 1(3), 181–210.

Rohrer, D. & Pashler, H. (2010) Recent research on human learning challenges conventional instructional strategies. *Educational Researcher*, 39(5), 406–412.

Schechter, R., Macaruso, P., Kazakoff, E.R. & Brooke, E. (2015) Exploration of a blended learning approach to reading instruction for low SES students in early elementary grades. *Computers in the Schools*, 32(3), 183–200.

Silver, H., Strong, R. & Perini, M. (1997) Integrating learning styles and multiple intelligences. *Teaching for Multiple Intelligences*, 55(1), 22–27.

Smithsonian Magazine (2013) You Can Visit the World's Oldest Zero at a Temple in India: Indian mathematicians were the first to treat zero as an equal. [Online] Available from: www.smithsonianmag.com/smart-news/you-can-visit-the-worlds-oldest-zero-at-a-temple-in-india-2120286/.

Stockley, D. (2003) E-learning Definition and Explanation (E-learning, Online Training, Online Learning). What is E-learning?: [Online] Available from: http://derekstockley.com.au/elearning-definition.html [Accessed 12 February 2018].

Sutton, J. & Krueger, A. (2002) *What We Know About Mathematics Teaching and Learning*. Mid-continent Research for Education and Learning, Aurora, CO, USA. Available from: http://eric.ed.gov/?id=ED465514 [Accessed 12 October 2017].

Tegmark, M. (2014) *Our Mathematical Universe: My Quest for the Ultimate Nature of Reality*. Penguin, New York, USA.

The Organisation for Economic Co-operation and Development (2016) OECD: Report on Australia's Education: [Online] Available from: www.oecd.org/australia/

The Sydney Morning Herald (2016) SMH: Federal election 2016: Prime Minister Malcolm Turnbull: Make maths and science compulsory. [Online] Available from: www.smh.com.au/federal-politics/federal-election-2016/federal-election-2016-prime-minister-malcolm-turnbull-make-maths-and-science-compulsory-20160619-gpmobe.html

Tularam, G. A and Amri, S (2011). Tertiary mathematics learning and performance in first year mathematics in the environmental sciences: a case of student preparedness for learning mathematics', in *Proceedings of Volcanic Delta 2011, the Eighth Southern Hemisphere Conference on Teaching and Learning Undergraduate Mathematics and Statistics*, held in Rotorua, NZ, November 16–21. University of Canterbury and University of Auckland, Auckland.

Tularam, G.A. & Kelson, N. (2003) Lecturing and tutoring critically in tertiary mathematics. *Effective Assessment at Queensland University Conference*. Available from: www.tedi.uq.edu.au/conferences/A_conf/program.html [Accessed 12 October 2016].

Tularam, G.A. & Kelson, N. (1998) Assessment in tertiary mathematics. In: *Proceedings of the Effective Assessment at University Conference*, held 4–5 Nov, in UQ Brisbane. Teaching and Educational Development Institute (TEDI), UQ. Available from: www.maths.qut.edu.au/pub/1998centerpubs.html [Accessed 12 October 2016].

Tularam, G.A. & Hulsman, K. (2015) A study of students' conceptual, procedural, logical thinking and creativity during first year mathematics. *International Journal of Mathematics Learning and Teaching*. Available from: www.cimt.plymouth.ac.uk/journal/ [Accessed 12 October 2016].

Tularam, G.A. (2015) *Essentials of Applied Mathematics for the Sciences*. John Wiley & Sons, Australia. ISBN: 9781119922629, Brisbane Australia.

Tularam, G.A. (1997a) The role of higher order thinking: Metacognition and critical thinking in algebraic word problem solving: First Year in Higher Education – Strategies for Success in Transition Years. 5–8 Jul 1997, Auckland, New Zealand, Auckland Institute of Technology, Auckland, New Zealand [Lecture].

Tularam, G.A. (1998) The role of algebraic knowledge, higher-order thinking and affective factors in word-problem solving. In: *Proceedings of International Conference-Transformation in Higher Education*, 7–10 Jul, Auckland, New Zealand. pp. 210.

Tularam, G.A. & Amri, S. (2011) Tertiary mathematics learning and performance in first year mathematics in the environmental sciences: A case of student preparedness for learning mathematics. In: Hannah, J. & Thomas, M. (eds) *Volcanic Delta 2011: Proceedings of The Eighth Southern Hemisphere Conference on Teaching and Learning Undergraduate Mathematics and Statistics*, 16–21 Nov, 2011, University of Canterbury NZ, Rotorua, NZ and University of Auckland, Auckland. pp. 386–398.

Tularam, G.A. & Hulsman, K. (2013) A study of first year tertiary students' mathematical knowledge-Conceptual and procedural knowledge, logical thinking and creativity. *Journal of Mathematics and Statistics*, 9(3), 219–237. Available from: http://thescipub.com/abstract/10.3844/jmssp.2013.219.237

Tularam, G.A. (1990) Higher-order thinking in mathematics. *Queensland Association of Mathematics Teachers Journal* 7(1), 10–12.

Tularam, G.A. (1992) Algebraic problem solving: Word-problems. In: *Proceeding of New Directions in Algebra Education Conference*, held in Brisbane, Australia. pp. 45–60, Brisbane, QUT.

Tularam, G.A. (1994a) Higher-order thinking and algebra. In: *Proceeding of Mathematics Education Research Group of Australasia*, Jul 1994, Lismore, Australia. pp. 41.

Tularam, G.A. (1994b) Early algebra users: Vedic Indians. *Third Annual Australian History of Mathematics Conference*, 16 Sep, Newcastle, Australia [Lecture].

Tularam, G.A. (1996) Critical thinking and algebraic problem solving. In: *Proceeding of Mathematics Education Research of Australasia*, 21 Jul 1996, Melbourne, Australia. pp. 25.

Tularam, G.A. (1997b) Interactions between algebraic knowledge, higher-order thinking and affective factors in word-problem solving. *Mathematics Education Research of Australasia*, 20 Jul 1997. Rotorua, New Zealand. pp. 15.

Tularam, G.A. (1997c) The importance of higher-order thinking in novel algebraic problem solving. *Queensland Association of Mathematics Teachers Residential Conference (QAMTAC'97)*, Sept, Bardon, Brisbane QLD [Lecture].

Tularam, G.A. (1997d) The role of algebraic knowledge, higher-order thinking and affective factors in word-problem solving. In: *Proceeding of Mathematics Education Research of Australasia*, 16 Jul 1997. Rotorua, New Zealand. pp. 25.

Tularam, G.A. (2013) Preparedness of students in first year mathematics: Importance of placing the focus and responsibility of learning on students. *Shaikshik Parisamvad, International Journal of Education*, 3(1), 24–33. Banaras University. Available from: http://spijebhu.in/SPIJ-January-13%20 1-27.pdf: http://spijebhu.in/SPIJ-January-13%201-27.pdf

Tularam, G.A. (2014) Mathematics in finance and economics: Importance of teaching higher order mathematical thinking skills in finance. *Journal of Business Education and Scholarship of Teaching*, 7(1), 43–73. Available from: www.ejbest.org/a.php?/content/issue/12

Tularam, G.A. (2016) Traditional and non-traditional teaching and learning strategies: The case of e-learning. *International Engineering Education Conference*, 21–14 Nov 2016, University of Western Sydney, Paramatta, Australia.

University of Oregon (2005) Math as a Language. [Online] Available from: http://pages.uoregon.edu/moursund/Math/language.htm

Winstead, L. (2004) Increasing academic motivation and cognition in reading, writing, and mathematics: Meaning-making strategies. *Educational Research Quarterly*, 28(2), 29–47.

Chapter 13

Curriculum development of a social hydrology course based on a blended learning approach

A. Rahman

School of Computing, Engineering and Mathematics, Western Sydney University, New South Wales, Australia

ABSTRACT: Water is a vital source of life on this planet. Understanding the interaction of water with society is vital for achieving sustainable water resources development. Many water engineering projects worldwide have provided significant benefits to the community; however, there are examples where ill-planned water projects have undermined the environment and have affected the community adversely due to an inadequate understanding of the interaction of water with the environment and society. This chapter presents an overview of hydrology courses in Australian universities. It has been found that most hydrology courses in Australia are biased towards mathematical computation, and little focus is placed on the interdisciplinary nature and social aspects of hydrology. A new hydrology course is proposed in this chapter to overcome some limitations of the current hydrology courses in Australia. The proposed course focuses on the interdisciplinary nature of hydrology, including its social aspects, in addition to mathematical aspects. It is proposed to initiate a forum of hydrology academics in Australia to design a hydrology course that can meet the need of water users, water industry and research sectors, as well as the need of the society as a whole. A blended learning approach is proposed to design such a hydrology course by developing and sharing resources among various disciplines across different institutions.

13.1 Introduction

Water is an essential resource for economic and societal development. Water is a vital component of life. Without water, no life can flourish or be sustained. Singh (2008) stated truly that "millions of people on this earth live without love and affection, but no one can survive without water even for a few days." Biswas (1970) stated that the history of humanity can be written in terms of human interactions and interrelations with water. In the quest for life in other planets, scientists search for water, which signifies its importance. Water-related problems are linked to the quantity, quality and spatial and temporal distribution of water on earth. Too much water causes flooding, too little water is linked to droughts, and deteriorated water quality undermines the health of living beings and the ecosystem. Uneven distribution of water over space and time requires water to be reserved and transported. Providing adequate water that is of acceptable quality is essential for societal benefits (Rahman, 2016a).

Water has played a major role in shaping human civilizations. For example, old Egyptian civilizations were built along the Nile, and old Indian civilizations were built along the

Euphrates and Tigris (Rahman, 2016b). Water problems contributed to the collapse of some ancient civilizations as well; for example, the Sumerian empire collapsed due to unmanageable irrigation salinity (Ponting, 1991). In modern times, numerous water issues affect our society. Annin (2006) stated, "water will be the oil of the 21st century."

Water is closely linked to society. It holds a special place in most world religions; for example, in Christianity, water is used in baptism; in Islam, water is used to make ablution (washing of hands, face and legs) before each prayer; in Hinduism, a dip in the Ganges River is regarded as a religious act. In the Qur'an (Muslim Holy Book), it is stated, "We made from water every living thing."

Water disasters can affect society negatively. For example, floods killed about 3 million people in China in 1931. In India, 6 million people died during 1896–1902 due to a famine that was linked to water shortage. Due to climate change, water-related disasters such as floods have been increasing (Ishak *et al.*, 2013).

Hydrology is a subject that studies water for the benefit of the society and the ecosystem under disciplines such as civil engineering, environmental engineering, earth science and geology. Hydrology covers the variability of water quantity and quality in space and time on earth, encompassing its lithosphere, atmosphere and hydrosphere. To plan and operate a water engineering project such as water supply, irrigation and flood control works, one needs to quantify water at a given location in future time (within the life span of the project). The estimation of water quantity needs the use of mathematical equations such as the water balance equation and rational formula (Mulvaney, 1851), and estimation and forecasting models of varying complexity such a regional flood frequency analysis (Haddad and Rahman, 2012). Generally, engineering hydrology courses focus on how these mathematical equations can be applied by students to solve given water-related problems; however, the courses provide little emphasis on societal aspects of water (Rahman, 2016a), e.g. the positive and negative impacts of water resources development projects on society. For example, the building of a dam can affect the livelihood of upstream residents by reducing the area of available agricultural lands. A well-designed and -managed irrigation project can reduce poverty of the rural community by increasing food production and generating new employment opportunities; however, factors such as water logging and groundwater-level rising need to be managed properly.

It is important to appreciate the dynamics of water resources as key determinants of physical development, human and environmental health, and conflict and sustainability (Gleick and Palaniappan, 2010; Postel and Wolf, 2001). Water shortage across the globe encourages universities to teach hydrology, providing a broader perspective in solving water issues that affect the society. Sustainable management of water resources is important to solve global water scarcity and to meet the needs of the environment and human beings. Hence, the teaching of hydrology needs a critical overhaul in terms of course content and the way the subject is taught (Uhlenbrook and de Jong, 2012; Wagener *et al.*, 2012). Hydrology graduates should have critical thinking skills, a collaborative spirit, interdisciplinary communication skills, intellectual confidence and an ethical foundation to deal with complex water issues (Thompson *et al.*, 2012).

The need for an interdisciplinary approach to hydrology education was pointed out by Nash *et al.* (1990) and Eagleson *et al.* (1991), among others. As noted by Wagener *et al.* (2007), hydrology is becoming "more interdisciplinary" due to new linkages of this subject with other disciplines (e.g. forest study, environmental science, ecology and geoscience) and many recent advances in computational and technical areas. Bourget (2006) noted that there is a great need for watershed hydrology and modeling courses in university from a survey of integrated water resources management in the USA.

Hydrology educators are heavily influenced by their own backgrounds in developing hydrology courses, rather than the needs of the industry and society. To develop a contemporary hydrology course, new educational tools and resources should be used in an interactive fashion, using new educational approaches such as blended learning. This is a learner-centered approach of teaching where the best aspects of face-to-face and online teaching approaches are combined to enhance student learning. Bloeschl (2006) noted that advancements in hydrological science are likely to arise from the synthesis of different approaches, from "collision" of theory and data, and from better communication. Aghakouchak and Habib (2010) stressed the importance of improving existing engineering hydrology curricula by incorporating more interactive modeling tools. As noted by Wagener *et al.* (2007), the first step in developing a new interdisciplinary hydrology course should be based on wider consultation among educators and practicing hydrologists and should "begin with examining the current state of hydrology education, who is teaching, in what department/disciplines, and with what materials."

Traditionally, all engineering courses were offered face to face. This follows a teacher-centered approach where students generally listen to lectures, take notes and complete examinations to demonstrate their understanding of the subject matter. In this approach, teachers actively deliver teaching materials, taking on the role of a subject matter expert (Pathirana *et al.*, 2012). In contrast, a learner-centered approach allows a student a degree of autonomy in his/her learning by taking more responsibility. In this case, the teacher serves as a guide or facilitator similar to postgraduate supervision (Bransford *et al.*, 1999). A learner-centered approach is generally based on practical problems. This allows students to see the relevance and applicability of theories introduced in lectures and then to connect his/her prior knowledge to the given practical problems (Herrington and Oliver, 2000; Prince and Felder, 2006). A learner-centered approach may consist of problem-based learning, project-based learning, inquiry-based learning, case-based learning and discovery learning (de Graaff and Kolmos, 2003; Herrington and Oliver, 2000). A teacher-centered approach may be more effective in gaining factual knowledge and its application to a given range of applications. In contrast, a learner-centered approach enhances student engagement, motivation and transferability of knowledge from lectures to novel problems (Light, 2004). However, a learner-centered approach is generally time and resource intensive, and too much student autonomy may result in the development of misconceptions (Yadav *et al.*, 2011). The best outcome could be achieved when both the teacher-centered and learner-centered approaches are combined via a "blended learning" approach. In this way, student learning experience is enhanced without sacrificing the core components of a traditional education system (Prince and Felder, 2006; Rahman, 2017; Smith *et al.*, 2005). Blended learning is regarded as one of the most efficient learning strategies (Bonk and Graham, 2012). Hence, it is recommended that social hydrology teaching should be implemented via a blended learning approach.

This chapter examines the hydrology course contents of a number of Australian universities and proposes a hydrology course that can meet the demands of all stakeholders, including the society and ecosystem. At the beginning, an overview of hydrology courses in Australian universities is presented. The importance of incorporating social hydrology into the hydrology course is then highlighted. Finally, a new hydrology course is designed that incorporates social hydrology based on a blended learning approach.

13.2 Overview of hydrology course in Australian universities

In Australian universities, hydrology is generally offered in engineering degrees as a third-year subject. Fluid mechanics, which covers the basic principles of fluid flow, including fluid properties, energy and momentum principles, basic open channel flow, flow measurement and dimensional analysis, is a prerequisite subject for hydrology. As shown in Table 13.1, a typical hydrology subject generally contains the following topics: water balance, hydrograph analysis, unit hydrograph, runoff routing, flood routing, flood frequency analysis and design rainfall estimation. In some universities, groundwater is included as a small chapter. Social hydrology is not included generally in the hydrology subject of Australian universities. It has the prerequisite of fluid mechanics or water engineering, which covers the basic principles of fluid flow including fluid properties, energy and momentum principles, basic open channel flow, flow measurement and dimensional analysis.

13.3 Importance of incorporating social hydrology in hydrology course

Water underpins sustainability of the environment. A hydrologist must know the implication of a water resource development project on the environment, such as how the environment and the society in the dam catchment will be affected if a dam is built across a river for irrigation or hydroelectricity generation (e.g. how many people could be displaced, how much agricultural land will be permanently inundated, how the drainage of the land will be impacted during flooding, how the fish population will be affected in the catchment system and how siltation problems will reduce the dam capacity slowly). The changing of landscape for a new water resources engineering project may require land clearing, which can exert notable impacts on water-cycle dynamics at local to regional and at decadal to century time scales. A hydrologist/water engineer should appreciate the impacts of such water resources engineering projects on the society at a local, regional and global scale.

Some water engineering projects are very large and can affect the society and ecosystem on a significant scale. One example is the Snowy Mountains engineering scheme in Australia that consists of 16 major dams, 7 power stations and 225 km of tunnels, pipelines and aqueducts. This project was implemented between 1949 and 1974 and is considered to be the largest engineering project ever undertaken in Australia. This project is regarded as an Australian identity, signifying its resources and potential as a powerful country. The project generated significant employment and benefited the Australian society after the Second World War. A hydrologist should appreciate the impacts of such a huge project on society and how its failure can affect many people in various ways. In this regard, social hydrology is receiving attention from a new generation of hydrology researchers and academics.

In the above context, Sivapalan *et al.* (2014) noted that social hydrology considers the impacts of decentralized human agents and institutions to water flows and storages, as well as their feedbacks. Zlinszky and Timar (2013) stated that social hydrology "is the science of human influence on hydrology and the influence of the water cycle on human social systems." Social hydrology deals with understanding the dynamics of coupled human-water systems over large spatial and temporal scales (Srinivasan, 2015). The above definitions of social hydrology indicate that it focuses on water over a large scale in space and time, and without

Table 13.1 Hydrology course outline in some Australian universities

University	*Course name*	*Course content*
University of New South Wales	CVEN9610: Surface Water Hydrology	Hydrologic cycle; climate and weather; meteorological and hydrologic measurement; evaporation and evapotranspiration; rainfall/runoff processes including loss models and hydrograph analysis; design rainfall; data and flood frequency analysis; flood estimation including regional methods and estimation of extremes.
University of Sydney	LWSC2002: Introductory Hydrology	This unit introduces students to hydrology and water management in the context of Australian integrated catchment management. It particularly focuses on the water balances, rainfall-runoff modeling, analysis and prediction of streamflow and environmental flows, water quality and sustainable practices in water management.
Monash University	CIV5882: Flood Hydraulics and Hydrology	This unit focuses on flood modeling for engineering design. Methods to estimate design flood magnitudes from experimental observations will be presented. Hydrologic and hydraulic routing models will be introduced along with software packages that apply these models.
Western Sydney University	300766: Surface Water Hydrology	The unit covers the principles of surface water hydrology. It will focus on catchment analysis, specifically focusing on rainfall-runoff relationships. Successful completion of this unit will enable hydrologic analysis of catchments to satisfy various regulatory requirements.
University of Technology Sydney	48362: Hydraulics and Hydrology	The objective of this subject is to give students knowledge of open channel hydraulics and hydrology, leading to understanding of the scientific foundations and basic principles of these fields, and the ability to apply hydraulic and hydrological methods to engineering applications in an integrated way covering the hydrological cycle, water balances, meteorology and climatology, data collection, statistics, hydrological models, design rainfalls, rainfall-runoff processes, flood estimation models and procedures, software packages, yield analysis, groundwater, environmental hydrology and integration of hydraulics and hydrology case studies.
Griffith University	2004 Engineering Hydrology	The course covers fundamental hydrologic processes such as rainfall, evaporation, infiltration, surface and groundwater, and hydrologic extremes, i.e. floods and droughts, flood estimation, flood frequency analysis, and flood routing.

a broad understanding of social hydrology concepts, hydrologists and water engineers may not be able to act appropriately to preserve and enhance water resources for a balanced and sustainable development for the present and future generations. Hence, the absence of social hydrology in hydrology courses is regarded as a significant weakness of the current hydrology course curriculum in Australian universities.

There are many examples where a poor understanding of the interaction of water with the environment and society has led to severe consequences. An example is the polder project in the coastal areas of Bangladesh, which attempted to build earthen embankments to create new lands for agricultural developments during the 1960s and 1970s (Rahman, 2016b). After approximately 20 years of construction, many of these polders failed and caused a serious water-logging problem. One of these polders is Beel Dakatia, located in southwestern Bangladesh. The polder has become an inland sea due to heavy siltation in the outfall river system that prevents natural drainage. This has caused a serious problem in the socioeconomic condition of the region by causing the collapse of agriculture, livestock and social infrastructure.

Furthermore, about 77 million people in Bangladesh have been reported to be at risk of arsenic poisoning due to contamination of drinking water caused by unplanned withdrawal of shallow groundwater. This has caused significant health and social problems in Bangladesh as highlighted by Rahman *et al.* (2018).

Sivapalan *et al.* (2014) noted examples of "wicked" water resources engineering projects causing great concern, such as trade-offs among ecosystems, hydropower and livelihoods in the transnational Mekong Basin (Ziv *et al.*, 2012), increased dry land salinity in Australia, expansion of hypoxic zones in the Gulf of Mexico due to higher nutrient loading in the agricultural lands of the Mississippi River (Turner and Rabalais, 2003) and "bad politics" around water-sharing issues for numerous national and international rivers (Gleick, 1993). Incorporation of the social hydrology component in the hydrology course will enable students to comprehend the possible negative impacts of these types of water engineering projects on the environment, ecosystem and society.

13.4 A proposed outline of hydrology syllabus

In this chapter, it is argued that hydrology courses in Australia should not focus merely on mathematical aspects of hydrology; rather, it should be designed as an interdisciplinary subject. Hydrology is linked with numerous disciplines and hence it has multiple dimensions of understanding (Vogel *et al.*, 2015; Wagener *et al.*, 2010). Hydrology syllabuses should contain a chapter on social hydrology to make students aware of the broader implications of water engineering projects on society at large, and how the spatial and temporal distribution of water quantity and quality can affect the ecosystem and human beings across local, regional, national and international domains. According to Sivapalan *et al.* (2012), social hydrology should include three fundamental sub-areas: (i) historical social hydrology (to know the history of hydrology, success and failure of past hydrology schemes); (ii) comparative social hydrology (comparative analysis of human-water interactions across socioeconomic and climatic gradients to map spatial and regional differences back to processes and their temporal dynamics) and (3) process social hydrology (focusing on a small number of human-water systems in greater depth, including data collection and modeling covering hydrological and social processes). Any new hydrology course that includes social hydrology should consider the above aspects as outlined in Table 13.2.

Table 13.2 Outline of a proposed hydrology course in Australian universities

Week	*Title*	*Topics*
1	Hydrologic cycle	Interaction among various components of the hydrologic cycle, water quantity and quality aspects, water balance modeling and examples of hydrologic projects such as Warragamba reservoir and Snowy Mountains engineering scheme. Introduction to *Australian Rainfall and Runoff.*
2	Rainfall and runoff measurement	Measurement of rainfall and streamflow data, relative accuracy of measured data, rating curve construction and associate uncertainty, and spatial and temporal measurements of rainfall.
3	Design rainfall estimation	Design rainfall estimation, areal rainfall and temporal patterns of rainfall. Use *Australian Rainfall and Runoff* to obtain design rainfalls for any given location of Australia.
4	Runoff estimation	Empirical methods (rational method and regional flood frequency analysis: RFFE Model) and loss modeling.
5	Unit hydrograph	Unit hydrograph calculation.
6	Runoff and flood routing	Flood routing and demonstration of RORB model.
7	Urban hydrology	Impacts of urbanization on water cycle. Demonstration of DRAINS model.
8	Flood frequency analysis 1	Probability concept, normal distribution, log normal distribution and LP3 distribution.
9	Flood frequency analysis 2	L moments and GEV distribution, demonstration of R program, and demonstration of FLIKE software.
10	Groundwater hydrology	Aquifers, basic groundwater flow, well hydraulics and interaction of surface and groundwater.
11	Social hydrology	History of hydrology, case studies on success and failure of large water engineering projects and the interaction of water and society through examples.
12	Impacts of climate change on water	Trend analysis and non-stationary frequency analysis.
13	Future of hydrology	Remote sensing and integrated hydrologic modeling. Climate change impact on water and society.

Table 13.1 shows that hydrology courses in Australian universities are heavily focused on mathematical aspects of the subject. A new hydrology course is proposed in Table 13.2 that focuses on the interdisciplinary nature of hydrology plus the demonstration of industry-based models and *Australian Rainfall and Runoff* (national guideline) (Ball *et al.*, 2016). It is proposed to initiate a forum of hydrology academics in Australia to prepare a hydrology course that can meet the needs of the water industry and research sectors. A blended learning approach can be adopted to develop the course materials using internet-based resources including video clips, recorded lectures, recorded tutorials and online discussion. Shareable resources can be developed by a collaborative approach among departments like civil engineering, earth sciences and geography across various universities.

Table 13.3 illustrates how the social hydrology component can be taught to students using a blended learning approach. It is assumed that social hydrology will be taught in a week, consisting of 3 hours of face-to-face sessions and 7 hours self-reading and practice by a student. Various tasks are proposed in Table 13.3, which can be developed and posted in the university learning system (such as vUWS in Western Sydney University) for student access. A learner-centered approach like blended learning will encourage a student to take some degree of autonomy in learning the social hydrology element in the hydrology course.

Table 13.3 Components of blended learning to teach social hydrology over a week in the semester

Self-reading/face-to-face in-class time	*Activities*	*Comments*
Self-reading by student:		
2 hours	Listen to pre-recorded lecture on social hydrology	
1 hour	Watch six VDO clips (10 minutes each)	Impacts of floods on economy and society, e.g. on poor, retired and vulnerable people. Impacts of droughts on Australian farming community, e.g. loss of income, family breakdown and dislocation. Impacts of dry land salinity on Australian farming communities. Impacts of arsenic pollution of the groundwater in Bangladesh. Rituals surrounding water in major world religions. Deaths caused by major floods and famines.
1 hour	Read two case studies (30 min each)	Water stress (highlighting issues such as 1.2 billion people lack access to adequate supply of safe water, half of the world's rivers and lakes are seriously polluted, 3900 children die every day due to dirty water and poor hygiene) and water sharing for international rivers.
3 hours	Write a reflective essay on social hydrology	Each student writes an essay (1000 words) on the importance of studying social hydrology.

Self-reading/face-to-face in-class time	*Activities*	*Comments*
Face-to-face session:		
1 hour	3-minute presentations by 15 randomly selected students from the class on randomly selected topic on social hydrology	Possible topics include water resources development, water resources management, integrated water resources management, climate change, environmental flows, room for river, virtual water (Allan, 2003), green water and blue water, water footprint, water sensitive urban design, atmospheric teleconnection.
1 hour	3-minute presentations by 15 randomly selected students from the class on their personal experience with water problems	Possible topics include visiting developing nations where water supply is intermittent, water restriction, drought affecting crops and livestock, salinity-affected region, polluted river, floods, polluted fish, waterborne disease.
30 minutes	Three student groups organized to debate the role of water to build civilization	History of hydrology, ancient civilization built around water, desertification, hydroelectricity, irrigation, Hurricane Katrina, flood insurance, 2010–11 Queensland flood.
30 minutes	Lecturer to share experience on social hydrology	Implication of failed water projects on society (e.g. arsenic pollution of drinking water in Bangladesh, negative impacts of dam, impacts of floods on vulnerable people of society).

13.5 Conclusion

This chapter presents an overview of social hydrology and the importance of incorporating social hydrology components in a hydrology syllabus. It has been found that most of the hydrology courses in Australia are biased towards mathematical aspects of the subject. There is little focus on the big picture and the interdisciplinary nature of hydrology connecting the broader society. In most Australian universities, social hydrology is absent in the hydrology syllabus. A new hydrology course is proposed in this chapter, which can overcome some of the limitations of the current hydrology courses in Australia. The new course focuses on mathematical aspects, demonstration of software and the interdisciplinary nature of hydrology, including social aspects of hydrology. It is proposed to initiate a forum of hydrology academics in Australia to prepare a hydrology course that can meet the need of the water industry and research sectors, as well as the society as a whole. A blended learning approach can be adopted to develop learning materials and to deliver the social hydrology component of the hydrology subject.

References

Aghakouchak, A. & Habib, E. (2010) Application of a conceptual hydrologic model in teaching hydrologic processes. *International Journal of Engineering Education*, 26, 4, 963-973.

Allan, J.A. (2003) Virtual water-the water, food, trade nexus, useful concept or misleading metaphor? *Water International*, 28(1), 4–10.

Annin, P. (2006) *The Great Lakes Water Wars*. Island Press, Washington, DC, USA.

Ball, J., Babister, M., Nathan, R., Weeks, W., Weinmann, E., Retallick, M. & Testoni, I. (2016) *Australian Rainfall and Runoff: A Guide to Flood Estimation*. Commonwealth of Australia, Canberra.

Biswas, A.K. (1970) *History of Hydrology*,. North-Holland Publishing Company, Amsterdam, The Netherlands.

Bloeschl, G. (2006) Hydrologic synthesis: Across processes, places, and scales. *Water Resources Research* 42: W03S02. doi:10·1029/2005WR-004319.

Bonk, C.J. & Graham, C.R. (2012) *The Handbook of Blended Learning: Global Perspectives, Local Designs*. John Wiley & Sons, New York City, USA.

Bourget, P.G. (2006) Integrated water resources management curriculum in the United States: Results of a recent survey. *Journal of Contemporary Water Research and Education*, 135, 107–114.

Bransford, J.D., Brown, A. & Cocking, R. (1999) *How People Learn: Mind, Brain, Experience, and School*. National Research Council, Washington, DC, USA.

de Graaf, E. & Kolmos, A. (2003) Characteristics of problem-based learning. *International Journal of Engineering Education*, 19(5), 657–662.

Eagleson, P.S., Brutsaert, W.H., Colbeck, S.C., Cummins, K.W., Dozier, J., Dunne, T., Edmond, J.M., Gupta, V.K., Jacoby, G.C., Manabe, S., Nicholson, S.E., Nielsen, D.R., Rodriguez-Iturbe, I., Rubin, J., Smith, J.L., Sposito, G., Swank, W.T. & Zipser, E.J. (1991) *Opportunities in the Hydrologic Sciences*. National Academy Press, Washington, DC, USA.

Gleick, P.H. (1993) Water and conflict: Fresh water resources and international security. *International Security*, 18(1), 79–112.

Gleick, P.H. & Palaniappan, M. (2010) Peak water limits to freshwater withdrawal and use. *Proceedings of the National Academy of Sciences*, 107(25), 11155–11162.

Haddad, K. & Rahman, A. (2012) Regional flood frequency analysis in eastern Australia: Bayesian GLS regression-based methods within fixed region and ROI framework – Quantile regression vs. Parameter regression technique. *Journal of Hydrology*, 430-431(2012), 142–161.

Herrington, J. & Oliver, R. (2000) An instructional design framework for authentic learning environments. *Educational Technology Research and Development*, 48(3), 23–48.

Ishak, E., Rahman, A., Westra, S., Sharma, A. & Kuczera, G. (2013) Evaluating the non-stationarity of Australian annual maximum floods, *Journal of Hydrology*, 494, 134–145.

Light, R.J. (2004) *Making the Most of College*. Harvard University Press, Boston, USA.

Mulvaney, T.J. (1851) On the use of self-registering rain and flood gauges in making observations of rainfall and flood discharges in a given catchment. *Transactions of the Institution of Civil Engineers of Ireland*, IV(II), 19–33.

Nash, J.E., Eagleson, P.S., Philip, J.R. & Van der Molen, W.H. (1990) The education of hydrologists. *Hydrological Sciences Journal*, 35(6), 597–607.

Pathirana, A., Koster, J.H., Jong, E.D. & Uhlenbrook, S. (2012) On teaching styles of water educators and the impact of didactic training. *Hydrology and Earth System Sciences*, 16(10), 3677–3688.

Ponting, C. (1991) *A Green History of the World: The Environment and the Collapse of Great Civilisations*. Sinclair-Stevenson, London, UK.

Postel, S.L. & Wolf, A.T. (2001) Dehydrating conflict. *Foreign Policy Magazine*, (126), 60–67.

Prince, M.J. & Felder, R.M. (2006) Inductive teaching and learning methods: Definitions, comparisons, and research bases. *Journal of Engineering Education*, 95(2), 123–138.

Rahman, A. (2016a) Social hydrology. In: Singh, V.P. (ed) *Handbook of Applied Hydrology*. McGraw-Hill, London, UK.

Rahman, A. (2016b) Updating hydrology course in Australian Universities: Incorporation of social hydrology using a blended learning approach. *Proceedings of the International Conference on Engineering Education and Research*, 21-24 Nov 2016, Sydney, Australia.

Rahman, A. (2017) A blended learning approach to teach fluid mechanics in engineering. *European Journal of Engineering Education*, 42, 3, 252–259.

Rahman, M.A., Rahman, A., Khan, M.Z. & Renzaho, A.M.N. (2018) Human health risks and socio-economic perspectives of arsenic exposure in Bangladesh: A scoping review. *Ecotoxicology and Environmental Safety*, 150, 335–343.

Singh, V.P. (2008) Water, environment, engineering, religion, and society. *Journal of Hydrologic Engineering*, ACSE, March 2008, 118–123.

Sivapalan, M., Savenije, H.H.G. & Bloeschl, G. (2012) Sociohydrology: A new science of people and water. *Hydrological Processes*, 26, 1270–1276.

Sivapalan, M., Konar, M., Srinivasan, V., Chhatre, A., Wutich, A., Scott, C.A. & Wescoat, J.L. (2014) Socio-hydrology: Use-inspired water sustainability science for the Anthropocene. *Earth's Future*, 2, 225–230.

Smith, K.A., Sheppard, S.D., Johnson, D.W. & Johnson, R.T. (2005) Pedagogies of engagement: Classroom-based practices. *Journal of Engineering Education*, 94(1), 87–101.

Srinivasan, V. (2015) Reimagining the past – use of counterfactual trajectories in socio-hydrological modelling: The case of Chennai, India. *Hydrology and Earth System Sciences*, 19, 785–801.

Thompson, S.E., Ngambeki, I., Troch, P.A., Sivapalan, M. & Evangelou, D. (2012) Incorporating student-centered approaches into catchment hydrology teaching: A review and synthesis. *Hydrology and Earth System Sciences*, 16(9), 3263–3278.

Turner, R.E. & Rabalais, N.N. (2003) Linking landscape and water quality in the Mississippi River Basin for 200 years. *BioScience*, 53(6), 563–572.

Uhlenbrook, S. & de Jong, E. (2012) T-shaped competency profile for water professionals of the future. *Hydrology and Earth System Sciences*, 16(10), 3475–3483.

Vogel, R.M., Lall, U., Cai, X., Rajagopalan, B., Weiskel, P.K., Hooper, R.P. & Matalas, N.C. (2015) Hydrology: The interdisciplinary science of water. *Water Resources Research*, 51(6), 4409–4430.

Wagener, T., Weiler, M., McGlynn, B., Gooseff, M., Meixner, T., Marshall, L., McGuire, K. & McHale, M. (2007) Taking the pulse of hydrology education. *Hydrological Processes*, 21, 1789–1792.

Wagener, T., Sivapalan, M., Troch, P.A., McGlynn, B.L., Harman, C.J., Gupta, H.V., Kumar, P., Rao, P.S.C., Basu, N.B. & Wilson, J.S. (2010) The future of hydrology: An evolving science for a changing world. *Water Resources Research*, 46(5).

Wagener, T., Kelleher, C., Weiler, M., McGlynn, B., Gooseff, M., Marshall, L., Meixner, T., McGuire, K., Gregg, S., Sharma, P. & Zappe, S. (2012) It takes a community to raise a hydrologist: The Modular Curriculum for Hydrologic Advancement (MOCHA). *Hydrology and Earth System Sciences*, 9, 2321–2356.

Yadav, A., Subedi, D., Lundeberg, M.A. & Bunting, C.F. (2011) Problem-based Learning: Influence on students' learning in an electrical engineering course. *Journal of Engineering Education*, 100(2), 253–280.

Ziv, G., Baran, E., Nam, S., Rodríguez-Iturbe, I. & Levin, S.A. (2012) Trading-off fish biodiversity, food security, and hydropower in the Mekong River Basin. *Proceedings of the National Academy of Sciences of the United States of America*, 109(15), 5609–5614.

Zlinszky, A. & Timar, G. (2013) Historic maps as a data source for Socio-hydrology: A case study of the Lake Balaton wetland system, Hungary. *Hydrology and Earth System Sciences*, 17, 4589–4606.

Chapter 14

Teaching science to engineering students

Application of student-centered and blended learning approaches

S. Rahman[1], R. Bhathal[2] and A. Rahman[3]

[1]School of Education, University of New South Wales, Sydney, Australia

[2]School of Computing, Engineering and Mathematics, Western Sydney University, Sydney, Australia

[3]School of Computing, Engineering and Mathematics, Western Sydney University, New South Wales, Australia

ABSTRACT: The study of engineering is highly dependent on mathematical ability and the understanding of scientific principles, since engineers apply mathematics and science to solve physical problems. In Australia, science, technology, engineering and mathematics (STEM) education has been reported to be declining for several years. This chapter shows that all the engineering courses at the University of New South Wales (UNSW) (having one of the largest engineering programs in Australia) have a significant science component, which means that high school leavers intending to undertake engineering studies at UNSW must have a sound background in science. The teaching experiences of two academics in engineering physics and fluid mechanics at Western Sydney University are also presented in this chapter. They note that poor mathematics and science backgrounds are a major obstacle in completing engineering physics and fluid mechanics successfully by many undergraduate engineering students in Western Sydney University. A few possible solutions to this problem are presented, *viz.* the use of student-centered and blended learning approaches, which include peer learning, development of interactive online tools, short courses, recorded and face-to-face lectures, online practice quizzes, face-to-face and recorded tutorials, case studies and remote labs.

14.1 Introduction

According to the Oxford dictionary, engineering is defined as "The branch of science and technology concerned with the design, building, and use of engines, machines, and structures." With this definition, the importance of engineering is apparent in virtually every built aspect of our physical environment, such as houses, roads, railways, irrigation systems, pipelines to transport oil, gas and water, cars, airplanes and ships. Engineering is traditionally divided into four branches: civil, mechanical, electrical and chemical. Regardless of which type of engineering a student specializes in, the duration of most Australian undergraduate engineering courses is four years. As a general overview, within these four years, the first year is spent mainly studying foundation subjects such as science, mathematics, introductory engineering and professional practice (Rahman *et al.*, 2016).

Engineers implement projects to preserve and enhance environmental and natural systems that support all living beings on this planet. They are concerned with the application of scientific laws and principles to build a better world. Therefore, it logically follows that engineering students need a strong grasp of science and mathematics to be able to understand their relevance and application in real-world problems, such as how to reduce the damaging force of flowing water to minimize river bank erosion (needs application of energy and momentum theories) and how to utilize gravitational force as much as possible to manage the water flow in an irrigation system with little/no use of mechanical pumps to save energy.

Students entering into engineering courses must have a high level of competence in high-school-level mathematics and science. However, statistics show that Australia is facing an emerging crisis in the standards of science and mathematics education in its schools, e.g. approximately 40% of mathematic classes in years 7–10 are taught by a teachers unqualified in mathematics, and the number of students choosing "science" subjects in schools are the lowest they have been in 20 years, with physics being studied by only 14% of students in 2010 (Chief Scientist, 2014a). Compared to 1992, there were 30,800 more students in year 12, but 8000 fewer students in physics and 4000 fewer in chemistry (Chief Scientist, 2014b). Physics has been reported to be the least common subject in Australian high schools (Chief Scientist, 2012). From tertiary education data in Australia, it has been found that only 16% of university students graduate with a science, technology, engineering or mathematics (STEM) degree. Although there has been an increase in undergraduate science enrollments in Australian universities, a significant number of students discontinue with science subjects after their first year (Chief Scientist, 2014c). This trend is also true for engineering courses. The dropping standards of STEM occurring in primary and secondary schools in Australia directly flows into the quality of students entering into engineering courses. McDowell (1995) mentions the importance of a solid understanding of foundational knowledge in science and mathematics in order to achieve success in engineering courses. Therefore, it is imperative to acknowledge and address the impact of poor science and mathematics education on engineering and implement strategies to effectively deal with the issue.

Educators, policy makers and business organizations are stressing the urgency for improving STEM skills to meet the future need of STEM-related jobs (Caprile *et al.*, 2015). STEM competencies have been drawing increasing global attention to meet the increasing demand not only within, but also beyond, specific STEM occupations (e.g. Commonwealth of Australia, 2015; European Parliament, 2015; National Science and Technology Council, 2013). To promote innovation, productivity and overall economic growth, STEM education is deemed to be of high importance globally due to perceived or real shortages in the current and future STEM workforce (e.g. Caprile *et al.*, 2015; The Royal Society Science Policy Centre, 2014). Assessments made by the Organisation for Economic Co-operation and Development (OECD) (OECD, 2016) have escalated the STEM issue. Views on how STEM education can be enhanced vary across school contexts, curricula and political spheres (English, 2017).

In an attempt to restore the decline in standards of STEM subjects in Australia and to prepare for its economic future, the Australian government has invested $17 million for STEM. Commencing in primary school, the STEM initiative aims to produce students with a solid understanding in STEM and its application in the real world and to ignite a curiosity in students to continue further tertiary study in STEM-related fields. Commonwealth Scientific and Industrial Research Organisation (CSIRO) in Australia has also taken the initiative to enhance science and mathematics education in Australian high schools with the aim of bringing real science, mathematics and information and communication technology (ICT) into the classroom through ongoing flexible partnerships between teachers (K–12) and scientists, mathematicians and ICT professionals.

In a similar manner to Australia, science education has declined in the USA. Freeman *et al.* (2014) noted that a 33% increase in the number of STEM bachelor's degrees completed per year is needed in the USA to cope with the forecasted demand. Drew (2015) stated that American students rank behind their international counterparts in the field of STEM education. In this regard, Bill Gates warned that this undesirable situation places the US at a serious disadvantage in the high-tech global marketplace of the 21st century (Drew, 2015). English (2016) noted that in the USA, the jobs of the future are STEM jobs; however, STEM education has been declining.

This chapter provides an overview of the problems associated with STEM education generally, identifies aspects of science and mathematics education in engineering studies, and suggests a few strategies to overcome these problems in the near future, which could assist in raising the standards of engineering education in Australia. The relationship between science and engineering is presented in Section 14.2. The science subjects in engineering courses are highlighted in Section 14.3. Problems faced by engineering students in learning science are discussed in Section 14.4. Reflection by the second and third authors of this chapter in teaching engineering subjects in Western Sydney University is presented in Section 14.5. Finally, possible solutions to enhance the learning of science by engineering students is discussed in Section 14.6.

14.2 Relationship between science and engineering

Science derives its root from the Latin word *scientia* meaning "knowledge; a knowing; expertise." Science focuses on studying the mechanisms and behaviors of the world with the aim of understanding and explaining natural phenomenon. Dealing almost entirely on collecting knowledge and proposing theories, science greatly differs from engineering in this regard, as engineers concentrate on the application of established and proven theories into the physical world. Dunn (1930) describes engineering to be "the art of the economic application of science to the purposes of man." However, science and engineering share an undeniable connection, as the basis of engineering developments are reliant on the knowledge of science. For example, an engineer required to develop a system to pump water from a river to an irrigation channel in the most energy-efficient manner utilizes scientific knowledge of atmospheric pressure, viscosity of water, the law of gravity and frictional losses. In solving this real-world problem, an engineer makes many simplified assumptions to achieve a solution that could be far from "perfect" in the view of a scientist.

14.3 Science subjects in the engineering curriculum

Every engineering degree includes science within its curriculum. Winkelman (2009) mentions how the first two years of contemporary US engineering degrees focus on teaching science before introducing engineering concepts, therefore highlighting the fundamental nature of science in engineering. The University of New South Wales (UNSW) has one of the largest engineering programs in Australia. UNSW offers 20 branches of engineering for their undergraduate bachelor's courses: aerospace, bioinformatics, chemical, civil, computer, electrical, environmental, geospatial, industrial, mechanical, mechanical and manufacturing, mechatronic, mining, naval, petroleum, photovoltaics, renewable energy, software, surveying and telecommunications. It is found that all of these engineering courses in UNSW have at least one compulsory mathematics course and, with the exception of software engineering, each of the engineering courses also requires a student to study a first-year physics course. In addition, bioinformatics, chemical, civil, environmental, geospatial and industrial engineering require students to study at least a first-year chemistry course. Bioinformatics is the only engineering specialization that necessitates the study of biology courses. Figure 14.1 shows that physics is the most commonly studied science subject in the engineering degrees at UNSW. These core science subjects are built upon higher-level courses, such as advanced thermofluids, which is a third-year subject that covers concepts such as heat transfer and exchange mechanism, steady-state and multi-dimensional conduction, structure of boundary layers, internal and external laminar and turbulent forced convection, chemical kinetics and emission control (UNSW, 2016). The scientific principles in these courses are applied in the real world by engineers in a myriad of ways. Table 14.1 lists a few of the engineering concepts that are dependent on scientific knowledge in the four major disciplines of engineering: civil, electrical, mechanical and chemical.

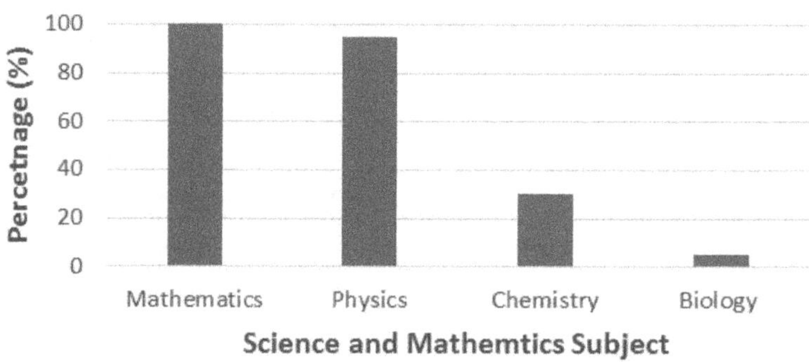

Figure 14.1 Percentage of individual mathematics/science subjects included in 20 different types of engineering degrees at University of New South Wales, Australia

Table 14.1 Examples of engineering topics founded directly on scientific principles

Branch of Engineering	*Science Subject*	*Application in Engineering courses*
Civil	Physics	Needed in fluid mechanics, hydraulics, statics, dynamics, mechanics of materials and structural analysis.
	Chemistry	Needed in environmental engineering (for example, treatment of waste water, drinking water, solid waste and storm water, remediation of contaminated soils).
	Geology	Needed for locating appropriate foundations in geotechnical engineering, waste disposal, landfill site selection, nuclear waste disposal, selection of materials for construction and examining weathering effects.

Branch of Engineering	*Science Subject*	*Application in Engineering courses*
	Material Science	Needed to know strength of materials, durability, creep, elasticity, plasticity, liquidity and viscosity.
Electrical	Physics	Needed to understand resistivity, conductivity and magnetism.
	Chemistry	Needed in corrosion control of electrical equipment and understanding of super conductors.
Mechanical	Physics	Needed in solid dynamics, fluid dynamics, heat transfer, gravity, friction, aerodynamics, noise, momentum and conservation of mass.
	Chemistry	Needed to control corrosion of moving parts, understand lubrication and fluid behavior under high temperatures.
	Material Science	Needed to know strength of materials, durability, creep, elasticity, plasticity, liquidity, reactivity and viscosity.
Chemical	Chemistry	Properties of solid, liquid and gases, reaction kinematics and mass balance.

14.4 Problems faced by engineering students in learning science

In contrast to Asian countries such as South Korea, Japan and Singapore, many engineering students in Australia are weaker in mathematics and hence they struggle to follow the concepts in physics in their first year of engineering study. In fact, physics courses at Australian universities have one of the highest failing rates. Low mastery of mathematics also plays a significant role in the inability of a student to excel in many engineering subjects such as fluid mechanics and also in applying mathematical models to solve real-world engineering problems (Mckagan *et al.*, 2007). Environmental engineering students generally struggle to learn environmental chemistry, which is needed in many real-life applications, such as the treatment of water and waste. As many engineering students in Australia do not take chemistry in high school, students have a mediocre or insufficient understanding of fundamental chemical principles needed to understand environmental chemistry.

Many engineering students find fluid mechanics to be a challenging subject because their inadequate understanding of the laws of physics, such as the principle of buoyancy, pressure, energy, momentum and boundary layer concepts. Some engineering disciplines have to use stochastic/probabilistic theories, for example the Bayesian theory (e.g. Haddad and Rahman, 2012), Monte Carlo simulation (e.g. Rahman *et al.*, 2002), bootstrapping (Mamoon and Rahman, 2017), joint probability approach (Caballero and Rahman, 2014; Loveridge and Rahman, 2014), generalized additive models (Rahman *et al.*, 2018) and artificial neural networks (Aziz *et al.*, 2015). However, for many engineering students these appear to be difficult topics, for example, in statistical hydrology (a civil engineering subject), students often struggle to utilize these principles in an attempt to solve water-related problems such as the forecasting of rainfall, wind, flood and drought.

14.5 Reflections of second and third authors in teaching science

The second author of this chapter (Ragbir Bhathal) coordinates and teaches engineering physics to over 450 first-year students at Western Sydney University (WSU). This is a compulsory foundational subject for all engineering students enrolled in the Bachelor of Engineering degree at WSU. Over the last few years, students enrolling into universities in Australia have been coming from a diverse range of educational backgrounds, mathematical and technical skills, prior knowledge and motivations. This is further compounded by the Australian government's policy to increase the participation rates of Higher School Certificate (HSC) students including those from low socioeconomic status (SES) (15.7%) backgrounds (Australian Government, 2016; Commonwealth of Australia, 2009). While this is an admirable initiative, it has created problems for universities, especially engineering departments, which rely on students having good backgrounds in mathematics and physics. In order to accommodate this change in the knowledge base of the student cohort, the author has adopted a multi-strategy approach to the teaching and learning of engineering physics. This includes an online Socratic tutorial system (Mastering Physics) in addition to face-to-face tutorials (Young *et al.*, 2011), using a just-in-time session (10–15 minutes) which revises the relevant mathematics before teaching the engineering topic for the day, an active lecturing system with question and answer and demonstrations, contextual physics, which shows the relevance of physics in an engineering context, and hands-on laboratories, which have a direct correlation to the theory taught in lectures (Bhathal, 2011; Bhathal *et al.*, 2010). Adopting the above multi-dimensional approach over 10 years has led to a decrease in the failure rate of students studying engineering physics at WSU from 32% in 2002 to 11% in 2014.

The third author of this chapter (Ataur Rahman) coordinates and teaches fluid mechanics at WSU, which is a second-year subject (having enrollments of over 200 students) in civil and mechanical engineering degrees. Fluid mechanics is strongly linked with physics. It had been a subject with a high failure rate (over 20%) for many years at WSU (Rahman and Al-Amin, 2014). He found that many students studying fluid mechanics had basic problems with fundamental concepts in mathematics and science, including difficulties in performing simple differentiations and integrations relevant to fluid movement, conversion of units, understanding momentum and energy equations and deriving basic equations to characterize simple fluid flows. Application of a blended learning technique in recent years involving recorded lectures, use of YouTube materials, online access of voice-recorded tutorial solutions and practice quizzes has reduced the failing rate from over 20% to around 10% (Rahman, 2017). Poor mathematical ability has been found to be the main problem in mastering the advanced concepts of fluid mechanics. This author also teaches statistical hydrology to third-year civil engineering students at WSU. In this subject, he found that a large number of students struggle in grasping the concepts of probability theory, probability distributions and stochastic simulation, which is linked to the weaknesses in mathematics and science by the enrolled students.

WSU College has been playing an important role in improving the science and mathematics skills for the students who do not meet the WSU admission criteria directly. These students take foundation courses including mathematics and science to improve their background knowledge before entering into the WSU engineering degrees. This program has been popular at WSU.

14.6 Application of student-centered and blended learning approaches

The issue of science education in Australia is a nation-wide problem that requires institutional-level intervention to steer it forward. The government has already taken action with the implementation of STEM, but the following suggestions could be of further benefit when applied to the university context, with specific focus on engineering students:

1. Introducing pre-university short courses on science before the commencement of formal first-year engineering classes. For example, a 20-hour course on physics covering the fundamentals of physics in a simplified manner would be notably useful to first-year engineering students. This would be especially helpful for students who have not completed physics in HSC. Although engineering students who did not study physics or chemistry in HSC are required to complete bridging courses in these subjects before the starting of their degrees, these bridging courses are often not emphasized enough or they are inefficient in conveying the necessary information.
2. Forming peer groups consisting of four/five students, two of whom have completed physics in HSC. Under the supervision of an experienced tutor, these students could be given assignments on the fundamental principles of physics that are needed in engineering subjects. Active learning that occurs in group work through discussions and questions are widely accepted to enhance learning. In addition, peer groups allow students to more comfortably seek help in problems that they would otherwise be reluctant to go to their teachers for (McDowell, 1995).
3. Adding a science course in the summer break between first and second year for students who achieve below-credit level. This course would allow these students to consolidate their science knowledge by reviewing the knowledge learned throughout the year.
4. Offering online tools and compulsory online classes to learn the principles of science subjects under the supervision of experienced academics.
5. Extra online interactive courses during being offered with science subjects, under the supervision of trained tutors.
6. Adoption of student-centered learning approaches in teaching science can increase the learning experience of students. The traditional teacher-centered science teaching methods include lecturing, reading and structured problem solving. In these approaches, teachers take the central role in delivering learning materials (Pathirana *et al.*, 2012) while students take notes, read and apply the theory to solve numerical problems. In contrast, student-centered approaches present learners a degree of autonomy in achieving the learning objectives, and learners take a greater responsibility in learning course materials (Bransford *et al.*, 1999). Here, teachers take the role of a guide or facilitator. Typical student-centered strategies include problem-based learning, project-based learning, inquiry-based learning, case-based learning and discovery learning (de Graaf and Kolmos, 2003; Herrington and Oliver, 2000). Freeman *et al.* (2014) noted that an application of student-centered learning of science in the USA has increased the examination performance significantly. A teacher-centered approach can be more useful in achieving factual knowledge for a given range of applications. In contrast, a student-centered approach increases student engagement, motivation and transferability of knowledge from teachers to novel problems (Light, 2004). However, a student-centered approach demands more time and resources to develop course materials. Also, excessive student

autonomy may result in the development of misconceptions (Yadav *et al.*, 2011). The best learning outcome can be gained when the teacher-centered and student-centered approaches are combined together in a "blended learning" approach, where students' learning experience is enhanced without sacrificing the core components of a traditional education system (Prince and Felder, 2006; Rahman, 2017; Smith *et al.*, 2005). Blended learning may be regarded as one of the most efficient learning strategies (Bonk and Graham, 2012). Hence, it is recommended that science teaching for engineering students should be implemented via a blended learning approach. As an example, Table 14.2 presents how a blended learning approach can be implemented to teach fluid mechanics for engineering students. This assumes that a student will spend 10 hours in a week (2 hours lecture, 1.5 hours tutorial and 1.5 hours lab and 5 hours self-study) to learn fluid mechanics for a 12-week semester. This method can be adapted to teach other science subjects for engineering students.

Table 14.2 Proposed weekly activities in teaching fluid mechanics by a blended learning approach

Study mode	*Activities*	*Comments*
Self-study by students:		
30 minutes	Listen to pre-recorded lecture on fluid mechanics	This will allow students to be familiar with new concepts (e.g. hydrostatic pressure, Bernoulli equation and energy equation) introduced in a given week and list possible questions to discuss with the lecturer and other fellow students during the face-to-face session.
1 hour	Watch six VDO clips (10 minutes each)	Relevant topics form each week's lecture to be selected, e.g. for Lecture 1 possible topics include how knowledge in fluid mechanics has contributed to design dams for hydroelectricity generation, irrigation canals to irrigate crops, pump-house to supply drinking water, flood protection levees to protect communities during floods and urban drainage systems to drain water quickly while reducing pollution of our waterways; why it is important to know fluid properties (e.g. density, specific weight, viscosity, surface tension and capillarity); how one can differentiate among solids, liquids and gases; and what the differences are between ideal and real fluids.
1 hour	Read two case studies (30 minutes each)	Relevant topics form each week's lecture to be selected, e.g. for Lecture 1 possible topics include Snowy Mountains engineering scheme in Australia, water supply system in Sydney and Murrumbidgee irrigation project in New South Wales, Australia.

Study mode	Activities	Comments
2 hours	Revise tutorial problems from recorded tutorial VDOs, hand-written solutions of selected tutorial problems and selected problems from textbook	This will assist students to achieve problem solving skills, e.g. to calculate pressure, velocity, discharge, energy and frictional resistance under different flow conditions and system configurations.
30 minutes	Complete online practice quiz	This will assist students to understand the concepts and theories presented in the recorded lecture of the week.
1.5 hours	Remote lab	A student will use remote lab to verify a theory or concept discussed during the weekly lecture.
Face-to-face session:		
10 minutes	Revision of last week's lecture	To highlight the main concepts presented in last week's lecture (e.g. key terms such as pressure head, velocity head and energy grade line and major equation such as energy equation).
1 hour	Presentation of lecture	Explain the key concepts and theories, derive equations, present possible applications and discuss assumptions and limitations.
50 minutes	3-minute presentations by 12 randomly selected students from the class on possible applications of the theories on the water-related problems experienced by the students themselves	This will assist students to relate the theory to personal experiences, e.g. experience of low water pressure in water tap, water hammer and hydraulic jump.
1.5 hours	Tutorial class	A student will solve given problems with the assistance of the tutor and other fellow students to apply the theory discussed during the lecture.

14.7 Conclusion

This chapter presents the importance of science in engineering studies. The poor background of students in science and mathematics in Australia is a major concern, since many engineering students do not have the necessary skills in science and mathematics to complete engineering courses successfully. It has been shown that all the engineering courses at the University of New

South Wales (UNSW) have a significant science component, which means that the high school leavers intending to undertake engineering studies in UNSW must have a sound background in science. The teaching experiences of two academics in engineering physics and fluid mechanics subjects in Western Sydney University are presented. They have noted that poor backgrounds in mathematics and science are a major obstacle in completing engineering physics and fluid mechanics subjects by many undergraduate engineering students in Western Sydney University. A few possible solutions are presented based on student-centered and blended learning approaches, which include peer learning, development of interactive online tools, custom-made short courses, face-to-face and recorded lectures, remote labs, online practice quizzes, face-to-face and recorded tutorials and case studies. A weekly activity is proposed to teach fluid mechanics through a blended learning approach which is likely to enhance student learning. A similar approach can be adopted to teach other science subjects to engineering students.

References

Australian Government (2016) *Driving Innovation, Fairness and Excellence in Australian Higher Education.* Department of Education, Canberra.

Aziz, K., Rai, S. & Rahman, A. (2015) Design flood estimation in ungauged catchments using genetic algorithm based artificial neural network (GAANN) technique for Australia. *Natural Hazards*, 77, 2, 805–821.

Bhathal, R., Sharma, M.D. & Mendez, A. (2010) Educational analysis of a first year engineering physics experiment on standing waves: Based on the ACELL approach. *European Journal of Physics*, 31, 23–35.

Bhathal, R. (2011) Retrospective perceptions and views of engineering students about physics and engineering practicals. *European Journal of Engineering Education*, 36(4), 403–411.

Bonk, C.J. & Graham, C.R. (2012) *The Handbook of Blended Learning: Global Perspectives, Local Designs.* John Wiley & Sons, New York City, USA.

Bransford, J.D., Brown, A. & Cocking, R. (1999) *How People Learn: Mind, Brain, Experience, and School.* National Research Council, Washington, DC, USA.

Caballero, W.L. & Rahman, A. (2014) Development of regionalized joint probability approach to flood estimation: A case study for New South Wales, Australia. *Hydrological Processes*, 28, 13, 4001–4010.

Caprile, M., Palmen, R., Sanz, P. & Dente, G. (2015) *Encouraging STEM Studies for the Labour Market.* Directorate-General for Internal Policies, European Union, Brussels, Belgium.

Chief Scientist (2014a) *Science, Technology, Engineering and Mathematics: Australia's Future.* Government of Australia, Canberra, Australia.

Chief Scientist (2014b) *Benchmarking Australian Science, Technology, Engineering and Mathematics.* Government of Australia, Canberra, Australia.

Chief Scientist (2014c) *Connecting the Science Pipeline.* Government of Australia, Canberra, Australia.

Chief Scientist (2012) *Mathematics, Engineering & Science in the National Interest.* Government of Australia, Canberra, Australia.

Commonwealth of Australia (2009) *Transforming Australia's Higher Education System.* Government of Australia, Canberra, Australia.

Commonwealth of Australia (2015) *Vision for a Science Nation: Responding to Science, Technology, Engineering, and Mathematics: Australia's Future.* Government of Australia, Canberra, Australia.

de Graaf, E. & Kolmos, A. (2003) Characteristics of problem-based learning. *International Journal of Engineering Education*, 19(5), 657–662.

Drew, D.E. (2015) *STEM the Tide: Reforming Science, Technology, Engineering, and Math Education in America.* JHU Press, Baltimore, MD, USA.

Dunn, G. (1930) The relationship between science and engineering. *Science*, 71(1837), 276–277.

English, L.D. (2016) STEM education K–12: Perspectives on integration. *International Journal of STEM Education*, 3(1), 3.

English, L.D. (2017) Advancing elementary and middle school STEM education. *International Journal of Science and Mathematics Education*, 15(1), 5–24.

European Parliament (2015) *Encouraging STEM Studies for the Labour Market*. Available from: www.europarl.europa.eu/studies [Accessed 21 April 2018].

Freeman, S., Eddy, S.L., McDonough, M., Smith, M.K., Okoroafor, N., Jordt, H. & Wenderoth, M.P. (2014) Active learning increases student performance in science, engineering, and mathematics. *Proceedings of the National Academy of Sciences*, 111(23), 8410–8415.

Haddad, K. & Rahman, A. (2012) Regional flood frequency analysis in eastern Australia: Bayesian GLS regression-based methods within fixed region and ROI framework – Quantile regression vs. Parameter regression technique. *Journal of Hydrology*, 430–431(2012), 142–161.

Herrington, J. & Oliver, R. (2000) An instructional design framework for authentic learning environments. *Educational Technology Research and Development*, 48(3), 23–48.

Lasry, N., Charles, E. & Whittaker, C. (2016) Effective variations of peer instruction: The effects of peer discussions, committing to an answer, and reaching a consensus. *American Journal of Physics*, 84, 639.

Light, R.J. (2004) *Making the Most of College*. Harvard University Press, Boston, MA, USA.

Loveridge, M. & Rahman, A. (2014) Quantifying uncertainty in rainfall-runoff models due to design losses using Monte Carlo simulation: A case study in New South Wales, Australia. *Stochastic Environment Research & Risk Assessment*, 28, 8, 2149–2159.

Mamoon, A.A. & Rahman, A. (2017) Selection of the best fit probability distribution in rainfall frequency analysis for Qatar. *Natural Hazards*, 86(1), 281–296.

McDowell, L. (1995) Effective teaching and learning on foundation and access courses in engineering, science and technology. *European Journal of Engineering Education*, 20(4), 417–425.

Mckagan, S.B., Perkins, K.K. & Wieman, C.E. (2007) Reforming a large lecture modern Physics course for engineering majors using a PER-based design. *AIP Conference Proceedings*, 883, 34–37.

National Science and Technology Council (2013) *A Report from the Committee on STEM Education*. National Science and Technology Council, Washington, DC, USA.

Organisation for Economic Co-operation and Development (OECD) (2016) Why access to mathematics matters and how it can be measured. In: *Equations and Inequalities: Making Mathematics Accessible to All*. OECD Publishing, Paris, France. doi:10.1787/9789264258495–4-en

Pathirana, A., Koster, J.H., Jong, E.D. & Uhlenbrook, S. (2012) On teaching styles of water educators and the impact of didactic training. *Hydrology and Earth System Sciences*, 16(10), 3677–3688.

Prince, M.J. & Felder, R.M. (2006) Inductive teaching and learning methods: Definitions, comparisons, and research bases. *Journal of Engineering Education*, 95(2), 123–138.

Rahman, A., Weinmann, P.E., Hoang, T.M.T. & Laurenson, E.M. (2002) Monte Carlo Simulation of flood frequency curves from rainfall. *Journal of Hydrology*, 256(3–4), 196–210.

Rahman, A. & Al-Amin, M. (2014) Teaching of fluid mechanics in engineering course: A student-centered blended learning approach, In: Alam, F. (eds) *Using Technology Tools to Innovate Assessment, Reporting, and Teaching Practices in Engineering Education*. IGI Global, Hershey, PA, USA.

Rahman, S., Bhathal, R. & Rahman, A. (2016) Challenges in teaching science to engineering students: An Australian perspective. In: Rahman, A. & Ilic, V. (eds) *Proceedings of the International Conference on Engineering Education and Research*, 21–24 Nov 2016, Sydney, Australia.

Rahman, A. (2017) A blended learning approach to teach fluid mechanics in engineering. *European Journal of Engineering Education*, 42, 3, 252–259.

Rahman, A., Charron, C., Ouarda, T.B.M.J. & Chebana, F. (2018) Development of regional flood frequency analysis techniques using generalized additive models for Australia. *Stochastic Environment Research & Risk Assessment*, 32, 123–139.

Smith, K.A., Sheppard, S.D., Johnson, D.W. & Johnson, R.T. (2005) Pedagogies of engagement: Classroom-based practices. *Journal of Engineering Education*, 94(1), 87–101.

The Royal Society Science Policy Centre (2014) *Vision for Science and Mathematics Education.* The Royal Society, London.

UNSW (2016) *UNSW Handbook, Course – Advanced Thermofluids – MECH3610.* University of New South Wales, Sydney.

Winkelman, P. (2009) Perceptions of mathematics in engineering. *European Journal of Engineering Education,* 34, 4, 305–316.

Yadav, A., Subedi, D., Lundeberg, M.A. & Bunting, C.F. (2011) Problem-based learning: Influence on students' learning in an electrical engineering course. *Journal of Engineering Education,* 100(2), 253–280.

Young, H.D., Freedman, R.A., Bhathal, R. & Ford, A.L. (2011) *University Physics with Modern Physics,* 1st Australian SI ed. Pearson, Australia.

Chapter 15

An analysis of the use of experiential learning principles for developing professional skills in postgraduate engineering students

D.S. Thorpe

School of Civil Engineering and Surveying, University of Southern Queensland, Australia

ABSTRACT: The engineering and built environment programs of the University of Southern Queensland include a number of coursework and research postgraduate engineering options. Programs offered through coursework include a number of engineering management courses aimed at developing the professional engineering knowledge and skills required for good engineering practice. These courses include advanced engineering project management, asset management in an engineering environment, management of technological risk and project requirements management. They are offered through both on-campus teaching and online in a blended learning environment, and they are normally completed through traditional approaches like written study materials, lectures and tutorials, which are supplemented by other study resources. These traditional teaching approaches are successful, but also have the potential to be improved in the way in which they develop those skills in learners that equip them to be successful professionals through approaches like student-centered learning, authentic assessment and experiential learning. Experiential learning is a particularly useful teaching method, as it not only teaches knowledge and skills, but also reinforces them through a cycle of concrete experience, reflective observation, abstract conceptualization and active experimentation. This approach is particularly successful when it is combined with other good teaching methods, such as practices designed to enhance and reinforce learning. An example of such practices is providing useful feedback from an early assignment with lower course weighting to a subsequent more difficult assignment, based on similar principles, that carries a considerably higher weighting. These practices can be supplemented by other teaching methods, such as embodied learning, which helps learners to better understand a principle through experiencing the application of theory, and the utilization of modern computer-based instruction material. While experiential learning is a positive approach to developing professional skills, there are also challenges in its implementation, with particular challenges in the online environment. The use of experiential learning approaches to develop professional skills in postgraduate engineering students is discussed, along with how this approach can be improved.

15.1 Introduction

Practicing engineers are expected to exhibit professional skills that require them to be knowledgeable in their discipline; understand their professional environment; have a good knowledge of areas like quality, risk and project management; work in a team; and exhibit good communication and interpersonal skills. They are also required to meet ethical standards.

While undergraduate courses provide basic education in these areas through courses in areas like sustainability, engineering management, professional practice and research projects, postgraduate courses have the potential to further develop their professional engineering skills. In the University of Southern Queensland, for example, coursework postgraduate programs, like the Master of Engineering Science and Master of Engineering Practice, are offered to engineering technologists to meet professional engineering requirements. Qualified professional engineers can undertake advanced programs, such as the Master of Advanced Engineering, the Master of Engineering Research, the Doctor of Professional Engineering (which has a one-third coursework component), and the Doctor of Philosophy (University of Southern Queensland, 2017a).

The courses offered in these programs include both technical engineering courses and engineering management courses and are focused on developing specific sets of professional knowledge and skills. A number of the courses, particularly in engineering management, are common to several of these programs. All of these programs contain a research component.

Some of the postgraduate engineering courses offered at the university include asset management in an engineering environment, advanced engineering project management, management of technological risk and project requirements management (University of Southern Queensland, 2017b). A blended learning environment, in which all of the postgraduate engineering courses at the university are delivered online through distance education (most are also available through classes at the Toowoomba campus of the university), is used to deliver these courses. The skills taught in these courses have been traditionally taught through standard educational approaches, such as the use of written study materials supplemented by lectures and tutorials, which are also made available to online learners. As a result of an ongoing continual improvement approach to the delivery of these courses and the

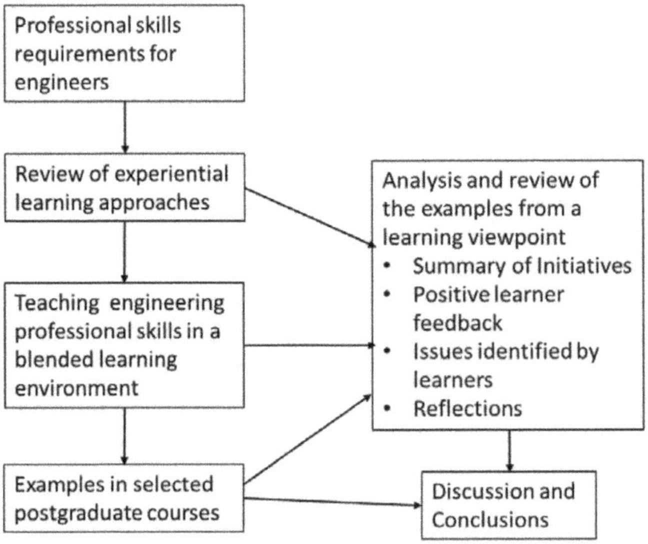

Figure 15.1 Conceptual framework of chapter discussion

desirability of embedding the concepts taught in them to learners, an increasing emphasis is placed on improved approaches to teaching and assessment, such as student-centered learning, authentic assessment and experiential learning.

Experiential learning, which can in its simplest form be considered as learning by doing or experiencing, can be described through the experiential learning cycle developed by Kolb (1984) of a concrete experience, reflective observation, abstract conceptualization and active experimentation. This chapter discusses the use of experiential learning approaches to developing professional skills in postgraduate engineering students, through outlining the professional skills required of engineers, reviewing experiential learning approaches, discussing their use in a blended learning environment to teach professional skills to engineers, providing examples of the use of this approach, analyzing and reviewing the examples from a learning viewpoint and developing conclusions. Figure 15.1 is a conceptual framework that outlines these processes and their connections to each other.

15.2 Professional skills requirements of engineers

Professional skills may be defined as the skills required for graduates to succeed in professional practice. They require graduates to be knowledgeable and skilled in their discipline, and include knowledge and skills relating to their discipline and the ability to apply knowledge and skills in a professional environment. As well as including discipline-specific skills (for example, civil engineering), these skills include generic skills and a range of personal skills required for workplace success, such as motivation and the ability to understand and apply ethical conduct, achieve goals, relate well and demonstrate initiative (Crebert *et al.*, 2011).

In Australia, professional engineers are required to achieve professional skills at two levels – initially at a graduate level (Engineers Australia, 2013), and then at the chartered professional engineer level, which may be achieved after the graduate has completed sufficient professional engineering experience and can demonstrate competencies in personal commitment, obligation to community, value in the workplace and technical proficiency (Engineers Australia, 2012). In addition, practicing engineers are required to comply with a code of ethics that defines the values and principles that shape the decisions that engineers make in professional practice (Engineers Australia, 2010). Engineers Australia have also more recently published guidelines for the principles and practice of implementing sustainability (Engineers Australia, 2017), which support the Sustainability Policy of Engineers Australia. These guidelines discus the principles of sustainability, including its relationship to risk management, and incorporate a number of practice notes, including design and project delivery considerations.

The professional skills of engineers at the graduate level are achieved through formal study that combines theoretical instruction with experiential tasks like work experience, laboratory practice and undertaking research projects. On the other hand, many of the competencies expected of more experienced engineers will be obtained through application of professional knowledge and skills. Postgraduate study that is designed to progressively develop the learner through successive learning tasks that build on previous tasks, and where possible is based on the learner's professional experience, can aid this process. Therefore, the requirements for engineers to develop sound practical professional skills and competencies underscore the importance of at having least some courses with an experiential learning approach in formal engineering education at both undergraduate and postgraduate levels.

15.3 Review of experiential learning approaches

Experiential learning, or learning by experience, has been developed by Kolb (1984) and other authors. The Kolb model consists of the stages of concrete experience (feeling), reflective observation (watching), abstract conceptualization (thinking) and active experimentation (doing) phases. Successive Kolb loops can be viewed as a spiral of continual development and improvement.

The experiential learning model has been extended into other applications. For example, Zwetsloot (2003) has applied experiential learning to the development of corporate social responsibility at the organizational level through a continuous improvement and innovation model, applied to a process of doing things right the first time and doing the right things. This model is used in the context of learning from experience, or experiential learning, which can be specific training, education or the supply of information.

Experiential learning, with its development of competencies through successive cycles of feeling, watching, thinking and doing, is well suited to the development of traditional engineering skills. An experiential learning approach can be used, for example, in teaching topics like asset management, which combine theoretical studies in strategic asset management with good engineering and business practice.

One teaching approach that can underpin experiential learning approaches is student-centered learning, which according to Biggs (2001) focuses on teaching that leads to learning and leads to high-quality learning. It is underpinned by the principles of good assessment in order to both certify and encourage learning (Boud, 1998). To meet the requirements of an experiential learning model, such good assessment should have constructive alignment and be criterion focused (Gulikers *et al.*, 2004). It will challenge learners and be relevant to their course of study.

A process that is linked with criterion-referenced assessment and therefore is a good basis on which to achieve professional skills development in engineers is authentic assessment, which is stated to be a form of performance assessment (Gulikers *et al.*, 2004). It is noted by these authors that to positively influence student learning, authentic assessment should be aligned to academic instruction, thus achieving constructive alignment. Good authentic assessment should also require learners to demonstrate competencies in a setting that is similar to professional practice. Authentic assessment is divided by Gulikers *et al.* (2004) into the five dimensions (task, physical context, social context, assessment form and criteria). Thorpe (2013) added to these dimensions a sixth dimension, professional issues, and has applied this approach to assessment in postgraduate engineering management courses, and in doing so was able to increase its range of applications.

Another approach that is related to experiential learning is embodied learning, or the development of relationship between the learner and the topic being studied. One description of it, which was supported by Dewey (1997) as cited in Nguyen and Larson (2015), is "learning that joins body and mind in a physical and mental act of knowledge construction." Such learning is stated to occur as senses are avenues of knowledge, not because external facts are somehow "conveyed" to the brain, but because they are used in doing something with a purpose (Nguyen and Larson, 2015).

Embodied learning in this sense can be considered to have three conceptual elements – bodily and spatial awareness of sensation and movement, unification of mind/body in learning and the body's role as socio-cultural context. It can be applied to activities like interdisciplinary collaboration, problem-posing instruction and learning space design (Nguyen and

Larson, 2015). It can be linked with good experiential learning. For example, Kolb and Kolb (2009) observe the importance of team learning, problem solving and learning space design when they discuss the application of experiential learning to management learning, education and development. Extensions of this approach to embodied learning can include computer simulation and games, particularly when they are designed to stimulate learning.

Because of its practical nature, experiential learning is well suited to the development of practical professional skills and competencies in professional areas like engineering, teaching, nursing and business management. For example, guest speakers from industry speaking to an asset management class can discuss real-world issues in planning, designing, delivering and maintaining assets and describe to the class approaches that they used, from a professional point of view, to solve real-world problems in achieving their goals. Similarly, work placements, workshops, simulations, site visits and role plays can provide insights into potential issues in practice and how they might be addressed. This approach, which can be applied at all levels of study, is well adapted to learners who, through work placements or other applied activities, are able to use it to understand and improve their performance of practical tasks through performing, reflecting on, modifying and applying their learning to the next time that they perform this task, or a similar task.

In addition to its application in practical sense, experiential learning is also suitable for linking with advanced educational practices like authentic assessment (Gulikers *et al.*, 2004). It is particularly suited to teaching professional skills. For example, experiential learning can be used to provide constructive feedback from an initial, relatively straightforward assignment in a professional engineering area that is undertaken fairly early in a study semester using basic skills in the field. This feedback can be used to aid the learner in the execution of a more detailed, complex assignment that is undertaken later in the semester and uses more advanced skills that build in the skills used for the first assignment. This approach has been used at the University of Southern Queensland in a postgraduate project management course. The main question on the first assignment in this course, which requires the use of a subset of project management skills, asks the learner to develop a solution for the ongoing management of a single project that is underperforming. Learners then use feedback from this assignment when undertaking a second assignment, which asks the learner to develop a process for the management of a program of projects and uses a much larger set of skills. If the feedback from the first assignment is applied well to the second, more complex assignment, learners will have developed a useful set of skills that apply the lessons learned from the initial assignment. A similar approach can be used for research students, who can apply feedback on their initial research in the more complex tasks of the later stages of their research.

While the traditional applications of experiential learning include problem-based learning, professional practice, work experience and similar activities, it has been shown above that it can have much wider application. For example, it can be used in areas like developing corporate social responsibility through a continuous improvement and innovation approach (Zwetsloot, 2003). Similarly, an experiential learning approach can be used for developing professional skills in postgraduate students, particularly in engineering management and research. The extent to which such development occurs varies with the course delivered and its objectives.

Therefore, experiential learning approaches have many applications and can be extended to a range of learning types. They are well suited to teaching the professional skills of advanced engineering, which are quite diverse and may have a strong practical element.

15.4 Teaching engineering professional skills in a blended learning environment

While ideally experiential learning approaches are best suited to a face-to-face learning environment, many courses are also delivered online, either to complement courses that are delivered on campus or as the only means of course delivery. Courses delivered online are suited to part-time learners, a number of whom would be studying professional skills or undertaking research while working in industry, such as practicing engineers. Learners undertaking these courses should, wherever possible, have a similar learning experience to learners attending a university campus. A blended learning environment can facilitate this process.

There are a number of definitions for blended learning. For example, Oliver and Trigwell (2005) have listed three definitions of it, which are:

- the integrated combination of traditional learning with web-based online approaches, which the authors state is the most common definition;
- the combination of media and tools employed in an e-learning environment and
- the combination of a number of pedagogic approaches, irrespective of learning technology use.

The University of Western Sydney (2013) defines blended learning in the university's context as referring to a strategic and systematic approach to combining times and modes of learning, integrating the best aspects of face-to-face and online interactions for each discipline and using appropriate information and communication technology (ICT).

While these definitions differ in detail, they tend have a common focus of a combination of a range of learning approaches that are normally comprised of face-to-face learning and one or more other approaches.

Singh (2003) observes that blended learning combines a number of delivery modes and can therefore offer more choices, such as self-paced learning and live collaborative learning. It can be used for a range of learning activities, including supplementing learning with practice. There are three main formats used – synchronous physical formats (such as instructor-led classrooms and lectures and field trips), synchronous online formats (such as online meetings, web seminars and conference calls) and self-paced, asynchronous formats (such as documents and web pages, online learning communities and discussion formats). He notes that the concept of blended learning is based on the premise that learning is a continuous process rather than a one-time event. Thus, it has the ability to extend the reach of a learning program and the potential to optimize the cost and time of developing and delivering a learning program. Singh (2003) recommends following Khan's octagonal framework consisting of institutional, pedagogical, technological, interface design, evaluation, management, resource support and ethical dimensions (Khan, no date, as cited in Singh, 2003). Khan (2007) further discusses the use of this approach in an open and distributed learning environment.

The University of Western Sydney (2013) lists a number of options for blended learning, including blogs, live internet streaming, flipped classrooms, video conferencing and discussion boards. Many of these options utilize online learning education in addition to face-to-face teaching. They are therefore suited to experiential learning.

Given its role in blended learning, the online education environment merits further examination from the point of view of its role in experiential learning. Some of the claimed benefits for online learning include:

- Online and on-campus courses are usually subject to the same criteria and meet the same academic standards.

- Courses can be accessed anywhere with an internet connection.
- Learners have the flexibility to study when it is most convenient for them and complete their course of study to meet their needs, subject to meeting the requirements of the course.
- Learners have the opportunity to become part of a wide, rich and varied community and have a similar level of access to university support services as on-campus learners.
- There are a range of options with respect to the delivery of online courses, including virtual learning environments, interaction with discussion forums, participation in online activities, linking with video streaming services and interaction with the virtual environment (University of Edinburgh, 2016).

Online learning can also provide greater opportunities to students, including those with disabilities, living in rural communities and with work and family commitments. It also allows shy, inhibited students, who may be intimidated by the conventional classroom environment, to voice their opinions, and it provides flexibility to learners. Society can also benefit from a more qualified workforce given the greater opportunities for learning offered to a wider range of potential learners (O'Donoghue *et al.*, 2004).

There are also some concerns about online learning. For example, online learners may not receive the benefit of face-to-face interaction and may feel isolated by online learning, particularly in asynchronous mode. In addition, while the price of technology continues to decrease, it may still be too expensive for some students, particularly when the cost of communication is considered. Learners also require to feel competent in using the technology, which in some cases may not always be reliable. Approaches for overcoming the disadvantages include balancing asynchronous and synchronous communication, increased use of multimedia, improving interaction between learners and teaching staff and developing instructor skills (O'Donoghue *et al.*, 2004).

Another perspective on online learning has been provided by Nedelko (2008), who surveyed undergraduate economics and finance students at the University of Slovenia and found that learners who had good prospects for success in an online environment had a number of characteristics, including positive attitudes towards information and communication technologies, high levels of skills and knowledge in it, a high motivation to study and a high level of self-discipline. Learners with lower prospects of succeeding in this medium had negative attitudes towards using information and communication technologies, low levels of skills and knowledge in it, a lower motivation for study and a lower level of self-discipline. Thus, the skills and knowledge that learners have in information and communication technologies, along with their motivation to study, have the potential to be factors in their ability to effectively participate in online study.

Therefore, a blended learning environment has considerable potential for experiential learning, provided the issues are considered. In particular, it is necessary to consider the type and location of the learner, how well the learner would benefit from a blended learning teaching environment and what learning style would be best suited to delivery of a course through blended learning tools. Matters to consider in using online study would include the suitability of a particular course to being studied online, the potential for learners to be located in a number of different time zones and the personal and work commitments of learners. From a teaching point of view, it is important to maintain the learner's interest, consider the learner's location and evaluate what components of the course can be taught using experiential learning approaches. For example, the most suitable course for online delivery may be one that primarily utilizes asynchronous communication (which allows learners to adjust their study to suit their time commitments), encourages open discussion on a course study forum, includes industry-based guest speakers and poses challenging tutorial questions that are based on authentic industry experience.

Not all courses are suitable for experiential learning in a blended learning environment. There are many courses that develop theoretical concepts, for example, which, while they can be delivered in an online environment, do not strongly lend themselves to experiential learning approaches. Possible courses in this category could include pure mathematics, economics, structural analysis, theories of information systems and other theoretically based courses with at best a limited practical component. Similarly, not all courses with a practical approach, while generally being suited to teaching by experiential learning approaches, are suitable for delivery by online means, particularly where the activity cannot be easily replicated electronically. Examples of such courses include engineering laboratory courses and presentation skills courses. Both of these activities require the learner to demonstrate knowledge and/or skills at a practical level in a physical environment, and are hence not easy to teach in an electronic environment. Developments in artificial intelligence, simulations and virtual reality may make such delivery a viable option in the future.

15.5 Examples of the use of experiential learning in developing professional skills

It has been shown that experiential learning has a range of applications in developing professional skills in engineers, particularly when combined with approaches like student-centered learning, authentic assessment and embodied learning.

To illustrate the application of experiential learning approaches in developing professional skills in postgraduate engineering students, the following three course groups are briefly reviewed:

- Postgraduate engineering management courses with one or more assignments and an examination;
- Postgraduate engineering management courses with multiple assignments and
- Research projects.

15.5.1 Postgraduate engineering management courses with one or more assignments and an examination

This group of courses is written from the point of view of developing professional skills in selected engineering management areas, such as asset management and risk management. Courses in this group have one or more assignments, followed by a written examination at the end of the semester. Two of these courses are "Asset Management in an Engineering Environment" and "Management of Technological Risk" (University of Southern Queensland, 2017b), both of which use a similar open approach to assignment questions. Asset Management in an Engineering Environment (University of Southern Queensland, 2017b), which has one major assignment and an examination, is typical of such courses. This course is divided into two main sections – asset management concepts and applications of the concepts.

As this course was written with the professional engineer in mind as a learner, it draws heavily on asset management practice. As part of this process, it illustrates its teaching by reflective examples that draw on the learner's experience and research. The assignment is divided into two components. The first question in this assignment is based on asset management strategic planning question and asks the learner, who has recently accepted a position as manager of an existing asset management network that has a number of issues, to write a submission to senior management that proposes a strategy to review and improve the organization's assets, their management and the standard of service that they provide, in order to achieve the goals of the organization's board with respect to asset management. There is also an engineering economics question that is based on an engineering asset decision-making scenario. This question is sometimes accompanied by a question on asset depreciation. While these more concrete questions are reasonably well defined, the first question on strategic asset management specifies the parameters of the problem to be solved, but leaves it open to the learner to construct the particular asset network being reviewed, using personal engineering experience or research. The examination draws on the course study material, and also achieves a measure of experiential learning approaches and authenticity through requiring candidates, in a number of questions, to illustrate their answer with an engineering example. Learners can also apply lessons learned from assignment feedback to it.

15.5.2 Postgraduate engineering management courses with multiple assignments

Some postgraduate engineering management courses offered at the university are assessed by two or more linked assignments. In such courses, assignments occurring later in the course tend to build on the assignments that precede them. Examples of such courses are "Advanced Engineering Project Management" and "Project Requirements Management" (University of Southern Queensland, 2017b). In both courses, the assessment consists of two assignments, the second of which builds on the first. Advanced Engineering Project Management (University of Southern Queensland, 2017b) teaches the concepts, processes and tools of advanced engineering project management through exploring the project life cycle, project management knowledge areas, professional issues in project management including the management of project sustainability, program management and current and future issues in engineering project management. It is supported by the *Guide to the Project Management Body of Knowledge* (PMBOK Guide) (Project Management Institute, 2013) and other project management references.

This course was written with the practicing professional engineer in mind as a learner and aims to further develop the learner, who would normally have some basic project management knowledge, to the program manager level through using strategic view of project management. It uses reflective, practical examples to illustrate its teaching. While each assignment has a minor question on specific aspects of the course, the main question in the first assignment challenges the learner, in the position of a project manager, to develop a process to complete, on time and to the required quality, a project with a number of issues, for which the learner has recently taken project management responsibility. The learner is charged with reporting on the revised project management plan and process to the project sponsor. In the second assignment, the learner is similarly required to address issues with three projects being managed at the higher-level position of program manager, and similarly provide a report to senior management. This task allows the learner to utilize learning from developing and receiving marking feedback from the first assignment to a similar, but more complex and higher-level task in the second assignment.

15.5.3 Research projects

All postgraduate engineering programs at the university have a research component. The extent of this research component varies with the program in which it is undertaken, and can be a relatively small proportion of the course in programs like the Master of Engineering Science to over 90% of the study program in the case of the Doctor of Philosophy.

The postgraduate research journey usually commences with the submission of a research proposal. On its acceptance, the research process moves through the literature review, research methodology development, confirmation of candidature, conducting the research, analyzing results of the research, developing conclusions and writing a research dissertation (or alternatively writing a set of suitable publications that can be used for award of a doctorate by publication). This journey can be challenging and can be subject to tight time pressures and regular progress reporting. Success in it usually comes after a period of considerable thought, experimentation, obtaining results, reflection, learning, further trials and resultant modification of the research until final outcomes are achieved. This process is similar to the experiential learning process of Kolb (1984) and the continuous improvement approach discussed by Zwetsloot (2003). In undertaking their research, many researchers become closely associated with their research topic, resulting in a synergy between research and research topic that impacts on the researcher's thinking processes and brings the researcher's personal approach to the research, resulting an embodied learning or similar experience.

15.5.4 Other initiatives

Guest lecturers from industry are now being used in a number of the university's courses, including postgraduate courses. Other initiatives are also being piloted. For example, in order to bridge the gap between lecturing staff and learners, regular video tutorials are being held for selected online classes, including the postgraduate professional skills course "Project Requirements Management," in which there are a series of interactive tutorials using the university's Zoom video conferencing system. Further initiatives, such as online simulations, are being considered.

It is also important to monitor the success of the experiential learning initiatives, particularly in an online environment where there may be no face-to-face contact with the learner. One approach to this process is through the use of learning analytics (for example Picciano, 2012; Van Barneveld *et al.*, 2012), which allows monitoring and analysis of online learner interaction during the time in which the course is being undertaken. Comments by learners can also be viewed and addressed during this period. The other main source of information is learner feedback on delivery of the course, which is surveyed by the university at the end of each teaching semester. The next section discusses examples of this type of feedback.

15.6 Analysis and review of examples

15.6.1 Summary of experiential learning initiatives

While a course like Asset Management in an Engineering Environment (University of Southern Queensland, 2017b) could be considered to have a typical course and assessment structure, it aims to focus on student-centered learning where possible, uses assessment procedures that draw on real practice as much as possible and allows learners to impart their professional engineering experience to assessment responses. While it does not strongly use the cycle developed by Kolb (1984), it has a number of components of experiential learning, including utilizing as much as possible the learner's existing expertise to further develop the learner's knowledge, skills and interest.

In Advanced Engineering Project Management (University of Southern Queensland, 2017b), the process of building the main question on the second assignment on the feedback of the first, simpler assignment question of a similar nature allows the learner the opportunity to have the concrete experience of developing the assignment and receiving feedback from it, reflect on the results, think about applying learning from the first assignment to the second assignment and write the second assignment using them. This process strongly utilizes the experiential learning cycle developed by Kolb (1984).

Postgraduate research has the potential to link all the elements of experiential learning. The research journey is very much based on what the learner wants, is normally quite challenging, will be assessed from the point of view of its authenticity and very much uses repeated cycles of the experiential learning process of Kolb (1984). It can also be argued that the postgraduate research journey contains the elements of embodied learning, as its consistent and challenging nature links the researcher with the research in a way that develops a synergy between them.

15.6.2 Positive learner feedback

There have been a number of positive comments from learners on the use of experiential learning environments in both the on-campus and online modes of delivery of the engineering professional skills development courses discussed in this chapter. Table 15.1 lists selected de-identified positive comments with respect to experiential learning initiatives from 2015 to 2017 from the university's survey of learner feedback on the delivery of each course at the end of its teaching semester. The research programs are not subject to student comments.

Table 15.1 Examples of anonymous student comments

Year	*Abbreviated title of course offer*	*Short description of comment type*	*Anonymous Comment*
2015	Asset Management in an Engineering Environment, Online	What were the best aspects of this course? Anything else about the course	Increasing my management skills and knowledge. I found it interesting, and think it will come in handy in my future career.
2015	Asset Management in an Engineering Environment, On-campus	What were the best aspects of this course?	Knowing the concepts and reality of engineering asset management. It's really a useful subject.
2015	Advanced Engineering Project Management, Online	What were the best aspects of this course?	It's practical and the assessment gives an opportunity to apply theory to real-life situations.

(Continued)

Table 15.1 (Continued)

Year	*Abbreviated title of course offer*	*Short description of comment type*	*Anonymous Comment*
2016	Asset Management in an Engineering Environment, On-campus	Has the assessment has contributed to my learning?	The assessment was a good task which made you think about the theoretical in a practical sense.
2016	Asset Management in an Engineering Environment, Online	What were the best aspects of this course?	The content of this course was directly related to my work environment.
2016	Management of Technological Risk, Online	What were the best aspects of this course?	Gaining a further, in-depth knowledge of risk management I may never have learned through experience.
2017	Asset Management in an Engineering Environment, On-campus	What were the best aspects of this course?	Understanding the technical aspects in asset management.
2017	Asset Management in an Engineering Environment, Online	What were the best aspects of this course?	The experience of the lecturers with real-life examples.
2017	Advanced Engineering Project Management, On-campus	What were the best aspects of this course?	Good materials and examples for me to master this course.
2017	Project Requirements Management, Online	What were the best aspects of this course?	Teaching style was a stand out along with the ability to utilize the Zoom function face to face.

The comments in this table are supportive of the experiential learning initiatives in these courses. The comments by students undertaking online study are of particular interest, as these courses are typically undertaken by experienced practicing personnel, whose view on the practicality and usefulness of the courses can be considered as reflecting an industry viewpoint.

15.6.3 Issues identified by learners

While there have been a number of positive comments from learners, there have also been some issues identified by learners. Paraphrased examples of these comments by learners include the following:

- There had been some reported difficulty in understanding recorded lecture material in an online environment.
- There have been requests for more detailed assignment information in some cases.
- Course material does not always meet the requirements of learners.
- In some cases, learners are concerned that assignments that have been designed for a range of engineering disciplines are too open and not sufficiently focused.

15.6.4 Reflections on the use of experiential learning approaches in teaching engineering professional skills

The University of Southern Queensland offers a number of postgraduate engineering management courses, all of which are available online, and many of which are available for on-campus delivery at the university's main campus at Toowoomba, Queensland, Australia. A number of these courses utilize experiential learning principles. In particular, the courses are designed to promote student-centered learning, adhere to authentic assessment principles where possible, utilize electronic study desks that are regularly updated by teaching staff and build sequential assessments on each other. Similar principles apply to research supervision. Further extensions of experiential learning principles are utilized through activities like using guest speakers and tutorials that are delivered through synchronous video conferencing.

Overall, the experiential learning approaches used for delivering postgraduate engineering professional skills have been received well by learners, who have gained a very good understanding of their course subjects through a conscious effort by teaching staff to use experiential approaches to learning, both in class and in the online environment. At the same time, it is recognized that some learners have indicated potential issues with the learning approach and process used. These issues, along with ongoing development of the program of courses offered, are being addressed by a program of continual improvement to course design and delivery, aided by regular interaction with learners on the online course study desk and improved monitoring of issues through the use of modern information tools like learning analytics.

15.7 Conclusions

Practicing engineers are required to possess those skills that enable them to meet their professional goals and objectives. In addition to their specialized technical skills, they are often required to develop and exercise engineering management skills and transferable skills as part of their professional practice. While many of these skills can be acquired through professional experience, academic study will facilitate their development. One approach to teaching such skills is through experiential learning, which can be provided either on a university campus or online, often as part of a blended learning delivery approach. A similar argument can be mounted for technologists desiring to upgrade their qualifications to qualified professional engineer.

The use of experiential learning approaches in postgraduate study offered by the University of Southern Queensland has been discussed, using the examples of different engineering management courses and postgraduate research. The assessment mode of the courses varies. Those courses that have been reviewed are offered either with one or more assignments and an examination, and/or with assignments, of which the second assignment builds on the development of and feedback from the first assignment. The discussions of these courses are supported by student feedback on delivery of the courses. Postgraduate research has been discussed as an example of how a range of good teaching practices, including experiential learning, can aid the research process and facilitate embodied learning in the form of the researcher and the research study developing a synergy following a series of experiential cycles.

Overall, the application of experiential learning approaches at the university in a blended learning environment, in conjunction with other good learning practices, has been considered to be positive. Such learning also has the potential to be further developed by other approaches like gaming, simulation, virtual reality and others that are facilitated by advanced information and communication technologies. While such approaches and other extensions to experiential learning have not been discussed in detail, they have the potential to add to the realism of learning experiences.

While experiential learning can be positive for many courses and has wide application in reinforcing theoretical engineering courses, it is unlikely to be suitable for all professional engineering development. For example, it may not be suited to courses with a high theoretical content. At the same time, it is considered a quite positive approach in professional development courses and in areas like research work, which have a clear cycle of experiencing, observation, thinking and experimentation.

One outcome of the discussion of the applications of experiential learning and related learning and teaching practices to the development of professional skills in postgraduate engineering students has been the necessity to design the experiential component of the course to meet the purpose and objectives of the course, and to consider the implications of offering them both on-campus and online in a blended learning environment. For example, an online environment may not be able to accurately replicate some activities, such as experimental work. This requirement requires consideration. On the other hand, courses that are designed to develop professional skills in postgraduate engineers and research skills development can normally be used effectively in an experiential learning environment. Research projects may require individual approaches to achieving the best outcomes. For these reasons, success in courses using experiential learning is only likely to be achieved through carefully considering and tailoring their experiential learning component to the requirements of learners. This position could change over time as advances in information systems technologies start to overcome some of the disadvantages of the online experiential learning environment.

It is concluded that experiential learning has considerable potential to be used in the development of professional skills in postgraduate engineering students in a blended learning environment, provided that its application is carefully considered and is aimed at meeting the requirements of learners. A range of options can be used to achieve this goal, particularly when courses and their assessment are structured to achieve good outcomes through combining experiential learning approaches with other good teaching practices, there is good interaction between teachers and learners via online systems and regular monitoring is undertaken.

References

Biggs, J. (2001) The reflective institution: Assuring and enhancing the quality of teaching and learning, *Higher Education*, 41(3), 221–238.

Boud, D. (1998) Assessment and learning – Unlearning bad habits of assessment. In: *Proceedings of Effective Assessment at University*, 4–5 Nov 1998, University of Queensland, Brisbane, Queensland, Australia.

Crebert, G., Patrick, C.J., Cragnolini, V., Smith, C., Worsfold, K. & Webb, F. (2011) *Professional Skills Toolkit*. Griffith University, Brisbane, Australia.

Dewey, J. (1997) *Democracy and Education*. Simon and Schuster, New York, USA. (Original work published 1916).

Engineers Australia (2010) *Our Code of Ethics*. ACT, Canberra, Australia.

Engineers Australia (2012) *Australian Engineering Competency Standards Stage 2 – Experienced Professional Engineer*. ACT, Canberra, Australia.

Engineers Australia (2013) *Guide to Assessment of Eligibility for Membership (Stage 1 Competency)*. ACT, Canberra, Australia.

Engineers Australia (2017) *Implementing Sustainability: Principles and Practice*. ACT, Canberra, Australia.

Gulikers, J.T.M., Bastiaens, Th. J. & Kirschner, P.A. (2004) Perceptions of authentic assessment: Five dimensions of authenticity. In: *Proceedings, Second Biannual Northumbria/EARLI SIG Assessment Conference. Journal of Vocational Education & Training*, Bergen.

Khan, B.H. (2007) Flexible learning in an open and distributed environment. In: Khan, B.L. (ed) *Flexible Learning in an Information Society*, Hershey, PA: Information Science Publishing.

Kolb, D.A. (1984) *Experiential Learning: Experience as the Source of Learning and Development*. Prentice-Hall, New Jersey, USA.

Kolb, A. & Kolb, D. (2009) Experiential learning theory: A dynamic, holistic approach to management learning, education and development. In: Armstrong, S. & Fukami, C. (eds) *The SAGE Handbook of Management Learning, Education and Development*. Sage, London.

Nedelko, Z. (2008) Participants Characteristics for e-Learning. In: *E-leader Conference*, Krakow, Poland.

Nguyen, D.J. & Larson, J.B. (2015) Don't forget about the body: Exploring the curricular possibilities of embodied pedagogy. *Innovation in Higher Education*, 40, 331–344.

O'Donoghue, J., Singh, G. & Green, C. (2004) A comparison of the advantages and disadvantages of IT. *Interactive Educational Multimedia*, 9, 63–76.

Oliver, M. & Trigwell, K. (2005) Can 'blended learning' be redeemed? *E-Learning*, (1), 17–26.

Picciano, A.G. (2012) The evolution of big data and learning analytics in American higher education. *Journal of Asynchronous Learning Networks*, 16(3), 9–20.

Project Management Institute (2013) *A Guide to the Project Management Body of Knowledge*, 5th ed. Project Management Institute, Pennsylvania.

Singh, H. (2003) Building effective blended learning programs. *Educational Technology*, 4396), 54–64.

Thorpe, D. (2013) Reflections on assessment: Comparison of assessment processes for postgraduate engineering management courses, In: *24th Annual Conference of the Australasian Association for Engineering Education (AAEE2013)*, Dec 8–11, 2013, Gold Coast, Queensland, Australia.

University of Edinburgh (2016) *What is Online Learning?* Available from: www.ed.ac.uk/studying/postgraduate/degree-guide/online-learning/about [Accessed 23 April 2017].

University of Western Sydney (2013) *Fundamentals of Blended Learning*. Western Sydney University, Parramatta, Australia.

University of Southern Queensland (2017a) *University of Southern Queensland Handbook*. Available from: www.usq.edu.au/handbook/current/filter-programs.html [Accessed 11 December 2017].

University of Southern Queensland (2017b) *Doctor of Professional Engineering*. Available from: www.usq.edu.au/study/degrees/doctor-of-professional-engineering [Accessed 11 December 2017].

Van Barneveld, A., Arnold, K.E. & Campbell, J.P. (2012) Analytics in higher education: Establishing a common language. *EDUCAUSE Learning Initiative*, 1, 1–11.

Zwetsloot, G.I.J.M. (2003) From management systems to corporate social responsibility. *Journal of Business Ethics*, 44, 201–207.

Chapter 16

Upgrading the Australian engineering curriculum to enhance communication skills of engineering students

T. Rahman[1], *C.C. Amos*[2,3] *and A. Rahman*[4]

[1]*School of Education, University of New South Wales, Sydney, New South Wales, Australia*

[2]*Civil Engineering, Western Sydney University, Sydney, New South Wales, Australia*

[3]*CSIRO Land and Water, Canberra, ACT 2601, Australia*

[4]*School of Computing, Engineering and Mathematics, Western Sydney University, New South Wales, Australia*

ABSTRACT: Engineers are regarded as problem solvers who use science, technology and mathematics to solve physical problems to build a better physical world. Poor communication skills have been regarded as a significant obstacle for many engineering students and junior engineers as they struggle to present their novel ideas and technical analyses effectively to others. On the other hand, it has been found that there are many engineers who have excellent communication skills. One possible solution to this problem is the introduction of a new subject in the engineering course curriculum on effective communication similar to some universities in the USA. A blended learning approach can be used to deliver this new subject that offers an enhanced flexibility in learning. It is suggested that Engineers Australia prepares a position paper on this issue and recommends a suitable roadmap for its implementation to help improving communication skills of Australian Engineering students.

16.1 Introduction

Engineers play a key role in building a better world by, among other things, building physical infrastructures, communication systems, modes of transport and many other items of service. They often use complex mathematics and physical and numerical modeling to study real system behaviors in providing an optimum sustainable design/solution to an engineering project. Apart from modeling, engineers need to present results of their analyses in a variety of ways such as technical reports, theses, project briefs, peer-reviewed papers, seminars and workshops to convey to the decision-makers, their peers and the general public their solution to many technical problems. This is also important for the opportunity of feedback between the societal needs and the manner of their engineering problem solutions. It is thus vital for a successful engineer to be a competent communicator as well as a skilled analytical problem solver. It has been reported that engineering students and many junior professional engineers – though very efficient in using communication devices (such as mobile phones, iPads and PCs) – often lack basic writing skills (Rahman *et al.*, 2016).

Communication skills are an integral aspect of every facet of interaction with people in every walk of life. This is especially true for professionals in the workforce; and though there are varying levels of importance concerning communication skills across the spectrum of professions, its fundamental necessity cannot be ignored. Kohn (2015) alludes to written communication having a high priority in the information age, contributing to success in the workplace and promotional opportunities. Employers therefore seek employees with strong communication skills and the ability to interact effectively with their customers. Many may have the impression that engineers are amongst the professions where communication is not important. Poor communication skills of junior engineers are often of concern to senior engineers, who note their various shortcomings in this respect.

Steiner (2011) mentioned that although engineers are required to write project proposals and research reports, they seldom receive formal training in effective writing of them. He also stated that engineers should take extra care to make sure that they avoid ambiguity in their writings so that others are not misinformed. Warnock and Kahn (2007) explained how engineering students often resist writing and stated that they misunderstand the linkage between the thinking process and writing skills. They argue that to implement ideas, engineers must express them in an effective manner so that they are well understood. Reave (2004) presented the results from a survey where she assessed 73 top-ranked US and Canadian engineering schools, which require a course in technical communication. Only 33% of these engineering schools used a dedicated elective course for improving technical writing of their students, while most other schools used an informal approach integrated with other subjects like professional practice.

Engineers are involved in the planning and completion of many large projects such as railways, bridges, power houses, ships, factories, dams and high-rise buildings. During the planning, design, implementation and operation of these projects, engineers must communicate in writing and verbally. The writing aspect includes writing technical reports when presenting specific results, analyses and investigations, while the verbal aspect focuses on a more direct communication with other engineers and stakeholders. In general, communication for engineers also includes e-drawings, meeting notes, project briefs, emails, text messages and slide (e.g. PowerPoint) presentations.

This paper addresses the underestimation and misunderstanding of the importance of communication and writing skills by identifying problems and proposing some solutions to enhance the communication skills of engineering students in Australia.

16.2 Issues in engineering communication

The root of the communication problem can be traced to engineers having an inherent focus and attachment to mathematics, computation and machines. This overwhelming focus leads to engineers neglecting the necessity of developing proficiency in communication skills. Engineers often lack basic social skills due to prolonged periods of seclusion and have tendencies to behave accordingly. This means they often become frustrated quickly when communicating with other professionals who possess wider social skills. Most engineering students do not get training in formal technical writing and thus underestimate the need of it in their professional career. However, based on the study of 243 engineers, Vest *et al.* (1996)

found that they spent more than 50% of their time communicating with others in their own or outside their organization.

The common problems that engineers have with their writing include:

- Having a poor command of grammar, such as incorrect tense, inappropriate or lack of conjunctions, poor spelling and incorrect choice of word selection/phraseology;
- Inadequate language expression: imprecise and poor sentence structure;
- Struggling to convey ideas eloquently;
- Poor interconnections between different sections of writing;
- Writing is often too long and convoluted;
- Inconsistency in writing, e.g. references, formatting and titles/headings and
- Poor conclusion in summing up the main ideas.

Another area engineers struggle with is their verbal communication. They are often lackluster in communicating ideas and instilling motivation, monotonous during presentations and unable to impart the big picture when explaining a concept. Often there is too much mathematics and too many equations without sufficient explanations. The same problem exists when writing emails in not being able to convey ideas adequately. Sentences often lack structure and have inappropriate word usage and tenor.

16.3 Successful engineers with exemplary communication skills

16.3.1 Founders of modern engineering: from the scientific revolution to the foundation of the first professional engineering society

The modern engineering disciplines are a product of the scientific revolution (Ahlström, 1978) and its emphasis on empirical sciences and experimentation rather than theory or pure logic. The *Encyclopedia Britannica* defines engineering as the "application of science to the optimum conversion of the resources of nature to the uses of humankind" (Academic, 2016). The Royal Society itself was founded along the lines of what was called The New Philosophy, or Experimental Philosophy, which had gained much popularity in England, Italy, France and Germany since the times of Galileo in Florence and Sir Francis Bacon in England (Syfret, 1948). So also the societies of the late 18th and early 19th centuries naturally followed on from the Royal Society founded 100 years or so earlier in the 17th century. Communication and literature particularly was a central feature of these societies that enabled the sharing and propagation of ideas, new scientific discoveries and engineering designs.

More than ever, becoming a good scientist or engineer meant that it was essential to be a good communicator by being able to read and comprehend reports and scientific literature and to be able to contribute to the profession by communicating new ideas. Engagement with the profession and also the public is another aspect of the skills required of the professional engineer that necessitate good verbal communication and written and presentation skills.

Engineers can be credited with the advancement of communication skills with the invention of the printing press, as this brought about widespread interest in literature. This is not to ignore the many ancient works of architecture that display engineering skill such as found in every age and as widespread as China and the Americas; for example, the ancient ziggurats ("tower of Babel"), the pyramids, Rome's coliseum and Taj Mahal in India (Hitchcock *et al.*, 1963). The fact that all these were achieved without a significant recorded history of discovery and access to computers and modern technology which we now have to hand is a credit to the ingenuity of the architects, chief builders, designers and "engineers" of those times. The word "engineer" was originally used as a military term related to engines of war, such as catapults and assault towers, and derived from the same Latin root as does "ingenious," "*ingenerare*," meaning "to create." Civil engineering is considered the second oldest of the engineering disciplines and was defined to distinguish it from military engineering when the first professional societies and schools of engineering were founded in the 18th century. Previously, the engineering of the built environment was in the domain of the architect. The word "architecture" has a longer history and comes via Latin from the Greek ἀρχι- (*arkhi*) "chief" and τέκτων (*tekton*) "builder." The architect's role is to design buildings and supervise their construction. This included simple structural calculations before the advances in science and technology and the introduction of new methods and materials during the scientific revolution. The etymology of the word chosen for the emerging "engineering" profession betrays the roots of the modern profession in the scientific revolution period, in which Latin was the *lingua franca*. During this time, many scientists published in Latin, which meant their works did not need to be translated into the various European languages and could be read immediately by the learned in a wide range of countries. A case in point would be Newton's *Principia Mathematica* published in English and in Latin. Many other works were translated into Latin, including those of Aristotle, who was subsequently debunked on many points through the scientific revolution's insistence on experimental evidence, notably his geocentric universe and the proposal that light and heavy objects fall at different rates. Today English, in its various forms, has in many respects taken over this position on a global scale. English is the *lingua franca* of trade, travel and tourism and for the foreseeable future is set to be the dominant global tongue (Gooden, 2009). It is the fourth most spoken first language after Chinese, Hindi and Spanish. However, it is the most spoken second language and is the primary language used for international communication in many fields like aviation, business internet and, most importantly in our context, science, technology and education (Nationsonline, 2017). Competent communication in English is therefore a great advantage to the engineer or scientist.

Johannes Gutenberg might not be a name regularly associated with engineering, but it would be negligent to consider the scientific revolution without reference to the printing press and the book, which inspired widespread literacy throughout Europe and indeed much of the world. The Royal Society right from the beginning employed printers whose duty was to print, vend, catalogue and check editions, and check the paper quality and the presentation of the figures. They not only sold copies but also ensured copies were distributed to key members and royalty and kept for the city's library (Rivington, 1984). This ability to print and distribute the literature so quickly and widely was made possible by the revolutionary invention of the movable type printing press by Gutenberg in the middle of the 15th century (Academic, 2017a). Although a goldsmith by profession, the design and assembling of the printing press should be considered a revelation and a work of engineering genius. Guttenberg's press was used without significant changes until the 20th century. The wooden press remained predominant for over 300 years at a rate of 250 sheets per hour (Academic, 2017b). Forty or more copies of the first book printed, the Gutenberg Bible, or the "Forty-two-line Bible" so-called because it was printed in 42-line columns, are still extant (Academic, 2017c). The Gutenberg

Bible was printed in Latin. By the middle of the 19th century, the impact of printing and the widespread literacy and distribution of literature is demonstrated by Edward Bulwer-Lytton coining his famous phrase, "The pen is mightier than the sword" for his play *Richelieu, or the Conspiracy*, in 1839. The engineer should be aware of the importance of communication through literature, the massive role it has played in the development of the engineering profession and the implication that represents for engineers to master good communication skills.

Sir Christopher Wren, one of the founders of the Royal Society and its president from 1680–1682, turned from pure science towards the built environment and architecture. After building a firm reputation as a formidable scientist, Wren combined his model making and physics to become known as the greatest English architect of his time, designing large and intricate structures. Wren designed St. Paul's Cathedral, 53 London churches and many other notable buildings including Hampton Court. He was in charge of the rebuilding of London after the great fire of 1666 and acted as the king's surveyor and head of the king's works. He also used his communication skills to act as a member of parliament and to publish scientific work which was held in high regard by Sir Isaac Newton and Blaise Pascal. He also wrote the preamble to the charter of the Royal Society, being one of its most active participants. He was knighted in 1673 (Britannica, 2017).

Sir Isaac Newton was born in the year of Galileo Galilei's death in 1642. Although considered a physicist and mathematician, Newton's engineering achievements cannot be neglected. Newton designed and built many of his own instruments, notably the reflecting telescope which was examined by Sir Christopher Wren and others at the Royal Society, as well as King Charles II. This earned him election as a Fellow of the Society (Mills and Turvey, 1979). Newton's work *Philosophiae Naturalis Principia Mathematica* (Newton, 1687), first written in Latin, is described as "one of the most important single works in the history of modern science" (Academic, 2017d). It is still the basis for many subjects in engineering at universities. Indeed, any engineering student has the constant reminder of Newton's efforts in the honorary naming of the unit of force after him. Edmund Halley is to be acknowledged, or rather thanked, for encouraging a reluctant Newton to write the book and for organizing the printing. If it wasn't for Halley's insistence that Newton publish his ideas, the *Principia* may never have been written and Newton would have most likely taken his knowledge to the grave. There is a good lesson here that engineering students may need encouragement to take time to communicate their ideas as well as help in the process. This process may include various kinds of editing, layout, spelling corrections, grammatical corrections and publishing. As part of the deal, Halley looked after the proofing and printing and also paid for the first printing (Iliffe and Mandelbrote, 2017). After *Principia*, Newton not only wrote scientific works but also theological works, such as the *Chronology of Ancient Kingdoms* and *Observations upon the Prophecies of Daniel* both published after his death. He also left more than a million words of notes on the Bible. In this he followed the principle of Sir Francis Bacon (1561–1626), one of the giants upon whose shoulders Newton said he stood and who is regarded by many as the father of modern science (Lennox, 2009). Bacon said that God has provided us with two books – the book of nature and the Bible – and that to be really educated you need to give your mind to both. The departure from this philosophy may be a contributing factor to the decline in literacy among science and engineering students.

John Smeaton founded the first specifically engineering society in 1771, The Society of Civil Engineers, later known as the Smeatonian Society of Civil Engineers. He is often referred to as the father of civil engineering. The society adopted a Latin motto in 1793, reflecting the pre-eminence of Latin at this time, "*Omnia in Numero, Pondere et Mensura*" adapted from Wisdom of Solomon 11:20 "(Thou hast ordered) all things by number, weight and measure." This society was established reflecting a time of increased civil engineering activities in Britain and not without influence from the Royal Society, with whom Smeaton had published many papers in their journal *Philosophical Transactions* (Skempton, 2004). It was also the Royal Society who recommended Smeaton for taking charge of the rebuilding of the timber Eddystone lighthouse in Plymouth, which had burned down. Smeaton designed a masonry lighthouse that could withstand the harsh marine environment. This became a prototype for all subsequent masonry lighthouses. Success was in part due to many hours of research that he put into hydraulic limes, thus beginning the chemistry of cements. This demonstrated an important application of research skills learned at university in the engineering field. The success at Eddystone established his reputation, and he went on to design a wide range of other works ranging from water mills, windmills and steam engines to bridges, harbors, river navigations, canals and fen drainage. He became the foremost civil engineer in Britain in his time and made many contributions to engineering science. He coined the term "resident engineer" in 1768 after years of working as a consulting engineer in conjunction with onsite assistants. Smeaton worked as a consulting engineer on many significant projects, including drainage projects that secured the foundations of London Bridge. He designed several water wheels with pump engines, including one for London Bridge waterworks, and pioneered the use of cast iron in millwork. He also worked on and designed many bridges, canals and harbors. Smeaton took literature and reports seriously. He made it a practice to have his reports copied by a clerk and supplied to clients and maintained in report books. All his drawing was by his own hand; he maintained a detailed diary and kept letters in a book format. Many of his reports are still kept in both manuscript and printed form in numerous libraries and record offices. The availability of his designs and the clear communication of his work in manuscripts enabled others to build on his work. For example, John Farey used Smeaton's work on steam engines, mills and hydraulics in Rees's *Cyclopedia.*

Thomas Telford was the first president of the world's first professional engineering body, the Institution of Civil Engineers (ICE). ICE had been founded in 1818, 47 years after the foundation of the Smeatonian Society of Civil Engineers in 1771 by three young engineers: Henry Robinson Palmer, James Jones and Joshua Field, the oldest of whom was only 32. This should serve as an encouragement to young engineers that they can have a big impact on the engineering profession. However, the institution only made significant progress when a mature Thomas Telford was appointed as its first president. Thomas was already greatly respected and well connected in both industry and government. As a result of his labors, ICE received a Royal Chartership in 1828, only 10 years after its foundation. This highlights the importance of good communication skills in bringing together professionals and making their worth recognized by governments and other stakeholders. Today, ICE is the foremost organization for engineers of all disciplines. Thomas's own works are formidable, boasting international reputation with involvement in numerous prestigious canal projects. These included one 60-mile-long canal constructed across the highlands of Scotland, as well as others abroad in Sweden, Russia, Canada and Panama (the eminent Panama Canal). His extensive involvement in bridge and road building earned him the nickname "Colossus of Roads" and "Pontifex Maximus." He was also responsible for 1200 miles of new or improved roads and 1100 bridges in the Scottish Highlands. In Russia, he advised on the 100-mile Warsaw to Brzesc major road towards Moscow completed in 1825 for Tsar Alexander I. Telford's architectural experience enabled him to impart grace and beauty to the appearance of many of the bridges in which he was involved, including the famous wrought iron suspension bridge across the Menai Straits in the UK. The bridge was the result of an ongoing experimental process, the results of which were widely propagated in leading textbooks. Telford also turned his hand to railways and steam carriages and was involved in early experiments with steam trains; most trains at the time were powered by horse. He was also involved in major drainage projects as large as 48,000 acres; over 100 harbors, docks and piers, water supply chains, reservoirs and dams; and highland churches as far north as the Shetland Islands. Telford had grown up in poverty, but had gained a good basic education at Westerkirk Parish School. He experienced farm work during this time and was at first an apprentice stonemason, which afforded him what would be described as an absolute necessity for an engineer: possessing both practical and theoretical knowledge. Students that work in manual labor in the building industry during their early years as student engineers should be encouraged by the advantage that this work can bring to their engineering competence and perhaps even incorporated into the curriculum somehow. Telford's success was often also attributed to his pleasant character and ability to communicate both orally and by the written word. It is said that he had a most benevolent character that made him approachable to anyone who came to him for advice. Telford also wrote 12 or more poems during his life, a number of which were published, including "Eskdale." Telford established the tradition of recording the society's proceedings, discussions and meetings and organized its library. He also encouraged people to present papers to the ICE. Telford took an important position advising on public works that were instigated in 1817 in a bid to break the severe economic depression in England after Waterloo. He had a large influence with the government due to his position as parliament's engineer that enabled him to obtain a Royal Chartership for the ICE in 1828. Telford is considered one of the greatest civil engineers of all time.

From the time of the establishment of the first professional engineering society in 1828, the role of an engineer had emerged and was being defined. Discussions of the ICE held on 2 January 1818, and their application for Royal Chartership the same year, gave some definitions.

Henry Palmer, co-founder of ICE, stated that,

> An Engineer is a mediator between the Philosopher and the working Mechanic; and like an interpreter between two foreigners must understand the language of both. The Philosopher searches into Nature and discovers her laws, and promulgates the principles and adapts them to our circumstances. The working Mechanic, governed by the superintendence of the Engineer, brings his ideas into reality. Hence the absolute necessity of possessing both practical and theoretical knowledge.
>
> (ICE, 1928)

With the explosion of natural philosophy or science, it was only expected that there should follow an explosion in its application, and so it did. By the early 19th century, the role of the engineer in relation to and in contrast with other disciplines had emerged to fill this new role of mediator between science and tradespeople. Even in this early definition, communication is a core skill; the engineer must not only be fluent in one language but must also be an interpreter able to mediate between the philosopher and the working mechanic.

The next 100 years or so saw the birth of a number of other new disciplines in science: notably mechanics, chemistry, electronics and computers. The birth of chemistry as a science can be dated to another co-founder of the Royal Society, Robert Boyle, and his book called

The Skeptical Chemist in 1661. In this book, he severely criticized the vague mystical ideas of the alchemists and again emphasized experimentation. His experimentations assisted him to reject the four element theory of the day (earth, air, fire and water) and establish the modern concept of elements. Evidence for this was also possible through experimentation because of the advance in clocks and optical instruments, telescopes, thermometers and compound microscopes in a way never before possible (Davies, 1980). According to Wankat (2009), the series of lectures in 1887 by George E. Davis at the Manchester Technical School in England and published in the chemical trade journal are usually considered the start of formal education in chemical engineering. James Watt's inventions saw the birth of mechanical engineering. The work of Michael Faraday and others saw the birth of electrical engineering, and that of James Clerk Maxwell and Heinrich Hertz that of electronics (Academic, 2016). Modern computing can also be seen as a product of engineering and the interactions between experimental science, logic and engineering practices as well as mathematics (De Mol and Primiero, 2015). Modern computing itself also holds an inherent aspect of communication within itself. Behind each of these societies and scientific discoveries there are a number of engineers and scientists with great communication and technical writing skills. Without their input and literature, engineering would not have progressed to its current level. These societies are addressing the need for scientists and engineers to clearly communicate their ideas to each other and to generations to come, so that we can stand on the shoulders of these giants.

16.3.2 Engineer novelists and play writers

The engineering profession, once born, has also worked to produce able communicators who have made a name for themselves in the world of literature. Examples include Fyodor Dostoyevsky, José Echegaray Eizaguirre and Nevil Shute Norway, whose novels often had an engineering setting.

Fyodor Dostoyevsky (1832–1916) was a lieutenant engineer who had attended the Nikolayev Military Engineering-Technical University, St. Petersburg, Russia. Dostoyevsky was contemporary with another famous Russian writer Leo Tolstoy, author of *War and Peace*, who is regarded as one of the greatest authors of all time. Dostoyevsky himself was not without influence. His novel *Poor Folk* earned him entry into St. Petersburg's literature circles in the 1840s, and his many other works, including *Notes from Underground, Crime and Punishment, The Idiot, The Brothers Karamazov*, made him one of the most highly regarded and widely read Russian writers. His books have been translated into over 170 languages. His own testimony is that it was his parents reading to him at night that gave life to his imagination. He also put his hand to politics, joining the socio-Christian Petrashevsky Circle which proposed social reforms in Russia, such as the abolition of serfdom and freedom from censorship. In 1880, he was elected vice president of the Slavic Benevolent Society and gave an impressive speech at the unveiling of the Pushkin memorial in Moscow.

José Echegaray y Eizaguirre (1832–1916) was born in Madrid, Spain. He was educated in engineering and also in economics. He worked as a civil engineer and became the Spanish minister of public works, and then the finance minister. José wrote a number of plays, his most famous being *El gran Galeoto*, about how unfounded gossip poisons a middle-aged man's happiness. Among his most famous plays are *La esposa del vengador* (1874) [The Avenger's Wife], *En el puño de la espada* (1875) [The Sword's Handle], *En el pilar y en la cruz* (1878) [The Stake and the Cross], and *Conflicto entre dos deberes* (1882) [Conflict of Duties]. Others include *O locura o santidad* (Saint or Madman?, 1877); *Mariana* (1892); *El estigma* (1895); *La duda* (The Calum, 1898); and *El loco Dios* (1900). José became a leading Spanish dramatist and was awarded the 1904 Nobel Prize for Literature for his many talented compositions which, "in an individual and original manner, have revived the great traditions of the Spanish drama" (Nobel Prize, 2014). José was the first Spaniard to win the prize and he was an engineer, which exemplifies how they can use their literary skills in a variety of ways to embellish society's experience.

Nevil Shute Norway (1899–1960) was born in London and immigrated to Australia in 1950. He was an aeronautical engineer, pilot and a co-founder of the aircraft construction company Airspeed Ltd, as well as a novelist. He had been chief engineer of the R100, a passenger-carrying airships project. After this, he teamed up with others and developed the "Airspeed Oxford," which became the standard advanced multi-engine trainer for the RAF and British Commonwealth. Shute was made a Fellow of the Royal Aeronautical Society for this work and also for designing a hydraulic retractable undercarriage for the Airspeed Courier. He also designed weapons during World War II, including a special anti-submarine missile. To avoid embarrassment, he had used "Nevil Shute" as his pen name in the novels he authored, and his full name in his engineering career. In the 1950s and 1960s, Nevil Shute was one of the world's best-selling novelists. His celebrity as a writer was later uncovered, and rather than having a negative effect on his career, it caused the Ministry of Information to send him to the Normandy landings and later to Burma as a correspondent. All 23 of his novels were reprinted in 2009. Many of them included an aviation and engineering setting, and he liked to highlight the role science and engineering could play in improving human life. It is important to impress this role upon the new generation of engineers, particularly today in the context of the sustainable development goals, which have been laid down as a gauntlet – Norway often quoted the anonymous quip, "It has been said an engineer is a man who can do for five shillings what any fool can do for a pound" (Anonymous, 1969).

16.3.3 An international icon

Alexandre Gustave Bonickhausen dit Eiffel (1832–1923) is renowned for his involvement in the Eiffel Tower in Paris and perhaps less so for his contribution to building the Statue of Liberty in New York. These projects were of great national and international significance and required great diplomatic skill to manage a project in the public eye. Gustave's symbolic intention for the tower, which he gave in a speech in presenting the design of the tower to the public, is still remembered and often quoted. Eiffel said,

> [N]ot only the art of the modern engineer, but also the century of Industry and Science in which we are living, and for which the way was prepared by the great scientific movement of the eighteenth century and by the Revolution of 1789, to which this monument will be built as an expression of France's gratitude.
>
> (Roumen and Mladjov, 2014)

Despite intense public criticism in the media during the construction, Eiffel's Tower stands today as a monument commanding international recognition. It is less well known that Eiffel conducted valuable research on the structural properties of cast iron and made significant contributions to the fields of meteorology and aerodynamics. Eiffel also personally instigated detailed meteorological measurements and record keeping at the Tower and later at 25 other locations in France. He also built wind tunnels to investigate the structural actions of wind on structures and later aircraft designs. Alexander Graham Bell, scientist-engineer and innovator who patented the telephone, said of Eiffel's written works that his writings upon the resistance of the air have already become classical. They have given engineers the data for designing and constructing flying machines upon sound, scientific principles. His contribution to the science of aerodynamics is probably of equal importance to his work as an engineer. All this was only achieved because of his competence in communicating to the public socially and the care he took to communicate his results clearly to other professionals through his technical papers.

16.3.4 An American president

Herbert Clark Hoover (1874–1964) started out as a mining engineer and became the 31st president of the USA. Hoover became a millionaire through his own private mining consulting business. Hoover was not only good at making money but also took an interest in more humanitarian efforts. After World War I, he was appointed head of the European Relief and Rehabilitation Administration and oversaw the moving of 34 million tons of American food, clothing and supplies to 20 nations in war-devastated Europe. During the war he had already organized the US Food Administration and aided the evacuation of Americans troops trapped in Europe. His negotiation and diplomacy can also be seen in his involvement in the Versailles Peace Conference when, as the director of the president's Supreme Economic Council, in 1918 President Woodrow Wilson relied on Hoover's counsel. His negotiation skills were again called upon in 1919, where he was to become instrumental in the negotiation of a treaty with Canada. His proven communication, negotiation and mediation skills and aptitude helped win him the American presidency from 1929 to 1933. The famous Hoover Dam in Colorado is named in his honor.

16.3.5 Australian engineers

There are a number of notable Australian engineers whose names will be familiar to most Australians through the various universities, companies, roads and other things named after them in their honor.

Sir John Monash (1865–1931) was one of Australia's greatest military commanders in World War I and took part in and survived the Gallipoli campaign. At the height of his military career, he was made a Knight Commander of the Bath, appointed commander of the Australian Corps and promoted to lieutenant-general. His non-military engineering achievements include contracting for bridge building, specializing in the then new reinforced concrete. His successful utilization of the Victoria brown coal deposits for efficient energy production, while serving as the general manager of the State Electricity Commission of Victoria, is regarded as one of his most difficult engineering successes. Monash, in association with others, also formed two businesses, one during the depression in 1894 and again in 1905 when he formed the "Reinforced Concrete and Monier Pipe Construction Co Ltd." Monash University in Melbourne is named in his honor. Monash began by studying engineering in conjunction with arts and law. He involved himself in student politics and co-founded the Melbourne University Union, which was aimed to be a source of encouragement for students to be involved in extracurricular activities at their university. In the 1920s, he represented return soldiers and became director-general of repatriation and demobilization. He continued his affiliation with and concern for Jewish affairs and became vice chancellor of the

University of Melbourne. Monash maintained a strong public profile and was respected for his careful and detailed planning, clear communication and ability to encourage his staff to give their best.

Dr John Jacob Crew Bradfield (1867–1943) studied civil engineering at the University of Sydney. He graduated with honors in 1889 and with a Master of Engineering in 1896. Bradfield is most remembered for being the engineer-in-charge of Sydney Harbour Bridge, which position he held for more than 20 years. He represented the government in dealings with the contractors; this mediatory role required competence in negotiation and communication. Bradfield prepared reports for parliament and represented the government in dealings with construction contractors of the Sydney Harbour Bridge. Bradfield was a foundation member of the Institution of Engineers Australia (EA), a position which shows his engagement and communication with the broader engineering profession. EA is the leading professional engineering organization in Australia and mirrors ICE in the UK. Chartership with EA is recognized internationally. Bradfield also served as deputy chancellor of the University of Sydney from 1942–43. He later became a consulting engineer for the Story Bridge and technical adviser to the constructors of the Hornibrook Highway near Brisbane and helped plan and design the St Lucia site of the University of Queensland.

Roger William Hercules Hawken (1878–1947) was chairman of the Queensland Institution of Engineers, president of the Institution of Local Government Engineers of Australia, and a member of the Australian Association for the Advancement of Science. He was chair of engineering and held the office of dean for 28 years at University of Queensland. Hawken's experience was deepened by working internationally in Malay for the British government as the acting executive engineer. He took a leading role in the formation of the Institution of Engineers from an amalgamation of various engineering societies then existing in Australia, including the Queensland Institution of Engineers whose part he represented. He was awarded the Peter Nichol Russell Medal by the institution in 1931. As well as being a member of the institution, he also wrote many technical papers, notably several on column design. He was a prolific writer, producing innumerable technical papers. His technical knowledge was supplemented by his wide experience of people and affairs and his ability to communicate effectively.

Sir Ove Nyquist Arup (1895–1988) was born in the UK, educated in Denmark and made his mark in Australia. He studied mathematics and philosophy and then engineering and is considered one of the foremost architectural structural engineers of his time (Ahm, 2015). He led the engineering design of the Sydney Opera House and made its construction possible when it seemed like its formation would not become a reality. The Opera House provided an opportunity for ground-breaking use of precast concrete, structural glue and computer analysis, giving Arup's firm a reputation, despite difficulties with the architect Jørn Utzon (Ahm, 2015). The Sydney Opera House is one of the most well-known buildings in the world – Sydney is the first major city to enter into the New Year, and so the Opera House receives international media coverage annually. Other projects he is known for include the Kingsgate foot-bridge over the River Wear at Durham and Coventry Cathedral in the UK, and in Paris the Centre Pompidou. Arup also took part in various designs for the war effort in World War II, including the design of air-raid shelters, which he published a number of papers on. He also played a significant part in the design of the Mulberry temporary harbors which helped make the Normandy landings possible on D-Day. Today he is perhaps more well known through the architectural structural engineering company which he founded in his own name and has come to be known simply as Arup, which is one of the largest and most important companies of its kind in the world. Arup's vision was to improve understanding between the professions and particularly between architects and engineers. Thus, Arup began as a practice of architects, engineers and quantity surveyors. He was a member of many professional engineering bodies and was vice president of the Institution of Civil Engineers from 1968 to 1971. He received prestigious medals in both architecture and structural engineering, namely the RIBA gold medal for architecture and the Institution of Structural Engineers' gold medal. He received a number of honorary doctorates from universities, and he was one of the original fellows of the Royal Academy of Engineering (Ahm, 2015). In 1965, he was knighted to the order of Order of the Dannebrog and 10 years later as Knight Commander First Class. His warmth and exactness in communication is reported to have made any conversation with him memorable. He was a visionary who hated compromise (Ahm, 2015). He stands as a lesson for us in interdisciplinary communication, particularly between architects and designers and engineers.

16.3.6 Contemporary engineers

Stuart Burgess (contemporary) is professor of engineering design, head of mechanical engineering and leader of the design engineering group at Bristol University in the UK. He taught engineering design at Cambridge and is a world expert on biomimetics, which involves imitating design in nature. His projects include copying insect wing motion (Burgess *et al.*, 2003), designing a prosthetic knee joint that imitates the human joint (Burgess *et al.*, 2014a; Etoundi *et al.*, 2013), examining the aerodynamic braking and flapping mechanism of birds (Burgess *et al.*, 2014b, 2015), analyzing the structural integrity of trees (Burgess and Pasini, 2004) and designing spacecraft for the European space agency (Burgess, 1995). Burgess holds seven patents and won the highly esteemed Company of Turners gold medal for his design of the solar array deployment system on the ENVISAT earth observation satellite, a project worth GBP 1.4 billion. His communication skills are demonstrated not only in his publication of many engineering and science journal articles, but also through several books in which he argues that the complex designs in nature which he frequently imitates through biomimetics is evidence of a designer. Thus, he lends his engineering mind and experience to the debate in broader society over the origins of life and various life forms in publications such as *Hallmarks of Design* (Burgess, 2000). In this he is continuing in the tradition of Sir Francis Bacon, Sir Isaac Newton, John Locke and others of the scientific revolution, who not only published scientific work but also wrote and thought deeply on ontological questions. This modern movement, which some call Intelligent Design, is as controversial as Galileo's heliocentric universe yet is gaining support. To write in this field takes confidence, diplomacy and advanced communication skills, as every work is liable to be highly scrutinized by critics. Burgess even had critics insisting that he be fired. Galileo, who advocated the heliocentric universe in the face of a majority of geocentric universe supporters, also had to be a good communicator to get his ideas across in the face of criticism. Dealing with criticism in a mature and comprehensive manner is an important talent for the engineer who may be involved in projects with various social and political importance.

Lizzie Brown (contemporary) was for three consecutive years named one of Engineers Australia's 100 Most Influential Engineers for her role as CEO of Engineers without Borders (EWB) in the community category (Kannegieter, 2013, 2014, 2015). EWB is a not-for-profit organization that seeks to create systemic changes in society through humanitarian engineering. The community category was added to better reflect the sectors of society and industry that engineers influence (Kannegieter, 2013). Brown says that her inspiration often comes from the passion of the volunteers. Brown joined EWB in 2004 and helped found the South East Queensland Chapter. She was CEO of EWB from 2010 to 2015, and then moved to lead their EWB connect section of the organization, which connects professional engineers with projects. With a strong emphasis on the millennium development goals and now the sustainable development goals (SDG), EWB is a voice for the potential of engineers' contributions to current global issues. It is worth noting that the SDG do not exclude the so-called developed world; Australia also falls short in some of the SDGs. During her time as CEO, EWB had 15,000 volunteers and engaged several engineering companies to work with over 40 community organizations across seven countries. EWB was able to bolster the engineering capacity of these organizations. EWB was started in Australia by a bunch of student engineers. This should be an encouragement for young engineers and also reflects on the very first professional engineering society, ICE, which was also started by three young engineers. EWB focuses on "human-centered design," which perhaps should be considered synonymous with "civil engineering," engineering for civilization and engineering for people. An organization such as EWB requires a CEO with not only technical knowledge but also great communication skills and a heart to care for developing countries. First, they need to listen and understand the needs of a community and then be able to match corresponding needs with engineering solutions. Brown has over 5 years technical experience in Australia's water sector as well as experience overseas with companies including OMV Porterra and WRM Water and Environment. Her communication skills were solidified through her involvement in education. She was EWB's director of education in 2006 and was awarded a Chief Executive Women Leadership Scholarship and a Churchill Fellowship centered on education, training and research programs for sustainable development in the UK and USA. In 2009, she became EWB's operations director before becoming CEO in 2010.

It is a pleasure to end this commendation of engineers with great communication skills with a reference to global needs, which are the pressing challenge of our time. They are not new, like the need for supply of clean drinking water. With the advent of mass media, everyone is aware of these needs. Knowledge means responsibility, and with such high levels of development in science and technology, it is the disgrace of our time that these needs have not been met. The UN in the SDGs has put together, into a readable quantitative format, what has always been known qualitatively: that globally there are basic needs of humankind, such as food and water, that are not being met. A quick look over the now 17 goals defined in the SDGs should make any engineer realize that many of the goals fall squarely into their domain. A little research will show that often the problems are tied in with political issues, and in this respect the engineer that desires to help must now deal with socioeconomic issues, ethical issues, war and politics. This will require an even higher degree of communication skills to be successful, as well as interdisciplinary cooperation, which again requires talent in communication. Interestingly, one of the two US presidents who studied engineering, Hoover, did exactly that in World War I, running the US Food Administration during the war and organizing the evacuation of trapped Americans in Europe. The needs of our time stand as a gauntlet for the engineer. Various communication skills will be crucial tools for the engineer to successfully instigate solutions. Jeremy Smith from the Australian National University (ANU) Research School of Engineering, who won the 2017 Australian Award for University Teaching, states that his philosophy is inspired by "the power of engineering to transform lives and create positive change for individuals and entire communities." He notes the urgency of these issues, stating that "The need for positive change is urgent, with hundreds of millions of people lacking access to water, shelter, energy and health care in every country, including Australia" (EA, 2017). Smith has pioneered a new approach to engineering education which involves students traveling overseas and applying their skills to solve real-life problems. In a conference held in Australia in 2016, it was reported that engineering educators who partook in 2-week placements in humanitarian design summits benefited substantially in their teaching practice and professional development, and particularly in their ability to teach the sustainable development goals (Brown *et al.*, 2016).

Table 16.1 lists the selected engineers discussed chronographically for reference and consideration. Some of their key positions and achievements are mentioned, but it should be noted that many of them held and achieved much more.

In summary, engineers through great communication skills have contributed to everything from the preservation of life, to improving the standard of living, to exploring the deeper questions in life and society's questions about the origins of our existence. They have all used their communication skills to great advantage to excel in their careers and/or advance the engineering profession in various ways. The list is only a selection of the many successful engineers who have developed advanced communication skills. This shows that many engineers had high-level communication skills which enabled them to serve at a very respectable

Table 16.1 Engineers holding important positions with great communication skills

Name	*Position and achievements*	*Comment on Communication Skill*
Johannes Guttenberg (1400–1468)	German inventor of the movable type printing press.	Enabled the mass production of literature.
Sir Christopher Wren (1632–1723)	Co-founder and president of the Royal Society.	Role in Royal Society's formation and running as its president.
Sir Isaac Newton (1642–1727)	President of the Royal Society.	Author of famous *Mathematical Principles of Natural Philosophy*.
John Smeaton (1724–1792)	Referred as father of civil engineering.	Well known for keeping records of practice.
(Sir) Thomas Telford (1757–1834)	First president of the Institution of Civil engineers, UK.	Famous for recording the proceedings and discussions of meetings.
Fyodor Dostoyevsky (1821–1881)	Lieutenant engineer.	One of the most highly regarded Russian writers.
José Echegaray y Eizaguirre (1832–1916)	Civil engineer, Spanish minister of Public Works.	Leading Spanish dramatist, awarded the 1904 Nobel Prize.
Alexandre Gustave Eiffel (1832–1923)	Built the Eiffel Tower in Paris.	Important paper on the design of Eiffel to the Société des Ingénieurs Civils.
Sir John Monash (1865–1931)	One of Australia's greatest military commanders.	Admired for his articulate communication and strong public profile.
Dr John Bradfield (1867–1943)	Engineer in charge of Sydney Harbour Bridge.	Represented the government in building the Sydney Harbour Bridge.

Name	*Position and achievements*	*Comment on Communication Skill*
Herbert Clark Hoover (1874–1964)	31st American president.	Ran the US Food Administration during the war.
Roger Hawken (1878–1947)	Chairman of the Queensland Institution of Engineers.	Served as dean at University of Queensland.
Sir Ove Nyquist Arup (1895–1988)	Led the engineering design of the Sydney Opera House.	Formed an international company for communication with governments.
Nevil Shute Norway (1899–1960)	Co-founder of the aircraft construction company Airspeed Ltd.	One of the world's best-selling novelists in the 1950s and 1960s.
Stuart Burgess (contemporary)	Spacecraft design for the European space agency.	Author of several books explaining the significance of design in nature.
Lizzie Brown (contemporary)	CEO of Engineers without Borders.	Named one of Australia's 100 Most Influential Engineers 2013.

position in society. This thus indicates that an engineer can also become an effective communicator. This information might encourage junior engineers and engineering students to enhance their communication skills and build their career.

16.4 Reflection on technical writing

This section presents the experience of Ataur Rahman (the third author of this paper) on technical writing as a student and academic. He completed his PhD from Monash University (Australia) in 1996 and then worked as a University Academic in Australia for over 20 years. Since English is not his first language, informal training in technical writing by his supervisory team (consisting of Professor Russell Mein, Mr Erwin Weinmann and Dr Bryson Bates in Monash University) assisted him in improving his technical writing standard from a "limited" to an "acceptable" level by the end of his PhD candidature. He wrote his first refereed conference paper in 1996, which needed significant guidance by his supervisory panel to reach the desired satisfactory level. However, his second conference paper, written in 1997, demonstrated a distinguishable improvement as noted by his supervisory team. To date, he has authored over 360 refereed publications, and his technical writing skills have been constantly improving since his PhD completion.

Ataur Rahman has supervised over 25 doctoral and 70 masters, honors and final-year engineering project students. These students required writing major/minor theses and/or project reports. According to his subjective judgment, only about 15% of these students had acceptable technical writing skills at the beginning of their candidatures. Most of his second-year fluid mechanics students had great difficulty in writing the literature review part of the laboratory report (Rahman, 2017; Rahman and Al-Amin, 2014). The generic problems in the technical writings of his students are summarized in Table 16.2.

Informal training on technical writing by him to his students mainly via meetings and feedback on written drafts have made "a difference" to their technical writings. A good number

Table 16.2 Common problems in technical writing among engineering students

Nature of communication problems		
Not thinking well enough before the start – what to write, what the main points are that need elaboration, and how to organize these points in the right order to convey the issue appropriately.	Lack of proofreading.	Do not enjoy their own or others' writings.
Not preparing structure of the report/thesis/paper in great detail before the start of writing.	Little effort in finding and correcting own mistakes.	Discussion is poor: e.g. lack of details of an observation, lack of comparison of results with similar studies and little on implications of assumptions.
Inconsistency in presentation: e.g. in line spacing, font usage, chapter headings and in-text and chapter-end referencing.	Little effort in becoming a good writer, tendency to write something for submission to meet cutoff dates.	Criticisms of others' work without proper justification.
Loose connection among different sections and sub-sections.	Not learning from the best examples.	Unnecessarily long reports.
Repetition of facts and figures unnecessarily.	Poor abstract of the report that lacks in important information from the body of the report.	Too long and complex sentences with different parts being poorly connected.
Absence of proper abbreviations at the right places.	Literature review is not critical, e.g. comparison and contrast among existing articles are generally absent.	Problem in use of articles "the," "a" and "an."
Improper use of italic fonts, e.g. symbols should be in italic font in a technical document.	There is poor flow of ideas, and readers often have to struggle in comprehending the main focus of the report.	Mixing of past, present and future tenses inappropriately.
Tables and figures are not mentioned/discussed in the body of the report.	The writing is not focused, e.g. no/little effort in proving/disproving a hypothesis/research question.	Use of capital letters, e.g. poor understanding of proper and common nouns.
Tables and figures are not numbered.	Leave it to the supervisor for finalization.	Use of semicolon and comma.
Copying from other sources without rewording and due references.	Take criticism negatively, e.g. "My supervisor always criticizes my poor writing, so what is the point in improving my writing."	Accepting supervisor's corrections automatically (without going through these); using auto editing of MS word.

of his honors students produced excellent theses, e.g. five State Awards were received by his honors students (including the "Best Engineering Thesis of the Year Award in NSW State" on two occasions). His research students have published over 110 referred papers based on their theses during the last 10 years; an exemplification of the sound technical writing skills of his former students. He recognized that some of his former students did not achieve acceptable technical writing skills. To overcome these shortcomings, these students would need active industry mentorship to enhance their writing skills.

16.5 Possible solution

There are no fast solutions to address this problem. The following suggestions, however, if followed and incorporated, can be a big step in alleviating some of the fundamental issues associated with the communication problems of engineering students. One important step this paper recommends is including, in the duration of a standard engineering degree, one subject per year focusing on communication enhancement techniques and strategies: English grammar and stylistic conventions appropriate for the workforce and communicating with professionals. This suggestion would likely cause a dilemma for universities with set syllabuses and subjects, which would potentially have to lengthen the degree program. To address this, it is worth considering either culling a few subjects whose relevance can be scrutinized, shortening and combining similar subjects, or introducing an elective subject on communication skills. The current imbalance in engineering curriculums in Australia calls for a serious reevaluation in incorporating some communication and literary-based subjects. In developing a new subject on communication skills, a blended learning approach can be adopted. Lectures and tutorials could be pre-recorded, which can be read by the students at a convenient time. This can save the lecturer's time and offer greater flexibility to student learning. Educational resources on communication skills can also be developed by a group of universities or by Engineers Australia for common use across universities.

The communication unit in engineering courses should focus on the following steps:

- Early intervention for new engineering students who are weak in writing and providing them with a proper learning methodology;
- Limit the use of electronic editing in MS Word by the teachers, which can be accepted by the students automatically without thinking; instead, do corrections on hard copy that allows students to think carefully before rewriting;
- Ensure a proper structure/outline before writing starts;
- Allow engineering students to interact with students of other professions, like law and sociology, who are traditionally good at writing;
- Allow engineering students to attend social/general presentations in other schools;
- Model the method of teaching English to the engineering mind where possible (e.g. engineers tend to relate to learning rules, such as rules of grammar and methods of writing);
- Make the engineering students aware of the value of communication, e.g. it can take one to the top of a company, and there is little career enhancement prospects without effective communication skills and
- Make the engineering students aware that they need to influence political and community leaders to initiate mega engineering projects like the Snowy Mountains engineering scheme in Australia.

16.6 Summary

This chapter presents communication issues of Australian engineering students. It has been found that many engineering students lack in communication skills and do not comprehend the importance of this in their studies. It is also noted that many engineers rose to a high level in their careers using their great communication and technical skills. To improve the situation of communication skills of engineering students in Australia, a proposal is presented in this paper, which involves the introduction of a new subject on formal communication similar to some universities in the USA. In developing a new subject on communication skills, a blended learning approach can be adopted, which provides greater flexibility in learning. It is suggested that Engineers Australia prepares a position paper on the technical writing issue to investigate the matter in order to recommend a suitable measure for implementation that can improve the communication skills of engineering students in Australia.

References

Academic, B. (2016) *Engineering, Encyclopedia Britannica*, viewed 3 Nov 2017. Available from: academic.eb.com.ezproxy.uws.edu.au/levels/collegiate/article/engineering/105842#.

Academic, B. (2017a) *Johannes Gutenberg*, viewed 28 Nov 2017. Available from: http://academic.eb.com.ezproxy.uws.edu.au/levels/collegiate/article/Johannes-Gutenberg/38592.

Academic, B. (2017b) *Printing press*, viewed 28 Nov 2017. Available from: http://academic.eb.com.ezproxy.uws.edu.au/levels/collegiate/article/printing-press/102489.

Academic, B. (2017c) *Gutenberg Bible*, viewed 28 Nov 2017. Available from: http://academic.eb.com.ezproxy.uws.edu.au/levels/collegiate/article/MTV/38593.

Academic, B. (2017d) *Sir Isaac Newton, Encyclopedia Britannica*, viewed 3 Nov. 2017. Available from: academic.eb.com.ezproxy.uws.edu.au/levels/collegiate/article/Sir-Isaac-Newton/108764.

Ahlström, G. (1978) Higher technical education and the engineering profession in France and Germany during the 19th century: A study on technological change and industrial performance. *Economy and History*, 21(2), 51–88.

Ahm, P. (2015) *Arup, Sir Ove Nyquist (1895–1988), H. C. G. Matthew Oxford Dictionary of National Biography. Ed. H. C. G. Matthew and Brian Harrison*. OUP, Oxford, 2004. Online ed. Ed. David Cannadine viewed 7 Nov. 2017 Available from: www.oxforddnb.com.ezproxy.uws.edu.au/view/article/40049.

Anonymous (1969) *Quoted in Nevil Shute Norway's Autobiography "Slide Rule"*, 2nd ed. Pan, London.

Britannica, E. (2017) *Sir Christopher Wren*, viewed 3 Nov. 2017. Available from: academic.eb.com.ezproxy.uws.edu.au/levels/collegiate/article/Sir-Christopher-Wren/77544.

Brown, N.J., Price, J., Turner, J.P. & Colley, A. (2016) Professional development within study abroad programs for engineering educators to gain confidence in preparing students to contribute to the Sustainable Development Goals. In: *27th Annual Conference of the Australasian Association for Engineering Education: AAEE* 2016. Engineers Australia, Coffs Harbour, NSW. pp. 96.

Burgess, S. (2000) *Hallmarks of Design*, Evidence of purposeful design and beauty in nature. Day One, Leominster, UK.

Burgess, S.C. & Etoundi, A.C. (2014a) Performance maps for a bio-inspired robotic condylar hinge joint. *Journal of Mechanical Design, Transactions of the ASME*, 136, 11.

Burgess, S., Lock, R., Wang, J., Sattler, G. & Oliver, J. (2014b) The effect of aerodynamic braking on the inertial power requirement of flapping flight: Case study of a gull. *International Journal of Micro Air Vehicles*, 6(2), 117–127.

Burgess, S.C. (1995) The design of an advanced primary deployment mechanism for a spacecraft solar array. *Journal of Engineering Design*, 6, 4, 291–307.

Burgess, S.C., Alemzadeh, K. & Zhang, L. (2003) The development of a miniature mechanism for producing insect wing motion. In: Cullis, A.G & Midgley, P.A (eds) *Microscopy of Semiconducting Materials 2003 – Proceedings of the Institute of Physics Conference*, Cambridge, 180. pp. 237–244.

Burgess, S.C., Lock, R.J., Wang, J., Sattler, G.D. & Oliver, J.D. (2015) The energy benefits of the pantograph wing mechanism in flapping flight: Case study of a gull. *International Journal of Micro Air Vehicles*, 7, 3, 275–284.

Burgess, S.C. & Pasini, D. (2004) Analysis of the structural efficiency of trees. *Journal of Engineering Design*, 15, 2, 177–193.

Davies, J.T. (1980) *Chemical Engineering: How Did It Begin and Develop?* ACS Publications, Oxford, UK.

De Mol, L. & Primiero, G. (2015) When logic meets engineering: Introduction to logical issues in the history and philosophy of computer science. *History and Philosophy of Logic*, 36, 3, 195–204.

EA (Engineers Australia) (2017) ANU teacher recognised for innovative approach to engineering education. Available from: www.engineersaustralia.org.au/portal/news/anu-teacher-recognised-innovative-approach-engineering-education.

Etoundi, A.C., Lock, R.J., Vaidyanathan, R. & Burgess, S.C. (2013) A bio-inspired condylar knee joint for knee prosthetics. *International Journal of Design and Nature and Ecodynamics*, 8, 3, 213–225.

Gooden, P. (2009) *The Story of English How the English Language Conquered The World*. Quercus Publishing Plc, London.

Hitchcock, H.R., Lloyd, S., Rice, D.T., Lynton, N., Boyd, A., Carden, A., Rawson, P. & Jacobus, J. (1963) *World Architecture an Illustrated History*. The Hamlyn Publishing Group Ltd, London, UK.

Iliffe, R. & Mandelbrote, S. (2017) The Life and Work of Isaac Newton at a Glance, The Newton Project, viewed 3 Nov 2017. Available from: www.newtonproject.ox.ac.uk/his-life-and-work-at-a-glance.

ICE (Institution of Civil Engineers) (1928) *A Brief History of the Institution of Civil Engineers, with an Account of the Charter Centenary Celebration*, June 1928. Institution of Civil Engineers, London.

Kannegieter, T. (2013) Australia's 100 most influential engineers. *Engineers Australia Magazine*.

Kannegieter, T. (2014) Australia's 100 most influential engineers. *Engineers Australia Magazine*.

Kannegieter, T. (2015) Australia's 100 most influential engineers. *Engineers Australia Magazine*.

Kohn, L. (2015) How professional writing pedagogy and university-workplace partnership can shape the mentoring of workplace writing. *Journal of Technical Writing and Communication*, 45(2), 166–188.

Lennox, J.C. (2009) *God's Undertaker Has Science Buried God?*, 2nd ed. Lion Hudson plc, Oxford.

Mills, A. & Turvey, P. (1979) Newton's telescope. An examination of the reflecting telescope attributed to Sir Isaac Newton in the possession of the royal society. *Notes and Records of the Royal Society*, 33, 2, 133–155.

Nationsonline (2017) Most widely spoken Languages in the World, viewed 12/1/17, Available from: www.nationsonline.org/oneworld/most_spoken_languages.htm.

Newton, I. (1687) *Philosophiæ Naturalis Principia Mathematica (Mathematical Principles of Natural Philosophy)*. Benjamin Motte, London.

Nobel Prize (2014) José Echegaray – Biographical, Nobel Media AB, viewed Web. 13 Jul 2016, Available from: www.nobelprize.org/nobel_prizes/literature/laureates/1904/eizaguirre-bio.html.

Rahman, A. (2017) A blended learning approach to teach fluid mechanics in engineering. *European Journal of Engineering Education*, 42, 3, 252–259.

Rahman, T., Amos, C.C. & Rahman, A. (2016) How to enhance technical writing skills of new generation engineering students? In the *Proceedings of International Conference on Engineering Education and Research*, 21–24 Nov 2016, Sydney, Australia.

Rahman, A. & Al-Amin, M. (2014) Teaching of fluid mechanics in engineering course: A student-centered blended learning approach. In: Alam, F. (ed) *Using Technology Tools to Innovate Assessment, Reporting, and Teaching Practices in Engineering Education*. IGI Global Publisher, Hershey, PA, USA. pp. 12–20.

Reave, L. (2004) Technical communication instruction in engineering schools: A survey of top-ranked U.S. and Canadian programs. *Journal of Business and Technical Communication*, 18, 4, 452–490.

Rivington, C.A. (1984) Early printers to the royal society 1663–1708. *Notes and Records of the Royal Society of London*, 39,1, 1–27.

Roumen, V. & Mladjov, S.E. (2014) Quoted in The Eiffel Tower at 125 years. *Structure Magazine: Historic Structures*.

Skempton, A.W. (2004) John Smeaton (1724–1792) viewed. Available from: www.oxforddnb.com.ezproxy.uws.edu.au/view/article/25746 [Accessed 3 November 2017].

Steiner, D.G. (2011) The communication habits of engineers: A study of how compositional style and the affect the production of oral and written communication of engineers. *Journal of Technical Writing and Communication*, 41(1), 33–58.

Syfret, R. (1948) The origins of the royal society. *Notes and Records of the Royal Society of London*, 5, 2, 75–137.

Vest, D., Long, M. & Anderson, T. (1996) Electrical engineers' perceptions of communication training and their recommendations for curricular change: Results of a national survey. *IEEE Transactions on Professional Communication*, 39(1), 38–42.

Wankat, P.C. (2009) The history of chemical engineering and pedagogy the paradox of tradition and innovation. *Chemical Engineering Education*, 43, 3, 216–224.

Warnock, S. & Kahn, M. (2007) Expressive/exploratory technical writing (XTW) in engineering: Shifting the technical writing curriculum. *Journal of Technical Writing and Communication*, 37(1), 37–57.

Chapter 17

Integration and adaptation of e-technology to deliver technical education in public and private universities of Bangladesh

A. Ahmed[1], *M.A. Rahman*[2], *M.S. Khan*[3], *F. Ferdousi*[4], *M. Haque*[5] *and A. Rahman*[6]

[1]*Nutrition and Food Engineering Department, Daffodil International University, Dhaka, Bangladesh*

[2]*School of Computing, Engineering and Mathematics, Western Sydney University, NSW, Australia*

[3]*Board of Trustees, Daffodil International University, Dhaka, Bangladesh*

[4]*Business Administration Department, East West University, Dhaka, Bangladesh*

[5]*EnviroWater Sydney, NSW, Australia*

[6]*School of Computing, Engineering and Mathematics, Western Sydney University, New South Wales, Australia*

ABSTRACT: Delivery of tertiary education is rapidly changing across the developed countries due to wider adoption of e-technology. However, most of the developing nations are not keeping up to date with the new education delivery systems such as blended learning and flipped classroom approaches. This chapter presents an overview of the technical education delivery systems in Bangladesh and presents the opportunities and challenges by two case studies: Bangladesh Open University and Daffodil International University. It has been found that Bangladesh is making some progress in delivering online technical education; however, it needs significant capacity building to develop an effective and viable online technical education system.

17.1 Introduction

&&The need for changing technological workforce that is becoming more computer-oriented than 20 years ago warrants a shift from the traditional "lecture – example – homework" format of course delivery to a more learner-centered approach (Ahmed *et al.*, 2016). The traditional "chalk and talk" approach is being phased away rapidly from the university education systems in developed countries. In this regard, several new education delivery methods are becoming popular in the developed countries. The use of learning management systems (e.g. Proprietary Blackboard and open-source Moodle software) has been playing a major role in these new education delivery methods (Martin, 2012). For example, the flipped classroom approach is emerging as an effective method in which lectures and tutorials are pre-recorded for students to listen to before coming to the classes, so that class time can be used for active learning.

Massive online courses (MOOCs) are attracting millions of students to a course at little or no cost. The shift toward MOOCs is growing and more "actors" have been entering into this market. This shift could put the traditional course offerings by many universities in danger, since MOOCs can be accessed by anybody in the world living anywhere with fast internet access at a very low price. As an example, one course offered by a Stanford faculty member in 2011 received 160,000 students; about 23,000 of these students completed the 10-week course (Lewin, 2012).

Another method becoming popular is known as the blended learning approach, where the best aspects of face-to-face and online deliveries are utilized to enhance learning experiences of the students. For example, Rahman (2017) showed that teaching and learning of fluid mechanics for civil and mechanical engineering students by a blended learning approach at the University of Western Sydney enhanced the learning experiences of the enrolled students. In this adopted blended learning approach, online recorded lectures, online recorded tutorials, hand-written tutorial solutions, a discussion board and online practice quizzes were made available to students via the vUWS system (online method), and the face-to-face lectures were used to provide more interactive discussion to enhance students' learning.

Engineering education in Bangladesh started in 1876 with the establishment of the then Survey School of Dhaka by the British Raj. In 1908, it was upgraded to Ahsanullah School of Engineering with the provision of technician-level education (diploma in engineering) and further upgraded to a college by offering degree courses in civil, electrical and mechanical engineering after the creation of East Pakistan in 1947. The initial intake capacity of 120 students in the college was increased to 240 students in 1960. With a view to organizing and developing both the undergraduate program and introduction of postgraduate study and research to meet the requirement of economic developments in the country, Ahsanullah Engineering College was upgraded to the status of a university (named the Bangladesh University of Engineering and Technology (BUET)) on 1 June 1962 by the East Pakistan government through promulgation of an ordinance.

To meet the growing demands for engineers and to provide higher educational facilities in different parts of Bangladesh, many new engineering and technical universities have been established in the country since 1962. As mentioned by Chowdhury and Alam (2012), there are about 7000 yearly intakes of engineering students in Bangladesh, i.e. 44 engineering students per million people, whereas there are 1344 and 214 engineering students per year per million people in South Korea and India, respectively.

In Bangladesh, three main types of universities (public, private and international) are providing tertiary education service to the students. There are 29 public, over 54 private and 2 international universities in Bangladesh (Chowdhury *et al.*, 2008). Besides the degree-awarding universities, there are a number of colleges affiliated with the National University of Bangladesh. Most of these colleges offer bachelor courses (e.g. Bachelor of Arts (BA), Bachelor of Science (BSc) and Bachelor of Commerce (BCom)) at the tertiary level. At present, the number of National University–affiliated colleges is around 1400. At the time of independence of Bangladesh in 1971, only four public universities were serving about 70 million people in Bangladesh.

The private universities started growing since 1992, when the Private University Act by the nation's parliament was enacted. Since then, the country has experienced a spectacular growth in private universities. Nevertheless, access to tertiary level education is still very poor in Bangladesh; only about 12% of high school finishers (i.e. year 12 graduates) can enter into higher education due to limited intake capacity of universities in Bangladesh.

Moreover, more than 80% of these students are in the National University–affiliated colleges; the remaining are fortunate enough to be in the public and private universities. This is due to the limited number of seats available in the universities. In a 2008 statistic presented by Chowdhury *et al.* (2008), it was found that only about 6000 students have the opportunity to be admitted into engineering universities in Bangladesh from a population of over 150 million.

Among all the engineering universities in Bangladesh, BUET is the most reputed one. Its predecessor was known as Ahsanullah Engineering College; the academic programs of this college were based on the British undergraduate curriculum (Rashid, 1958). The engineering curriculum went through notable changes in 1953 through a linkage program with the Texas Agriculture and Mechanical College (presently known as Texas A&M University) (Rahman and Haseeb, 2007). Since then, BUET has maintained its academic relationship with many foreign universities through different linkage programs. All the engineering universities in Bangladesh are providing major engineering courses. For example, BUET provides 11 branches of engineering degrees to the students: civil, mechanical, chemical, water resources, material and metallurgical, naval architect and marine, industrial and production, electrical and electronic, computer science, biomedical, and glass and ceramic.

The current mode of teaching in engineering and technological universities in Bangladesh is generally "classroom-based/teacher-centered" where students get their lessons face to face during the lectures and tutorial classes. Laboratory classes are also conducted face to face, where students do experiments under the supervision of the lab teachers. Currently, little online or blended learning (i.e. mixture of face-to-face and online) has been implemented by the public universities in Bangladesh. In addition, use of computers/projectors in classroom is still limited. Only a few teachers have adopted the PowerPoint presentation along with black/whiteboard teaching in the classroom; most of them are still following the "chalk and talk" approach of teaching. Evaluation of the students is based only on a few traditional written examinations (e.g. class tests, assignments and final examination). Industry-university relationships are also not significant for most of the universities. In Bangladesh, most of the teachers follow a deductive approach (i.e. teacher-centered) in teaching, in which they deliver and explain the concepts and then expect students to complete given tasks to practice the concepts.

This chapter presents an overview of online learning opportunities in technical education in Bangladesh. Firstly, it provides a brief overview of online technical education delivery in India, the neighboring country of Bangladesh. It then presents the challenges and opportunities of implementing technical online education in Bangladesh. Thereafter, it discusses how the technical education delivery method in Bangladesh can be upgraded. Finally, two case studies are presented in which recent developments in and future opportunities for online technical education are discussed.

17.2 An overview of the current online technical education delivery situation in India

Over the last two decades, India has made substantial reforms in the delivery of technical education through adopting new teaching approaches blended with information and communication technologies. Bangladesh, being a neighboring country and possessing a similar educational system, could benefit by looking into its current state of online technical education delivery methods. Bhattacharya (2008) elucidates how India has enhanced the use of e-technology in engineering education through implementing the National Programme on Technology Enhanced Learning (NPTEL), launching an educational satellite and adopting some significant approaches such as the use of "virtual classrooms" and "virtual laboratories."

Bhattacharya (2008) reports that within the NPTEL, 120 web-based and 120 video-based engineering courses containing high-quality resource materials have already been developed. Some of the leading engineering institutes like the Indian Institute of Technology (IIT), Madras have already offered such web-based courses. Besides, IIT, Kharagpur has developed more than 100 full-semester video courses which have been digitized and are stored in a central server. The video lectures can be accessed at any time and from any place within the intranet through "unicast technology" (Bhattacharya, 2008). Many of the video lectures are also uploaded and accessed through 'YouTube.' Through IIT-Kharagpur's local area network (LAN) students can also join a live and virtual classroom which allows them to ask questions and watch and listen to the responses of the teachers (Ray and Bhattacharya, 2000). Similarly, at IIT-Kharagpur, students can also join a virtual laboratory, which allows an engineering student to simulate and conduct laboratory experiments on a computer (Bhattacharya, 2008).

In 2004, India launched a satellite named EDUSAT solely for spreading education at all levels of the curriculum in the country (ISRO, 2004). All forms of educational institutes in India, including their technical and vocational training institutes, are utilizing the potential benefits of EDUSAT through a collaborative knowledge network to educational servers over the internet, including the possibility of having a "virtual campus" (Bhattacharya, 2008). The satellite has opened the opportunity of interactions between students and teachers through two-way video and audio links from distant locations. For example, IIT-Bombay has been establishing state-of-the-art classroom cum satellite transmission units in their remote destinations (hub), which has already facilitated to offer at least 60 semester-long and short-term courses for about 6000 students (Bhattacharya, 2008). Another successful instance of a collaborative group learning model through e-technology in India is the off-campus distance education program at IIT-Kharagpur named "Electronically Networked Lifelong-Learning," where more than 12,000 students have studied (Stone, 1987). Positive applications of online delivery of technical education in India have taken place possibly because of several significant steps taken by the Indian government to fulfill the prerequisite for effective implementation of technology-enhanced learning initiatives, such as high-bandwidth networks, internet access in users' institutions, installation of intranet within institutions and video-on-demand setups for multi-user access of online video courses (Bhattacharya, 2008).

17.3 Opportunities and challenges for online technical education in Bangladesh

Online teaching, in general, presents many challenges. Teaching and learning engineering and other technical subjects present additional challenges unique to programs with highly technical content. Two major concerns of delivering such technical programs through an online or distance learning model are (a) how to deliver effectively the technical contents at distance and (b) how to use laboratories as part of online courses. In addition, a few common constraints involved with online education include (a) enrollment and payment issues, (b) quality of learning outcomes and (c) security and authenticity. However, engineering and other technical education programs through an online approach are getting increasingly popular in developed countries due to the availability of advanced information technology.

The growth rate of online higher education in the USA is disproportionately higher than conventional higher education. Studies also reveal that an online engineering program is as effective as on-campus models, if not better (Abdellah *et al.*, 2008; Cook *et al.*, 2008). Although the online education concept is widely accepted in developing countries, the adoption rate is not very significant (Belangera and Lui, 2008). While e-learning has become a norm in the United States, it has just reached adolescence in the Pacific due to academic isolation, struggling economies and limited access to technology (Heine, 2007). Evidence of online technical education, especially engineering courses, in developing countries is limited. The underlying reasons that developing countries are lagging in the race of online technical education fall into five main categories:

1 Inadequate technical infrastructure due to high initial cost involvement (Abdellah *et al.*, 2008).
2 Lack of specific merit of students to participate in full online module of technical education (ABET, 2009; Pitchian and Churchill, 2002; Sarangi, 2004).
3 Insufficient competent instructors to deliver online technical/engineering programs at tertiary level.
4 Absence of an effective simulation method to integrate virtual and remote laboratories in the education delivery architecture of full/partial online technical/engineering programs (ABET, 2009).
5 Motivations of policy planners towards the approach of full/partial online delivery of technical/engineering programs.

17.3.1. Technical infrastructure

Bangladesh was ranked 133 for "infrastructure" and 126 for "technological readiness" in global competitiveness rankings out of 139 countries in 2010–2011 (CPD, 2010). However, integration of e-technology in both government and private business sectors in Bangladesh has attained a remarkable success in the last two decades. The adoption rate of e-technology towards the development of socio-cultural well-being has also been increased rapidly in Bangladesh over the last 10 years. On the other hand, significant technological reformation of the education sector in Bangladesh has been observed at all levels, particularly in the tertiary education delivery system provided by private universities. The use of information and communications technology (ICT) in both education delivery method and administrative activities has been adopted widely and effectively in most of the private universities, including some top-end public universities. The use of enterprise resource planning (ERP) for academic and administrative management, online enrollment and mobile money transfer are now a common practice in most of the universities in Bangladesh. It is evident, however, that providing engineering or technical education through a fully online model compared to a blended approach is prohibitively costly in some cases, as the initial investment is significantly higher than for a traditional approach. Promisingly, some top-ranked public and private universities, e.g. Bangladesh University of Engineering and Technology, Bangladesh Agricultural University, North-South University, East-West University, BRAC University, Daffodil International University, Ahsanullah University of Science and Technology, Islamic University of Technology, Independent University Bangladesh and American International University Bangladesh have established state-of-the-art technical infrastructure that has the potential to effectively deliver online engineering programs.

17.3.2. Ability of students

Transformation of the engineering curriculum from a traditional classroom-based approach to a full or partial online model has been apprehended as a paradigm shift. However, 126 universities worldwide successfully implemented such curriculum (Ibrahim and Morsi, 2005). One of the key success factors is probably the students' technical compatibility to cope with such changes (ABET, 2009). In developed countries, students enjoy exposure to a technology-driven academic learning environment from the school level, whereas in developing countries like Bangladesh, students get similar exposure at the tertiary level of education at only a few institutions. Hypothetically, this could be a key challenging factor against implementing full online delivery of any technical course at the tertiary level in Bangladesh. However, generally vivid challenges of e-learning present themselves with different magnitudes from one educational environment to another and from one country to another. Unpublished data and personal observation confirms that initial partial application of e-learning in engineering courses in a few top-end universities in Bangladesh has been successful. Students of civil engineering at BUET scored higher marks in online-based coursework (where a 3D model of column structure was used) compared to a traditional classroom test. The "10 Minute School" channel on YouTube is another brilliant example of e-technology integration in teaching basic mathematics, geometry and physics for elementary-level students in their native language. Use of "content loaded SD Card" in cell phones was first introduced by Bangladesh Open University (BOU) to facilitate offline study of different courses offered at tertiary-level education. BRAC University is the first educational institute in Bangladesh to launch its own nano-satellite, which was developed by three of its students. The satellite "BRAC Onnesha" was designed and manufactured by the SpaceX Falcon 9 rocket (Prothom Alo, 2017).

17.3.3. Role of instructors

As stated in the previous section, the adoption rate of e-technology in education delivery at the tertiary level in Bangladesh has been rapidly increasing over the last 10 years. A great deal of credit goes to the university teachers for their relentless effort behind this phenomenal transformation. Peterson and Feisel (2002) confirmed from the international experience of adopting e-learning that scholars' technological orientation, technical competency and strong enthusiasm play a vital role in overcoming general challenges of incorporating a full/partial online model in traditional curriculum. However, facing inevitable teaching loads and the trade-offs between online and on-campus teaching time are still controversial in the teachers' community (Abdellah *et al.*, 2008). In contrast, changes in students' advising protocol, efficacy assessing students' high-end technical work and specific software requirements with adequate training for using them are still debatable. Use of Google Extension as an alternative tool to enhance off-campus learning is being practiced in a few private engineering universities in Bangladesh. Unpublished data of Ahsanullah University of Science and Technology (AUST) reported that the rate of assignment submission on the due date using Google Extension is relatively higher than that of on-campus submission. However, it is inspirational that a few private universities (e.g. Daffodil International University) in Bangladesh have introduced "Google Classroom" as a compulsory component for tertiary education delivery. Undoubtedly, this accelerates students' interaction with the teacher on a virtual platform and provides a scope for students' orientation towards a unified goal of e-learning.

17.3.4. Engineering laboratory

One of the major challenges of incorporating a fully online curriculum for engineering courses is student's laboratory work, which would be fundamentally different in online courses from that in a campus-based program. Although the concept of using virtual and remote laboratories can functionally solve the problem, the accreditation of such use is still controversial (Abdellah *et al.*, 2008). However, it is argued that the key objectives set by the Accreditation Board for Engineering and Technology (ABET) for engineering laboratory classes can effectively be met by using virtual or remote laboratories (ABET, 2009). In a laboratory simulation model, engineering experiments can be exercised on computers with an optimum level of interactivity. Examples are found in National Instruments DAQ (data acquisition), Field Point, Measurement Studio and LabVIEW. This practice is relatively low in cost and significantly increases student access (Gowdy *et al.*, 2002; Hung *et al.*, 2007). The highest level of this technique is classified as "virtual reality" (Klett, 2002).

Another effective approach is remotely controlled physical laboratories. Students can get convenient access and use physical laboratory equipment located in a convenient place to perform experiments (Bing and Voon, 2010). Through a resource-sharing approach, many institutions may share common physical laboratories. Examples of successful online laboratories can be found in Massachusetts Institute of Technology, Johns Hopkins University and the University of Texas.

It is promising that a good number of engineering graduates have returned to Bangladesh after completing their degrees from world-renowned engineering universities in last two decades. These graduates are now working in different public and private universities in Bangladesh, and it is found that many of them have been implementing these modern technologies at their universities. A few faculty members of BRAC University and Ahsanullah University of Science and Technology reported that 3D images of laboratory equipment using a 3D printer significantly help undergraduate students to practice items before examination when physical equipment is hard to reach.

17.3.5. Policy planners

In Bangladesh, all public and private universities are overseen by the University Grants Commission (UGC) under the Ministry of Education. Although the present government has made remarkable achievements in telecommunication sectors through rapid digitalization under the political manifesto "Digital Bangladesh," the concept of full digitalization of the tertiary education sector (especially engineering and technical programs) may face intense debate from the experts and the policy planners. An informal interview of a UGC representative confirms that the accreditation issue of an e-learning-based degree would be the major barrier towards online engineering and technical education in Bangladesh, along with a few other issues as noted below:

1. Setting regulations for credit transfer from an online engineering to a campus-based program;
2. Assessment of learning outcomes of students under an e-learning program;
3. Monitoring of laboratory work of students through virtual labs and
4. Training of regulatory staff for such technology-driven programs.

17.4 Government approach towards gradual integration of e-technology

Tables 17.1 and 17.2 provide some key ICT features and capabilities of Bangladesh (BTRC, 2017; ETMA, 2015). About 74.43% of people own a mobile phone in Bangladesh, which is a good sign; however, only 3.9% of households have computers. The literacy rate is only 57.7%. Only 6.5% are broadband users. These data show that mobile phone technology may assist in delivering online education in a more effective way than other means in Bangladesh.

17.4.1 Vision 2021 of Digital Bangladesh

The present Bangladesh government outlines its plans for the progress and development of Bangladesh in the Vision 2021 program. Ample focus has been placed on technological development (especially in the ICT sector) in Vision 2021. It pledges to combat poverty by building a Digital Bangladesh and joining the ranks of middle-income countries. The government also recognizes the huge and increasing demand for a skilled work force at home and abroad. Vision 2021 emphasizes the production of more skilled workers for work abroad. It envisions that formal training will enhance the knowledge, skills and creativity of all new entrants to the workforce, allowing Bangladesh to achieve nearly full employment by 2021. The government also envisions that both the public and private sectors will expand the network of quality vocational training institutes, thus ensuring full-scale vocational training

Table 17.1 ICT profile of Bangladesh (ETMA, 2015)

Item	%
Literacy rate	57.7
Fixed telephone	0.69
Mobile/Cellular phone	74.43
Active mobile broadband subscribers	6.5
Fixed (wired) broadband subscribers	0.97
Households with computers	3.9

Table 17.2 Telephone network characteristics in Bangladesh (BTRC, 2017)

Item	*Status*
Wireless Broadband Internet Service Provider	3(4G)
Mobile Service Provider	5(3G), 1(2G)
Mobile/Cellular Phone User	130 million
International Internet Gateway	6
International Internet Bandwidth	900 Gbps
Nationwide Transmission Backbone	DWDM
Optical Fiber Transmission Line	60,000 km
Optical Fiber Connectivity	up to Union Level
Local Internet Service Provider	483
Nationwide ADLS Connectivity	GPON

with adequate practical exposure. The expanding scope of Vision 2021 clearly indicates an evolution of an e-learning environment through a massive digitalization in the near future.

17.4.2 a2i project of Bangladesh

a2i is the world's first Innovation Lab+ introduced and successfully implemented by the prime minister's office of the Bangladesh government that primarily ensures easy, affordable and reliable access to quality public services for all citizens of Bangladesh. One of the prime focuses of the a2i project is rapid digitalization of the education sector. Currently, 8.9 million children are learning from multimedia content developed by over 100,000 teachers. The project has introduced 105 digital talking textbooks for all visually impaired students as well.

17.4.3 National OER policy of Bangladesh

Bangladesh is on the verge of introducing its national plan for an open educational resource (OER) policy in association with Commonwealth of Learning (COL), Canada. The term OER was coined in 2002 at the Forum on the Impact of Open Courseware for Higher Education in Developing Countries organized by UNESCO (Atkins *et al.*, 2007)

17.4.4 TVET Bangladesh

Technical and Vocational Education and Training (TVET) stated its strategies in the Education Policy 2010. The comprehensive education policy emphasizes new ideas about strengthening and scaling up TVET in terms of access to information. Improving access to and the quality of TVET, harnessing the potential of ICT and maximizing the contribution of the private sector are some focus areas of the policy. TVET has comprehensive plans to promote public-private partnership to foster e-learning into its curriculum.

17.5 Upgrading technical education delivery method in Bangladesh

There are plans to improve the course delivery methods in Bangladesh to improve the quality of graduates from engineering universities by taking up recent developments in engineering education. One of the important concepts to be implemented is "blended learning" in engineering education in Bangladesh. Blended learning is an education delivery method that integrates both online and face-to-face learning, which can improve learning experience of the students (Khan *et al.*, 2012). This model is enriched with online educational materials available to students for 24/7. The students can use the internet to access these rich multimedia contents at anytime and anywhere, at the university or at home, which increases their flexibility and understanding in learning. Therefore, when blended learning is designed and applied properly depending on the online materials, students' need and teachers' requirements, it can offer great advantage to students and teachers. The major differences between traditional classroom learning and blended learning are presented in Table 17.3.

The online component of the blended learning model can be delivered through "internet-based learning" or "web-based learning" (Yigit *et al.*, 2014). In internet-based learning, a virtual communication channel is established between the teacher and the students, in which the teacher lectures at one end and students join the class at another end. The students not only

Table 17.3 Major differences between classroom learning and blended learning approaches

Features	*Traditional classroom method*	*Blended learning method*
Flexibility	Not flexible as students need to attend physical classes on a fixed time schedule	Flexible as students can access the online materials (i.e. lectures, tutorials and other supplementary materials) anywhere and anytime with internet access
Learning technique	Only face to face	Both face-to-face and online delivery methods are integrated
Use of technology	Not much use of technology in teaching and learning, as teachers mostly use chalk and talk approach	Technology-enabled learning (e.g. use of multimedia, virtual classroom and recordings of lectures and tutorials)
Evaluation	Normally evaluation of students' performance is done through year-end results	Real-time evaluation can be done through the online quizzes; also, progressive evaluation is done and reported through the online grading system along with year-end final evaluation
IT skill requirement	Not much IT skill is required	IT knowledge is required for both teachers and students to enable them to use blended learning approach

receive the information but also contribute and interact in the online classes that enable effective learning. One of the major challenges of internet-based learning is the current bandwidth and speed limitations. In developing countries like Bangladesh, those limitations would be the major drawback in using internet-based learning, especially for students who reside in rural areas.

Web-based learning is a type of computer-based learning in which the uploaded course materials on the World Wide Web (WWW) are used to support the learning process. This model provides flexibility of time and location. Students are provided with login details containing user name and password. They can access the materials at their own time and place and can go at their own pace.

Adoption of blended learning is the emerging trend in most universities across the USA, Canada, Australia, the EU and China (Gudimetla and Mahalinga, 2006), whereas it is not the case in the engineering universities in developing countries like Bangladesh. A blended learning model is proposed in this chapter for the engineering universities in Bangladesh to improve the teaching and learning outcomes.

The proposed blended learning model consists of the following steps:

1 Deliver the lecture face to face and record it through university systems, and then make it available to the students through hosting a website on the university's intranet.
2 Along with recorded audio lectures, teachers should upload the PowerPoint slides of the lecture with various directives to students for further readings and references, so that students can review the lecture slides and listen to audio lectures as many times as required to capture the essence of the concepts discussed.

3 In the engineering universities in Bangladesh, the concept of tutorial has not been well implemented, whereas in the developed countries it is an integral part of a course where students can practice and solve the engineering problems on the lecture topic with the help of tutors. A tutorial class is proposed here under the blended learning approach for each of the lectures delivered, and some of the videos of the solutions to the tutorial problems are proposed to upload online after the tutorial classes. This will allow the students to review the tutorial solutions as many times as needed to achieve the full understanding of the problems.
4 Practice online quizzes are proposed to implement after each of the lectures through adopting a commercial learning management system, in which a student can access the quizzes and can get immediate feedback on the results. Moreover, they should be able to attempt the online quizzes as many times as they want. The quizzes may be multiple choice or fill in the blank, based on the lecture topic delivered. This will help and encourage students to understand the lecture topics in a better way.

All these efforts will simultaneously lead to the creation of a knowledge base that is expected to enhance the learning experience of the students. However, implementation of blended learning in engineering courses is challenging. It can have an adverse effect on the students' learning if it is not designed properly. Moreover, students' lack of computer skills, poor self-discipline and deficiency in patience and time management can lead to poorer results for the potentially slow-learning students. Therefore, teachers must design the blended learning approach properly and prepare online materials that are suitable and interesting to the students, which could enable the delivery of a true learning product rather than serve as a mere repository for reading materials somewhere in cyberspace.

17.6 Case study 1: Bangladesh Open University

17.6.1 General

Bangladesh Open University (BOU) was established in October 1992 as a public university. The university started its journey with very few students or staff. BOU has now 26,625 academic staff and 574,000 students, and it is the seventh largest university in the world according to enrollment (Hossain & Islam, 2015).

Immediately after independence in 1971, when mass education became the priority of the nation, the School Broadcasting Program (SBP) was established as a project to facilitate mass education, which in due course formed as BOU (Forhad and Kamal, 2013). This is the only university in Bangladesh which provides online and distance education at both undergraduate and postgraduate levels in business administration in addition to other disciplines, including science, agriculture, humanities and social science. An interview with the participants of BOU suggest that, along with 21 formal academic programs, it also offers 19 non-formal programs (environmental protection, basic science, elementary mathematics, agriculture, bank service, marketing management, health nutrition, population and gender issues) (Forhad and Kamal, 2013) (Table 17.4). Under these disciplines, the university uses a blended learning approach in which the instructional system comprises mainly the development of instructional materials in modular form, delivery of printed materials to the learners, face-to-face tutoring of students at selected tutorial centers and delivery of learning materials using various ICT modes (Hossain and Islam, 2015).

Table 17.4 Number of programs being offered in different disciplines at Bangladesh Open University

Discipline	*Programs*
Humanities	14
Education	3
Engineering and Technology	—
Social Science	10
Management	7
Computer Science & Informatics	2
Commerce	—
Basic Sciences	—
Medical, Nursing & Health Science	1
Legal Studies	2
Agricultural Veterinary Science	3

17.6.2 Current facilities

The university is designed with state-of-the-art technology and is fully equipped with most modern technological devices like silicon graphics, digital editing suites, electronic preview theaters, microwave communication link and full-fledged audio-video studios with chroma key facilities. In this regard, one participant from BOU stated:

> Our university is fully equipped to provide quality education using the Information and Communication Technology (ICT) . . . we extensively use different technologies such as Learning Management System (LMS), Interactive Virtual Classroom (IVCR), Mobile Technology, Pre-recorded Video Programs and Live Program (Live Streaming) through using National TV Channel (BTV) as well as Satellite TV Channel and Radio Program.

In addition, BOU is dedicated to facilitating learning for those having limited/no internet connection. Accordingly, it introduces micro-SD cards containing video and audio programs inserted into mobile sets of students. These assist students even when they are offline. The university also uses YouTube and BouTube (www.boutube.com) to facilitate students by providing the necessary textbooks/modules. Moreover, various technology-based systems and application software are in use that ensure smooth administration and provide better online service to the students. In this regard, the participant from BOU stated,

> [F]or better access to resource and information, we have Online Admission and Result Management Systems, Online Student Complaint System, Online Service and Payment System, OER Repository, Enterprise Resource Planning (ERP).

When the participant was asked about their future initiatives, it was mentioned that BOU was underway to launch "podcast" and "vodcast" in the campuses. The participant said that along with the implementation of state-of-the-art technologies, BOU has specific plans in the next five years to implement e-learning programs at all levels and establish a Center for Research and Training. In respect to research, the university has already implemented some research projects under the sponsorship of the British Council and Commonwealth of Learning. To further the research activities, some collaboration with other universities is underway.

Furthermore, for the very first time, this university has offered PhD degrees to interested students via an online mode.

However, while quality becomes the key concern for higher educational institutes, BOU established a Quality Assurance Cell with the financial and technical assistance from the University Grants Commission's Higher Education Quality Enhancement Project. The key focus is to undertake situational assessment of the academic programs and management which will eventually develop into a quality assurance manual to serve as the framework of providing control on quality education.

17.6.3 Future plans

In respect to future plans, the participant from BOU unveiled the interest for online engineering program. The participant mentioned that while there is an increasing demand for online engineering education in the developed countries due to the advantages of utilizing modern technologies, evidence of online engineering courses in developing countries is rare. The fundamental reason that developing countries are lagging in the race for online education is the lack of infrastructure facilities. The participant also mentioned that in Bangladesh, most of the public universities, including engineering universities, are in urban areas and very few in number. Students from rural areas have to incur a huge living cost and face significant relocation difficulties to attend an on-campus program. On the other hand, these insubstantial tertiary education facilities exert extensive socioeconomic pressures on urban lifestyle as well. Accordingly, considering the growing need and market potential for online engineering programs in Bangladesh, BOU is also planning to introduce online engineering as a separate discipline. In this regard, the participant stated,

> We firmly believe that soon we will be able to execute online engineering program. This program requires more sophisticated laboratories and technological facilities. We are underway to identify the required resources and technical supports.

In addition, BOU is currently in progress of developing an e-learning center and interactive virtual classroom in collaboration with the Korean International Cooperation Agency. This will increase the capacity of the university to deliver technical courses through an online module. However, in order to meet the increasing demand of online education in Bangladesh, BOU is dedicated to increasing and improving its existing programs, thereby establishing strong linkage with national and international organizations. BOU is looking for the opportunities to become one of the best distance education providers not only in terms of enrollment numbers but also in terms of quality education.

17.7 Case study 2: Daffodil International University

17.7.1 Introduction

Daffodil International University (DIU), one of the pioneers in the field of IT education in Bangladesh, has implemented e-technology right after the birth of this university in the year 2002. Since the beginning, DIU has been using the Learning Feedback System (LFS), which was developed by the university to support its students via online services such as submission of assignments, sharing of class contents and storage of educational resources. It began an era of supplementary education in online using internet as a primary resource. However, the integration and adaptation of e-technology to deliver technical education was not that effective initially because of the factors such as inefficiency of the software, limitation of the storing capacity and user experience. Afterwards, the university decided to use Google Classroom, which has already been popular. This again proved the university's commitment to provide the best IT education for the nation. Using this technology, previous issues with the LFS has been reduced and the user experience has since become positive.

17.7.2 History of usage

In the spring of 2015, during the beginning of the year, faculty members of DIU started to use the Google Classroom. It was a transition from the use of LFS to the use of Google Classroom. Many of the faculty members started to use Google Classroom to a different extent, however, a few took it seriously and later most of the teachers took part willingly. Google Classroom became the salvation to the problem during natural calamities, such as floods and heavy rainfall, which are very common in Bangladesh. During the period of strike for consecutive days, while regular classes could not be arranged, some of the faculty members started teaching classes using Google Classroom. They instructed their students to be logged in at a particular time from wherever they want and started communicating in the class stream. Teachers uploaded the class lectures and contents prior to the class and after watching the contents and lectures, students started to comment and discuss, and then teachers replied back to students' comments. Some of the teachers started taking live audio classes using another service from Google named Google Talk. These steps taken by the teachers created a new era in Bangladesh in the field of education using e-technology. The use of this technology seemed the way out in case of emergency situations; however, it showed a path towards a new education system in Bangladesh where this online classroom can be used as a supplementary method to provide education service in parallel with the regular in-class education. Thus, the method of blended learning was introduced in DIU where benefits of in-class education and online education were blended together to enhance student learning.

17.7.3 A real scenario

The following example has been taken from a faculty member's Google Classroom. That faculty member has been teaching in the Department of Software Engineering under the Faculty of Science and Information Technology. This teacher has been teaching in DIU since 2013, and in January 2015, the teacher willingly took the opportunity to use Google Classroom as a supplementary method in parallel to the regular in-class teaching. Since then, the teacher has been using this tool for every course the teacher taught. The snapshot of this particular teacher's classroom is shown in Figure 17.1.

17.7.4 Integration of Google Classroom technology

First of all, to integrate Google Classroom e-technology, the teacher had to change and modify the existing course outline, where the teacher had to mention the weighting of the assessments, which indicates the percentage of mark assessed in-class and the percentage of marks assessed online. Table 17.5 shows assessment types with weighting for a course in DIU (Titled "SWE112: Computer Fundamentals with Lab") taught by the participant teacher.

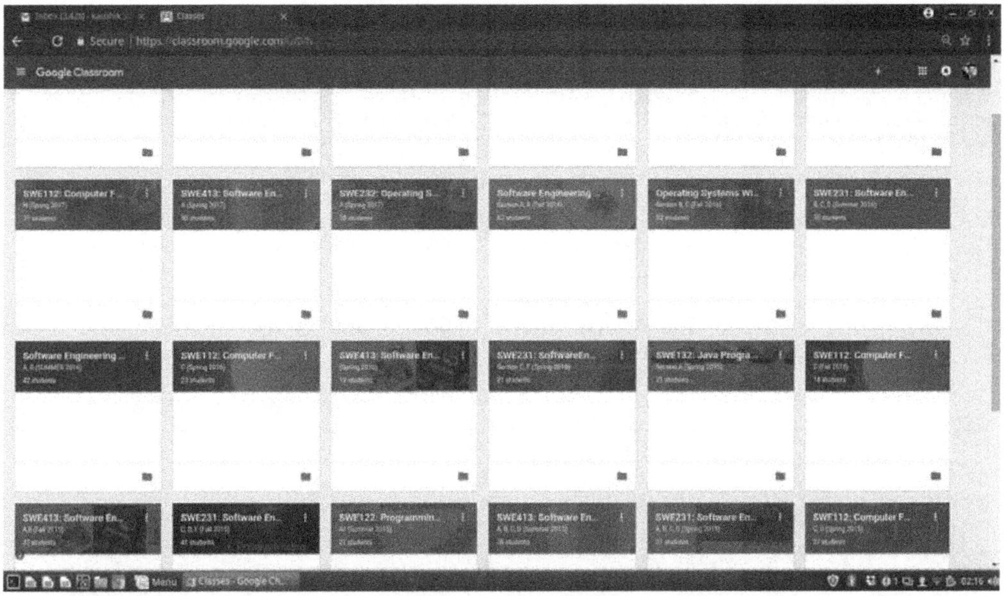

Figure 17.1 Screenshot of Google Classroom of a teacher at Daffodil International University

Table 17.5 Assessment methods with weighting for a lab-based course indicating the percentage of marks assessed online (using Google Classroom)

Serial	*Assessment Methods*	*Weighting*			*Weighting*		*Remarks*
					In class	*Online*	
1	Continuous Assessment	50%	5%	Attendance	5%	0%	To measure how well students have learned throughout the semester
			15%	Quiz (Min 3 Quizzes)	10%	5%	
			5%	Assignment	0%	5%	
			25%	Lab	15%	10%	
2	Examination	50%	20%	Mid-term exam	50%	0%	To measure how well students have achieved the learning outcomes
			30%	Final Exam			

Further, Table 17.6 shows assessment types with weighting for a course (Titled "SWE413: Software Engineering and Cyber Law") taught by the participant teacher.

From the above two examples in DIU, it is found that there are several assessment methods which can be implemented using Google Classroom. The assessment methods are broadly two types, continuous assessment and examination, with several sub-types as shown in Tables 17.5 and 17.6. Besides sharing contents and group/individual discussions, several sub-types of above assessments under continuous assessment can be performed using Google Classroom. So far, the examinations are not performed using this technology, and the ratio of in-class weighting and online weighting may vary from department to department and may also vary from course to course. To integrate this technology further, the lesson plan has been

Table 17.6 Assessment methods with weighting for a non-lab course indicating the percentage of marks assessed online (using Google Classroom) at Daffodil International University

Serial	*Assessment Methods*	*Weighting*			*Weighting*		*Remarks*
					In Class	*Online*	
1	Continuous Assessment	35%	7%	Attendance	5%	2%	To measure how well students have learned throughout the semester.
			15%	Quiz (Min 3 Quizzes)	10%	5%	
			8%	Presentation	0%	8%	
			5%	Assignment	0%	5%	
2	Examination	65%	25%	Mid-term exam	25%	0%	To measure how well students have achieved the learning outcomes.
			40%	Final Exam	40%	0%	

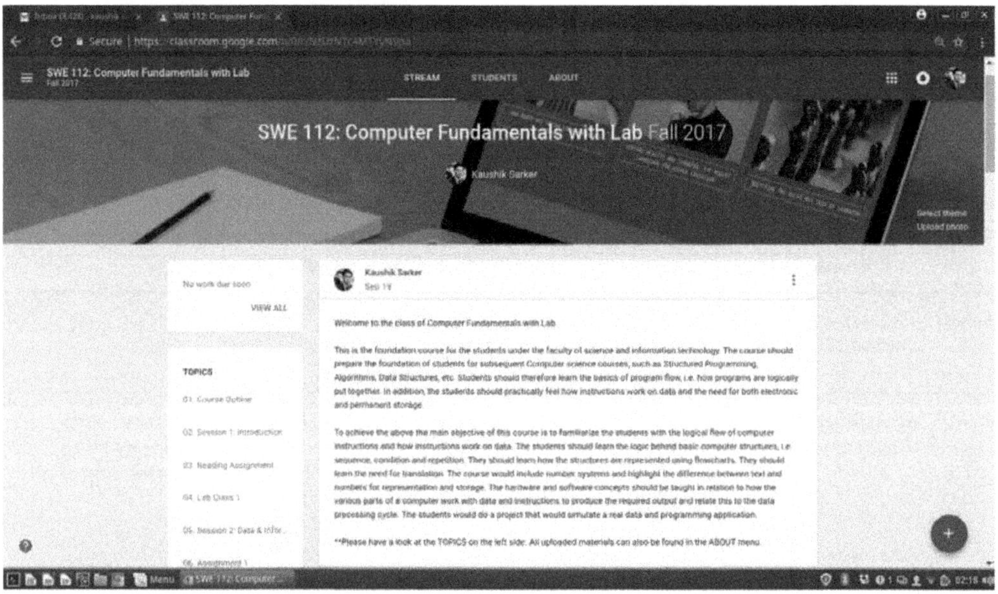

Figure 17.2 Screenshot of a class with course code SWE112 of the participant teacher at Daffodil International University

modified too, so that all these assessments along with the contents are shown and arranged session-wise in the classroom. Figure 17.2 shows the class with course code SWE112: Computer Fundamentals with Lab, which has the topics arranged at the left panel of the classroom according to the designed lesson plan.

17.7.5 Steps taken to adapt Google Classroom technology in DIU

Adapting Google Classroom technology by the teachers and students of DIU was not easy. It took numerous seminars, workshops and group discussions to popularize it among the teachers and students. When new teachers are recruited, the necessary training is arranged for them to learn this new tool to plan, design and deliver the course. Similar steps are taken for the students at orientation day at DIU. DIU encourages its faculty members to use this tool by providing awards and certificates to the innovative users of Google Classroom. The award is regularly given under the banner "Innovation in Teaching and Learning." To motivate the students, regular awards with other benefits are also given based on effective interactions using Google Classroom.

17.7.6 Future opportunities and challenges

The Google Classroom has a great future, in which its adoption can change the way education is served to a developing nation like Bangladesh. In the future, as far as technical education is concerned, laboratory classes can also be taken and assessed by this online tool. For example, a course in software engineering, which has a programming language lab associated with it, may need to integrate the compiler, editor and executioner with the Google Classroom so that the teacher can verify and assess the code written and submitted online by the students. Therefore, this opportunity may then create a new challenge to Google and to DIU at the same time. In the future, like many other universities in the developed world, DIU may run several paid online courses for registered users or the public, and that may further create future employability and revenue of DIU. Courses for distant learners can also be developed in which students will not be required to attend the university physically for education, but rather the university will come to the students to serve at one click away. However, besides infrastructural challenges, DIU may face social challenges, as students, their families and the society still believe education should be served in a classroom, not online or over the internet. Fraudulent activity may increase while completely serving the courses online because of the incapability to properly verify the identification of the distant learner. However, at the end, DIU believes that with the advancement of technology, these challenges can be met to create the positive opportunities for all in Bangladesh.

17.8 Conclusion

This chapter presents an overview of technical education delivery methods in Bangladesh. It has been found that Bangladesh still uses the traditional "chalk and talk" approach in delivering engineering education in most of the public universities. Bangladesh Open University (BAU) has been using different technologies such as learning management system (LMS), interactive virtual classroom (IVCR), mobile technology, pre-recorded video programs and live program (live streaming) through the national TV channel (BTV) as well as satellite TV channels and radio programs to deliver online education. Daffodil International University (DIU) has been using innovative online approaches such as Google Classroom in delivering some of its courses. Both BOU and DIU have the capability to develop new engineering and technological courses based on online and blended approaches. Public engineering universities in Bangladesh need to catch up in this regard to make their courses more flexible and effective by adopting various blended and online course delivery systems.

References

Abdellah, G.A.H., Taher, S. & Abdel-Rahman, M. (2008) Recent developments in Egyptian engineering education through competitive projects. *National Committee Meeting, End of Term of HEEPF Phase 7, Egypt*. Government of Egypt, Cairo.

ABET (2009) *Accreditation Board for Engineering and Technology (ABET): Accreditation Policy and Procedure Manual*. Engineering Accreditation Commission Publication, USA.

Ahmed, A., Haque, M.M., Rahman, M.A. & Rahman, A. (2016) Engineering education in Bangladesh: Implementation of online delivery approach. *Proceedings of International Conference on Engineering Education and Research*, 21–24 Nov 2016, Sydney, Australia.

Atkins, D.E., Brown, J.S. & Hammond, A.L. (2007) A review of the open educational resources (OER) movement: Achievements, challenges, and new opportunities. Creative common. The William and Flora Hewlett Foundation, Menlo Park, CA, USA. pp. 1–84.

Belanger, D. & Liu, J. (2008) Education and inequalities in rural Vietnam in the 1990s. *Asia Pacific Journal of Education*, 28(1), 51–65.

Bhattacharya, B. (2008) Engineering education in India – The role of ICT. *Innovations in Education and Teaching International*, 45(2), 93–101.

Bing, D. & Voon, L.K. (2010) *Developing a Scalable and Reliable Online Laboratory for Engineering Education Using NI Products*. National Instruments Company, Austin, TX, USA.

BTRC (2017) Bangladesh Telecommunication Regulatory Commission. Available from: www.btrc.gov.bd/journals/information-and-communication-technology-ict-status-issues-and-future-development-plans [Accessed 10 November 2017].

Chowdhury, H. & Alam, F. (2012) Engineering education in Bangladesh – An indicator of economic development. *European Journal of Engineering Education*, 37(2), 217–228.

Chowdhury, H., Alam, F., Biswas, S.K. & Islam, M.T. (2008) Engineering education in Bangladesh. In: *Proceedings of the 4th BSME-ASME International Thermal Engineering Conference (ICTE2008)*, 933–946, Dhaka.

Cook, D.A., Levinson, A.J., Garside, S., Dupras, D.M., Erwin, P.F. & Montori, V.M. (2008) Internet-based learning in the health professions: A meta-analysis. *Journal of the American Medical Association*, 300(10), 1181–1196.

CPD (Center for Policy Dialogue) (2010) Global Competitiveness Report 2010–2011 and Bangladesh Business Environment Study 2010, Press Advisory, Dhaka [Online]. Available from: www.cpd.org.bd/pub_attach/GCR_2010.pdf [Accessed 7 November 2017].

Education Technology Management Academy (ETMA) (2015) *A Baseline Study on Technology Enabled Learning in The Asian Commonwealth*. Commonwealth of Learning, Canada.

Forhad, M.A.R. & Kamal, M.A. (2013) Inclusion and Access to Technology in ODL programs – A Case of Commonwealth Executive MBA Program in Bangladesh, MPRA Paper No. 39242, 15:38 UTC. Available from: http://mpra.ub.uni-muenchen.de/39242/ [Accessed 28 March 2016].

Gowdy, J.N., Patterson, E.K., Wu, D. & Niska, S. (2002) Development and use of simulation modules for teaching a distance-learning course on digital processing of speech signals. *Proceedings of ECI Conference on e-Technologies in Engineering Education: Learning Outcomes Providing Future Possibilities*, Davos, Switzerland.

Gudimetla, P. & Mahalinga, I.R. (2006) The role for e-learning in engineering education: Creating quality support structures to complement traditional learning, *Proceedings of 17th Annual Conference of the Australasian Association for Engineering Education*, Auckland, New Zealand.

Heine, H.C. (2007) Preparing and licensing high quality teachers in Pacific Region jurisdictions. National Center for Educational Evaluation and Regional Assistance. Available from: http://ies.ed.gov/ncee/edlabs [Accessed 2 July 2016].

Hossain, M.F. & Islam, M.A. (2015) Bangladesh Open University educating people through distance mode. *International Journal of Educational Studies*, 2(1), 37–43.

Hung, T.C., Liu, C.C., Hung, C., Ku, H. & Lin, Y. (2007) The establishment of an interactive e-learning system for engineering fluid flow and heat transfer. *Proceedings of the International Conference on Engineering Education (ICEE)*, Sep 3–7, 2007, Coimbra, Portugal.

Ibrahim, W. & Morsi, R. (2005) Online engineering education: A comprehensive review. *Proceedings of the 2005 American Society for Engineering Education Annual Conference & Exposition*, 12 June 2005. American Society for Engineering Education, Washington, DC, USA.

ISRO (Indian Space Research Organisation) (2004) *Conference on EDUSAT – Press Release* [Online]. Available from: www.isro.org/pressrelease/Jul23_2004.htm [Accessed 22 July 2007].

Khan, A.I., Shaik, M.S., Ali, A.M. & Bebi, C.V. (2012) Study of blended learning process in education context. *International Journal of Modern Education and Computer Science*, 4(9), 23.

Klett, F. (2002) Improving engineering student learning in a web-based learning space due to virtual reality techniques and advanced interactivity. *ECI Conference on e-Technologies in Engineering Education: Learning Outcomes Providing Future Possibilities*, Davos, Switzerland.

Lewin, T. (2012) Instruction for masses knocks down campus walls. *New York Times*, 4 Mar 2012.

Martin, F.G. (2012) Will massive open online courses change how we teach? *Communications of the ACM*, Aug 2012, 55(8), Pulman PR.

Peterson, G.D. & Feisel, L.D. (2002) E-learning: The challenge for engineering education. *Proceedings of ECI Conference on e-Technologies in Engineering Education: Learning Outcomes Providing Future Possibilities*, Davos, Switzerland.

Pitchian, A.S. & Churchill, D. (2002) E-learning design for engineering education. *Proceedings of Fourth Asia-Pacific Conference on Problem-Based Learning*, Hatyai, Songkola, Thailand.

Prothom Alo (2017) BRAC Onnesha launched into space, Newspaper article, June 4 [Online]. Available from: http://en.prothom-alo.com/science-technology/news/150171/BRAC-Onnesha-launched-into-space [Accessed 15 November 2017].

Rahman, A. (2017) A blended learning approach to teach fluid mechanics in engineering. *European Journal of Engineering Education*, 42(3), 252–259.

Rahman, M.A. & Haseeb, A.S.M.A. (2007) Challenges of research and graduate studies in engineering at BUET. *Proceedings National Symposium on Engineering and Technological Education*, Dhaka.

Rashid, M.A. (1958) Engineering education in East Pakistan in the context of modern development. *Journal of the Ahsanullah Engineering College*, 8(1), 4–9.

Ray, A.K. & Bhattacharya, B. (2000) Individualised and virtual group learning via networked computers. *Proceedings International Conference on Distance Education – An Open Question?* Sep, University of South Australia, Dunedin.

Sarangi, S.K. (2004) Continuing engineering education through e-learning: A strategy for development. *International Association for Continuing Engineering Education News Letter*. European Journal of Engineering Education, England, UK.

Stone, H. (1987) Comparative performance analysis of on- and off-campus video-based engineering graduate students. *IEEE Transactions on Education*, E-30, 254–258.

Yigit, T., Koyun, A., Asim, Yuksel, A.S. & Cankaya, I.A. (2014) Evaluation of blended learning approach in computer engineering education. *Procedia – Social and Behavioral Sciences*, 141, 807–812.

Chapter 18

Recent advances in the use of remote labs in fluid mechanics

A review

S. Noor[1] and A. Rahman[2]

[1]EnviroWater Sydney, Sydney, New South Wales, Australia

[2]School of Computing, Engineering and Mathematics, Western Sydney University, Sydney, New South Wales, Australia

ABSTRACT: The evolution of digital computers along with the advancement of simulation technologies, remote controlling of instruments, automated data acquisition and rapid data analysis methods have opened a new door of delivering engineering education via online and blended learning methods. Distance delivery of engineering degrees often needs the use of virtual and remote laboratories in order to facilitate online teaching of intrinsic laboratory components. Remote laboratories present an opportunity for a student to complete the required laboratory components of his/her degree, usually from a distant location, and more importantly at their convenient time. The experimentation is performed by controlling the laboratory equipment using web-based means. This chapter presents the state of the art of recent research and developments of online learning via virtual and remote laboratories, and in particular makes a comparison among three different types of laboratories: hands-on/physical, virtual and remote. Virtual/remote laboratories could be useful to many developing countries where setting up of physical laboratories may be more expensive and hence establishing virtual/remote laboratories at a few central locations may cater to a large number of students across the country.

18.1 Introduction

The main objective of engineering education is to enable students to deal with real-world engineering problems. An engineering degree generally embodies a significant component of instructional laboratory sessions. Laboratory experimentation is one of the core requirements for undergraduate engineering studies, since it allows them to reinforce and improve their understanding of the theoretical knowledge gained during lectures and reading of other learning materials, such as textbooks and lecture notes. The engineering graduates are expected to develop hands-on skills during their period of education. Engineering students should develop necessary practical skills including collaborative work, conceptual knowledge testing, equipment handling, experimental data generation and analysis and trial-and-error-based learning. There are numerous online means to enhance theoretical knowledge in a given engineering subject. However, the necessary practical knowledge can be achieved via laboratory practice. Feisel and Rosa (2005) discussed three different types of hands-on/real laboratories. These include developmental laboratory practice, which focuses on design, development and determination of performance parameters in relation to a given set of specifications for a given project. Another type is research laboratory practice which is used to test new research hypotheses or improve a current industrial practice. The third type is educational laboratory practice, where the classroom theoretical knowledge is applied to improve students' practical experience. It is an established fact that practical experimentation in the laboratory plays a significant role in engineering education.

According to Nersessian (1991), hands-on or practical experience is the heart of science education. In this regard, Clough (2002) stated, "laboratory experiences make science come alive." Therefore, the importance of laboratory experimentation for engineering students is irrefutable regardless of the objective focus, e.g. design, research or education. Despite the important role of practical training of students in science and engineering education, teachers and students face various problems, which include the limitation of space and laboratory resources due to high cost of equipment and prescribed/limited time for a student to perform experimentations. Due to numerous branching of an engineering discipline, students also face problems of performing experiments which are relevant to their study. In addition to this, undergraduate students who intend to pursue a research career after the completion of their degree cannot yet get exposed to the ongoing research-related experiments. Therefore, provision of appropriate, meaningful and sufficient laboratory experimentations for students remains a critical component in engineering education, as discussed by Tiernan (2010). More recently, online education is making its way at a considerable pace in order to meet the increasing demand of students who cannot attend the regular university courses due to work and other commitments.

It is comparatively easy to provide theoretical knowledge in engineering subjects via online means; however, integration of relevant laboratory experiments in online engineering courses continues to be one of the major challenges in distant learning methods. An e-learning solution to this problem is possible via development of web-based experimentation using today's high-speed internet technology (Macedo-Rouet *et al.*, 2009). Virtual/simulation laboratory environment, remote controlling of laboratory equipment, automated data acquisition – all of these provide a new opportunity for online delivery of laboratory components in engineering education. Controlling instruments remotely is on the increase with the rapid improvement of communication technologies.

Virtual laboratory can be used when the laboratory equipment is too expensive or unavailable or the experimental procedure is not safe (Hercog *et al.*, 2007). One of the most important features of the virtual laboratory is the non-destructive nature of an experiment; for example, a student can learn from his/her failure without any damage to the instrument involved. According to the Accreditation Board for Engineering and Technology (ABET), learning from failure is one of the nine important aspects of engineering education. Virtual laboratory allows the teachers and students to perform a large number of experiments using the same system and does not limit the number of simultaneous users. It may be very difficult to change some of the parameters of the experiment in a real laboratory, but it can easily be done in a virtual laboratory. A simulation study before the real experiment in the hands-on laboratory allows the students to have a deeper understanding of experimental procedure and the corresponding theory (Ma and Nickerson, 2006). Virtual laboratory involves different software like LabVIEW, MATLAB/Simulink, ANSYS, Java Applet and Flash, which enable simulation of a real laboratory experiment. However, creating a 3D dynamic model of a complex experimental setup may be difficult as well as time consuming in a simulation software system. This is also dependent on the processing capability of computers.

Remote laboratory may be defined as the experimental procedure where the laboratory instruments are controlled from a distant location. Users do not need to be physically present in the laboratory to perform remote experimentation. For engineering educational institutions, the remote laboratory increases the efficiency of the classroom learning process. Remote laboratory is not only used for educational purposes but also in industry and research centers. Scientists can carry out the experiment instantly whenever they need to confirm any of their hypotheses, which enables them to undertake a high level of research. Remote laboratory can be useful for the experiments which are time consuming and need continuous monitoring. Use of remote laboratories along with hands-on laboratories improves students' learning in the classroom. Research experiments can be shared with the undergraduate students, which helps them to be familiarized with research.

Nowadays, most of the remote laboratories allow the users to control the experiment parameters as well as obtain a tabular or graphical result from the linked software. The amount of laboratory equipment which can be controlled remotely is increasing with the rapid development of network technology and microprocessors. Remote laboratory may be suitable for cases in which the number of enrolled students in a subject is too high and when there is a limitation in the availability of certain laboratory equipment (Chen *et al.*, 2010). Another form of remote laboratory is "distributed laboratory." This type of infrastructure is formed by the sharing of resources among a number of laboratories (Muller and Ferreira, 2004). This type of laboratory will not only be a cost-effective solution for collaborative research projects among different universities around the world, but also will open a new door for researchers in the developing countries where there is a lack of physical laboratories. High standard remote laboratory facilities can be a source of income for the hosting institution as well.

Ma and Nickerson (2006) presented a comparative study of hands-on, virtual and remote laboratories. Virtual laboratories were defined as the "replications of real experiments" where the structure required for experiments are not real, but rather simulated by computer software. They defined remote labs as the "mediation of reality," which is similar to real or hands-on laboratories since they also require real laboratory equipment as well as space. The main feature which distinguishes remote labs from hands-on laboratories is the distance of the laboratory equipment and the proximity of the person carrying out the experiment (Nedic *et al.*, 2003). In real or hands-on laboratories, the equipment may be controlled using a computer; however, in remote laboratories, the experimenter is located geographically at a certain distance from the experimental facility. In many of the hands-on experiments, users may not get an adequate internal view of the experimental setup. However, in virtual and remote laboratories, students may be able to see the experimental process in slow motion to achieve greater understanding of the experiment. Some of the well-developed remote laboratories also allow students to visualize more technical details of the experiment.

This chapter presents a review of recent literature on virtual and remote laboratories. It also presents a comparative assessment of different types of laboratories used by engineering education providers.

18.2 Remote and virtual laboratory systems

In engineering education, the application of remote laboratories depends on the flexibility, cost and effectiveness of the system. The idea of remote laboratories first emerged when internet technology started during the 1970s. Since the 1970s, different online learning programs have been developed and applied by educational institutions such as PROLEARN and MIT's OpenCourseWare (Gomes and Zubía, 2008 and Gustavsson *et al.*, 2009). Recent developments in the field of remote laboratory include LabVIEW, which is based on all three laboratory types: remote, virtual and hands-on (Abdulwahed and Nagy, 2013). LabVIEW is designed to allow multi-user experimentation.

18.2.1 Virtual laboratory

Computer programming languages can be used to create the simulation models mentioned above. These languages include Visual Basic, C++, FORTRAN and Java. Developing a simulation model from first principles gives the user a great deal of flexibility, but this becomes time consuming and reduces the ease of use. Java is mainly used to develop a simulation model that can be run on the internet. C++ supports object-oriented simulation modeling (Pidd, 1998). Examples of some of the widely used systems of virtual laboratory adopted by engineering educational institutions are presented below.

1. C++: Originally C and C++ were developed for numerical analysis. In time, C and C++ both became popular for developing simulation models. Currently, much high-level commercial and open source engineering as well as scientific simulation software use C++ because of its ability to process complex data structures (Stroustrup, 1995).
2. Flash: Flash is compatible with any browser since it can be viewed with a plug-in only. A number of virtual laboratory platforms have been designed using Flash. It has been widely used by many engineering schools around the world. For example, the University of Delaware has developed a virtual microscope using Flash (Chen *et al.*, 2009).
3. Java: Comparing to C++, Java is very much object oriented to code a simulation program. Java supports distributed computation of programs which can be accessed by multiple clients at the same time (Kortright, 1997). A Java application can easily be converted to Java applet to upload it into an HTML page. This allows a standalone simulation program to be accessed by other clients on their browser (Hamilton, 1996). Java applets and applications can be run in any platform having a Java Virtual Machine. Java has been used by Chen *et al.* (2009) to create a virtual laboratory system for teaching the resistor color code. The clients can use the combo box to select different combinations of color bands in this model. The result is then calculated by the Java applet. Java has been adopted in many other remote and virtual laboratories.
4. GNU Octave: GNU Octave was designed for numerical computations. Both linear and non-linear problems can be solved numerically using an interactive command-line interface. This is an open source redistributable software. Anyone using it can contribute to the development of GNU Octave by modifying it (Eaton *et al.*, 1997). GNU Octave has been used by many researchers to design mathematical models for high-level numerical simulation.
5. OpenFOAM: OpenFOAM is a C++ based object-oriented software which has been used for numerical simulation in continuum mechanics and computational fluid dynamics. OpenFOAM uses polyhedral mesh where a list of cells is represented by the faces of the mesh volume. Many different types of fluids and structural analysis models can be designed by OpenFOAM, in which coupled and non-linear equations are emphasized (Jasak *et al.*, 2007).

6 COMSOL Multiphysics: COMSOL Multiphysics is a finite element analysis–based multiphysics simulation software. This software has a built-in model builder which can extend one type of physics model into multiphysics models and vice versa. This software allows the user to design a model by defining relevant physical quantities such as loads, constraints, material properties, sources and fluxes, which eliminate the need for the underlying equations. These variables can be applied to solid or fluid domains independently of the computational mesh. COMSOL can be used as a standalone black box simulation software or users can program their own simulation by coding with Java or MATLAB. The simulation can be carried out in various fields of science and engineering including fluid dynamics, heat transfer, microfluidics, photonics and quantum mechanics (COMSOL, 2012).

7 MATLAB/Simulink: This software is used in many engineering universities today to carry out simulation and modeling of different engineering problems. Many universities have included this software in their curriculum. Users can focus more on designing and implementing the technical side of simulation, unlike traditional programming language where the simulation is created from scratch and a lot of time is spent for coding. This is also widely used for research purposes. For example, Schmid (2001) presented a virtual laboratory using MATLAB/Simulink. Casini *et al.* (2001) developed an automatic control telelab using Java servlet and MATLAB/Simulink. MATLAB has been used by Sánchez *et al.* (2004) to build a remote laboratory in order to demonstrate an inverted pendulum experiment.

8 LabVIEW: This commercial software is developed by National Instruments. It is a popular software used by many universities and educational institutions. LabVIEW is based on graphical programming language. Unlike text-based code, the graphical interface makes it easier and more efficient for engineers and scientists to design a virtual experiment in their computer. Many function blocks known as "virtual instruments" are used to design an experiment. Users only need to drop the function blocks on the graphical interface, which saves time. It is relatively easy to program and has high performance while dealing with complicated and advanced applications. Currently, many laboratories use LabVIEW for its strong data acquisition and instrument control system. In addition to the applications for virtual labs, this is also widely used in the remote laboratory system due to its ability to interface the sensors and transducers of the laboratory equipment. The graphical programming of LabVIEW allows nonprogrammers to build their own simulation model and experiment (Tiernan, 2010).

9 ANSYS: ANSYS is one of the most comprehensive engineering simulation software in the analysis industry. This software has a strong platform called ANSYS Workbench, which combines physical modeling of fluid flow, heat transfer, turbulence behavior of fluid, furnace combustion and many other applications ranging from bio-fluid simulation (e.g. blood flow) to semiconductor physics and waste water treatment plant modeling to clean room. ANSYS proved its capability to bring a robust and scalable solution for design industry. Optimization of product design can be achieved by parametric simulation on Workbench. The fluid-structure interaction module of ANSYS has been proven to simulate efficiently fluid flow through turbines and valves and over airplane wings and many other structures. ANSYS is being widely used in various industries for product design and development. This software has successfully proved its capability not only for industries but also for engineering educational institutions. These days, students' learning is much enhanced by prior simulation before performing the hands-on experiment (Madenci and Guven, 2015).

Many researchers are using virtual laboratory or simulation software to carry out their research. There is a comprehensive literature published in journals, books and conference proceedings which are based on simulations (e.g. Lee *et al.*, 2007; Manseur, 2005; Sampio *et al.*, 2011). This demonstrates the validity of results obtained by simulation.

18.2.2 Remote laboratory

Many innovations are currently being introduced in remote laboratory research (Olmi *et al.*, 2011). For example, LabVIEW has introduced a combined system of all three laboratories: hands-on, virtual and remote. The remote laboratory in this system has multi-user capabilities. A brief discussion is given below for some of the most well-known remote laboratory systems.

1 iCampus iLabs: iCampus iLab project was started by Prof. J.A. del Alamo of the Electrical Engineering and Computer Science Department at Massachusetts Institute of Technology (MIT) in 1998. This system is based on a distributed software system and a middleware web service infrastructure. iLab allows users from different educational institutions around the world to access shared laboratory instruments via the internet. Users can log into this remote laboratory system thorough a simple administrative interface. iLab supports both "batched" and "interactive" type of experiments. In batched experiments, all the parameters of the entire experiment can be specified at the very beginning using a graphical editor. Interactive experiments are those which allow the users to control single or multiple parameters of the experiment while executing it. Since the interactive experiments involve student interaction, they typically require more time than batched experiments. For this reason, a scheduling process is necessary for interactive experiments. Recently, iLab released a LabVIEW integrated interactive lab server since National Instruments' LabVIEW is widely used by many engineering educational institutions (Harward *et al.*, 2008). During the last few years, iLab has been adopted by many educational institutions all over the world. iLab is partnering with universities in the developing countries wherever internet connection is available. This is enabling the students to access high-level laboratory experimentation which they would not possibly get at their own university.
2 Labshare: Labshare is an Australian government-funded project that aims to create a national network of shared remote laboratories among five different universities: University of Technology Sydney, Curtin University of Technology, RMIT University, University of South Australia and Queensland University. The Australian government has funded $3.8 million for 3 years for this project. Labshare has created a shared network of the laboratories of these universities via remote laboratory system so that students can access a high-quality experimental facility from anywhere in Australia and the world. A highly sophisticated remote laboratory system called "Sahara" has been developed to allow the users to access the remote experimental setup. Sahara is based on client-server architecture which is built with the SOAP interface using extensible markup language (XML). Simple Object Access Protocol (SOAP) is a simple information exchanging protocol among various programs in a distributed network. Sahara runs on Java framework. The student feedback survey from the first national sharing trial has identified a high level (over 90%) of student satisfaction with potential improvement in different areas of study (Lowe *et al.*, 2012).

3 LiLa (Library of Labs): LiLa was developed in collaboration with eight different European universities and three business organizations funded by the European Commission. The objective of this remote laboratory system is to provide a common platform for the mutual exchange of laboratory facilities among European institutions, especially for the students of undergraduate engineering schools (Mateos *et al.*, 2012). LiLa fulfills the purposes of both remote and virtual laboratories. Teachers can identify the experiment relevant to their course material from the LiLa web portal and download these experiments. The LiLa architecture consists of a web server which controls the LiLa portal. A database server is utilized for booking and reservation time slots. In case of the remote laboratory system, a plug-in for the browsers needs to be installed in the users' computers to gain access to the laboratory equipment. Exercises with online experiments using LiLa have become very good supporting material and have improved the efficiency of delivering lectures. In addition to this, students are very positive about using the exercises to prepare for their exams, which has improved their academic performance as well (Richter *et al.*, 2011).
4 WebLab-Deusto: This is an open source remote laboratory system, which was designed mainly for the students of University of Deusto. It was first launched in February 2005, with the main objective of supporting the laboratory experiments for engineering subjects. The platform used for this system is client to server architecture software. It enabled the students to perform real laboratory experiments, e.g. CPLD, PIC microcontrollers and FPGA from a distant location via computer network (Orduña *et al.*, 2011).
5 Virtual and Remote Laboratory (VR-Lab): This remote laboratory system has been developed by Texas Southern University (TSU) with the collaboration of the University of Houston (UH). To use this remote laboratory, there is no need to install an extra plug-in in the browsers. The platform is based on browser-server architecture software. It was designed to carry out the experiments of data communication systems and digital signal processing.

18.3 Application of virtual and remote laboratories in robotics

Some of the first successful projects of remotely driven industrial robotic components were the Telegarden project (Goldberg *et al.*, 2000), Mercury project (Goldberg *et al.*, 2000) and the remote laboratory developed at the University of Western Australia (Taylor and Dalton, 2000). These systems were developed to enable the students to move the robot and operate the objects in the workspace via internet. These internet-based remote laboratories opened a new trail for robotics e-learning during the past few years. Among the widely used remote and virtual laboratory platforms used for robotics these days, four important ones include the ARITI project (Marín *et al.*, 2005), the UJI Robot (Marín *et al.*, 2003), RLab (Safaric *et al.*, 2001) and Robolab (Candelas *et al.*, 2005). Students can control a Cartesian robot using ARITI's telerobotic features, which consist of an interface based on amplified reality. UJI Robot, having a multirobot architecture system, allows the users to access both industrial and educational robots through the internet. In addition to this, the robot arm can be controlled by using an amplified reality. RLab is a combination of internet-based prototype laboratories for engineering robotics subjects. It provides the users with online access to real hardware for remote control of the robotic components. Robolab is an open platform used for simulation and controlling of robotic arms via the internet. The main limitation of these systems is that they can only be used for reaching the kinematics of robots. The control of the dynamics and programming issues related to many subjects of robotics and automation is absent in these systems. Casini *et al.* (2008) discussed another successful virtual and remote laboratory system for the robotics and computer vision fields: RACT. RobUALab is a complete software platform for simulation, teleoperation and programming of a robotic plant (Jara *et al.*, 2010). Using this platform, the clients can perform experiments with different concepts of robotics and automation, such as kinematics, dynamics and path planning of a robot manipulator. In addition, the programming of different components of the robot can also be carried out using this package.

18.4 Example of remote experiments in fluid mechanics

18.4.1 Flow rate measurement

The laboratory experiment on flow through Venturi was developed at Washington State University's thermal fluids lab (Ellis *et al.*, 2014). Both satellite campus and on-campus students can access this lab experiment. The remote experiment was built by using an online data acquisition system, automatic valve control and a web camera. The remote users apply a voltage to the solenoid valve to control the flow of colored water. Then a calibrated multimeter sends the voltage readings from the pressure transducers along the Venturi and orifice plate to the user via web portal. The web camera can be rotated to visualize every part of the experimental setup, including the level of water in the manometer.

18.4.2 Introduction to aerodynamics using wind tunnel

This remote experiment was set up at the thermo-fluids laboratory at the Rochester Institute of Technology (Rock *et al.*, 2011). The wind tunnels in this experiment were modified using an airfoil stepper motor controller and a data acquisition board which can record data from 16 channels of pressure transducers. The experiment was performed to evaluate the performance of an airfoil at different velocities of wind and angle of attack. LabVIEW software has been used to control the angle of attack of airfoil using the motor controller as well as the velocity of air in the wind tunnel. The same software has been used for data acquisition from the sensors. A web camera installed in the laboratory transmitted a live streaming video during the execution of the experiment. The audio output provided a sense of realism for this remote experiment. Rock *et al.* (2011) found a significant decrease in the experimentation time compared to the local laboratory.

18.4.3 Pumps and linear momentum experiment

Sundararajan and Dautremont (2014) developed this remote experiment at the Mechanical Engineering Lab of Iowa State University. In this experiment, the water is pumped out of the tank through a Coriolis flow meter and a flow control valve and back into the tank again. The rise in head and flow rate measurements are taken. A torque sensor, placed between the pump and motor, measures the horsepower. Keeping the pump speed constant, the parameters are measured at various flow rates. A load cell senses the reaction force. The horizontal momentum is then calculated from the vertical momentum. A compact data acquisition system as well as National Instrument's software, LabVIEW, is used for creating a virtual control panel in the computer. Bridge excitation voltage is given to the load cell by a universal analog input. This also converts the signal from the load cell and torque sensor. The flow meter and pressure transducer is connected with another analog input module. The control signals to the control valve and variable frequency drive operating the pump motor are sent by using an analog output module. A web camera is used in the system to visualize the experimental procedure. LabVIEW is used for data acquisition and creating a virtual control panel in the computer.

18.5 Learning outcomes from hands-on, remote and virtual laboratories

In recent studies, it has been found that fewer literatures (less than 70%) published before 2002 favored e-learning in engineering education. This statistic increased to 84% after 2003 (Shachar and Neumann, 2010). In most cases, the outcome is dependent on the prior knowledge of the students. Simulation of abstract concept increases the conceptual learning in cases where the underlying mechanism of the problem is comparatively simple. This applies only with the students having low prior knowledge. Students with higher prior knowledge benefitted more from the simulation of a problem understanding the complex underlying mechanism (Olympiou *et al*., 2008). The Colorado Department of Higher Education (Colorado DOHE, 2012) carried out a study to find out the effectiveness of three different types of laboratories. It was shown in this study that students who completed the course in a non-traditional learning system (virtual and remote) achieved slightly lower grades than the students who completed their study in traditional learning system. After the completion of virtual laboratory experiments, the students were engaged in face-to-face laboratory experimentation. It was found that they performed better than those who already carried out the experiment in a hands-on laboratory (NRC, 2006). In some cases, the virtual and remote laboratories are more effective than hands-on laboratories because they allow sufficient time for experimentation. Education providers can design the delivery system of practical lessons by taking the advantages of the affordability of these laboratories. Ensuring the highest benefit from the designed instruction of laboratories is the most crucial variable in determining the success of engineering education (De Jong *et al*., 2013).

18.6 Possible applications of remote laboratory in the fluid mechanics subject of Western Sydney University

Western Sydney University (WSU) has been encouraging a blended learning approach in Engineering courses. Blended learning is a concept that utilizes the best aspects of face-to-face and online delivery methods. In order to make an engineering course fully online, one of the main problems encountered by the students and teachers at WSU is the laboratory component of the course. Fluid mechanics is the core subject for both mechanical and civil engineering undergraduate students in WSU. Blended learning has been applied to fluid mechanics in WSU as noted by Rahman and Al-Amin (2014). In this study, the tutorials in fluid mechanics were uploaded to YouTube, lectures were recorded and posted to the website and online practice quizzes were introduced. In another study by Rahman (2017), it was found that this blended learning increased the satisfaction level of the enrolled students and improved the learning outcomes overall. As a part of the future enhancement of this subject in WSU, the virtual and remote laboratory facilities should be incorporated with the current learning materials.

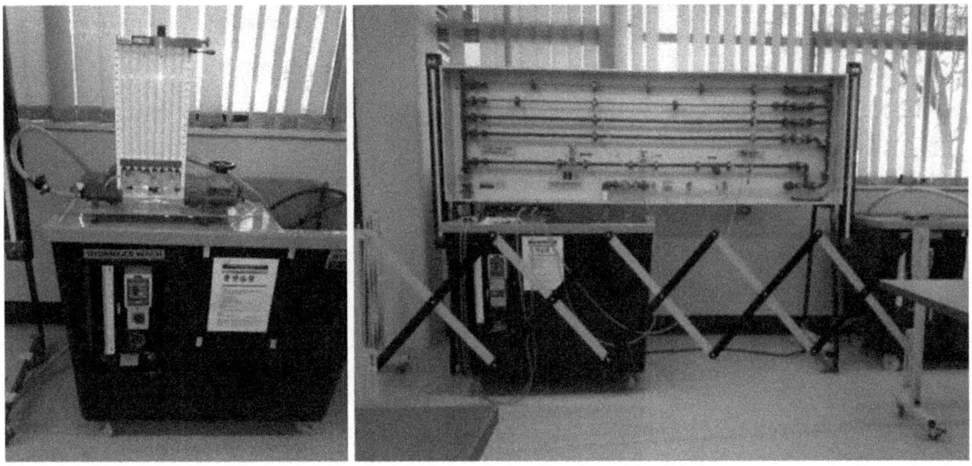

Figure 18.1 Experimental setup in WSU to demonstrate (i) Bernoulli's principle and (ii) friction in pipes

Two of the most important experiments of fluid mechanics are (i) demonstration of Bernoulli's principle and (ii) analysis of friction in pipes. The experimental setup of these two experiments is shown in Figure 18.1. Dai *et al.* (2010) demonstrated the application of virtual and remote laboratory systems in three different laboratory experiments of fluid mechanics. In this study, pressure readings were considered as the output of the experiments. The remote laboratory consisted of a server, client and control-side interface and the virtual laboratory consisted of a 3D model of experimental setup, input parameters and output results.

At Western Sydney University (WSU), the two experiments mentioned above can be incorporated within a virtual system using commercially available simulation software such as ANSYS Fluent, SimScale and NEi Nastran. For a remote laboratory, the diameter of the tubes used in the experiment can be obtained from pre-calculated values since they are fixed. The flow rate can be controlled by a solenoid valve wired to a power supply. The valve position solenoid or the current flowing through it needs calibration for the flow measurement. The pressure gauge reading is converted to a current reading. The computer can receive data from a web-connected multimeter. A data logger can be used for control and data acquisition, which is connected to the computer. The liquid level sensor, which in turn transmits a current or resistance reading to the computer, gives the volume of water collected in a given time. The manometer reading as well as visual access to the experiment can be obtained by using a wireless network camera. This camera should have sufficient optical zoom and autofocus technology. The remote users should be able to control the camera position via a web interface. The above procedure could allow WSU fluid mechanics laboratories to be accessible to the enrolled students (about 250 students in a year) remotely.

18.7 Conclusion

Recent advancement in e-technology has opened many new opportunities in the field of virtual and remote laboratories. More recently, many universities in the developed countries are adopting online education methods for numerous courses. In engineering courses, the laboratory component has been regarded as the main obstacle in online delivery; however, virtual and remote laboratories can provide a solution to this problem. Furthermore, it is possible to improve the learning outcomes of the on-campus students by combining physical, virtual and remote laboratories. With the increase in number of students and having limited laboratory equipment facility, education providers are showing interest in virtual and remote laboratories. This chapter presents an overview to help developing virtual/remote laboratories for engineering education providers. These laboratories may be useful to many developing countries where the setting up of physical laboratories may be more expensive and hence the establishment of virtual/remote laboratories at convenient locations may cater to a large number of students across the country or several countries in a region.

References

Abdulwahed, M. & Nagy, Z.K. (2013) Developing the TriLab, a triple access mode (hands-on, virtual, remote) laboratory, of a process control rig using LabVIEW and Joomla. *Computer Applications in Engineering Education*, 21(4), 614–626.

Candelas, F.A., Puente, S.T., Torres, F., Segarra, V. & Navarrete, J. (2005) Flexible system for simulating and tele-operating robots through the internet. *Journal of Robotic Systems*, 22(3), 157–166.

Casini, M., Chinello, F., Prattichizzo, D. & Vicino, A. (2008) RACT: A remote laboratory for robotics experiments. *IFAC Proceedings Volumes*, 41(2), 8153–8158.

Casini, M., Prattichizzo, D. & Vicino, A. (2001) The automatic control telelab: A remote control engineering laboratory. *Proceedings of the 40th IEEE Conference*, 4, 3242–3247.

Chen, X., Jiang, L., Shahryar, D., Kehinde, L. & Olowokere, D. (2009) Technologies for development of virtual and remote laboratories – A case study. *Proceedings of American Society for Engineering Education Annual Conference and Exposition, ASEE*, Austin, TX.

Chen, X., Song, G. & Zhang, Y. (2010) Virtual and remote laboratory development: A review. *Proceedings of Earth and Space*, 55, 3843–3852.

Clough, M.P. (2002) Using the laboratory to enhance student learning. *Learning Science and the Science of Learning*, 85–94.

Colorado DOHE (Department of Higher Education) (2012) *Online versus Traditional Learning: A Comparison Study of Colorado Community College Science Classes*. [online] Available from: http://wcet.wiche.edu/wcet/docs/blog/1622CCCSOnlinevsTraditionalScienceStudyReportJune 20 2update.docx

COMSOL (2012) *COMSOL Multiphysics User Guide (version 4.3 a)*, COMSOL, MA, USA. pp. 39–40.

Dai, S., Song, Z. & Jia, R. (2010) Web based fluid mechanics experimental system. In: *Proceedings of Electrical and Control Engineering (ICECE), 2010 International Conference*. pp. 3134–3137, IEEE.

De Jong, T., Linn, M.C. & Zacharia, Z.C. (2013) Physical and virtual laboratories in science and engineering education. *Science*, 340(6130), 305–308.

Ellis, G.E., Richards, C.D. & Thompson, B.R. (2014) A remote access laboratory for fluids education in mechanical engineering. In: *American Society for Engineering Education, Proceedings of 122nd ASEE Annual Conference & Exposition*, 14–17, Jun 2015, Seattle, WA, USA, 26.96.1–7.

Goldberg, K., Gentner, S., Sutter, C. & Wiegley, J. (2000) The mercury PROJECT: A feasibility study for Internet robots. *IEEE Robotics & Automation Magazine*, 7, 35–40.

Hercog, D., Gergic, B., Uran, S. & Jezernik, K. (2007) A DSP-based remote control laboratory. *IEEE Transactions on Industrial Electronics*, 54(6), 3057–3068.

Eaton, J.W., Bateman, D. & Hauberg, S. (1997) *Gnu Octave Manual*, 3rd ed. Network Thoery, London.

Feisel, L. & Rosa, A. (2005) The role of the laboratory in undergraduate engineering education. *Journal of Engineering Education*, 94, 121–130.

Goldberg, K., Gentner, S., Sutter, C. & Wiegley, J. (2000) The mercury PROJECT: A feasibility study for Internet robots. *IEEE Robotics & Automation Magazine*, 7, 35–40.

Gomes, L. & Zubía, J.G. (2008) Advances on remote laboratories and e-learning experiences. *Engineering*, 6, University of Deusto Publications, Madrid, Spain.

Gustavsson, I., Nilsson, K., Zackrisson, J., Garcia-Zubia, J., Hernandez-Jayo, U., Nafalski, A. & Hkansson, L. (2009) On objectives of instructional laboratories, individual assessment, and use of collaborative remote laboratories, learning technologies. *IEEE Transactions*, 2(4), 263–274.

Hamilton, M.A. (1996) Java and the shift to net-centric computing. *IEEE Computer*, 29(8), 31–39.

Harward, V.J., Del Alamo, J.A., Lerman, S.R., Bailey, P.H., Carpenter, J., DeLong, K., Felknor, C., Hardison, J., Harrison, B., Jabbour, I., Long, P.D., Mao, T., Naamani, L., Northridge, J., Schulz, M., Talavera, D., Varadharajan, C., Wang, S., Yehia, K., Zbib, R. & Zych, D. (2008) The iLab shared architecture: A Web Services infrastructure to build communities of Internet accessible laboratories. *Proceedings of the IEEE Conference*, 96(6), 931–950.

Jara, C.A., Candelas, F.A., Puente, S.T., Pomares, J. & Torres, F. (2010) Practical experiences using RobUALab, ejs: A virtual and remote laboratory for Robotics e-learning. *IFAC Proceedings*, 42(24), 1–6.

Jasak, H., Jemcov, A. & Tukovic, Z. (2007) OpenFOAM: A C++ library for complex physics simulations. *In International Workshop on Coupled Methods in Numerical Dynamics, IUC Dubrovnik, Croatia*, 1000, 1–20.

Kortright, E.V. (1997) Modeling and simulation with UML and Java. In: *Simulation Symposium. Proceedings of 30th Annual*, IEEE, New York, NY, USA. pp. 43–48.

Lee, C.H., Liu, A., Del Castillo, S., Bowyer, M., Alverson, D., Muniz, G. & Caudell, T.P., (2007) Towards an immersive virtual environment for medical team training. *Studies in Health Technology and Informatics*, 125, 274–279.

Lowe, D., Machet, T. & Kostulski, T. (2012) UTS remote labs, labshare, and the Sahara architecture. *Using Remote Labs in Education: Two Little Ducks in Remote Experimentation*, 8, 403.

Ma, J. & Nickerson, J.V. (2006) Hands-on, simulated, and remote laboratories: A comparative literature review. *ACM Computing Surveys (CSUR)*, 38(3), 7.

Manseur, R. (2005) Virtual reality in science and engineering education. *ASEE/IEEE Frontiers in Education Conference*, Indianapolis, IN.

Macedo-Rouet, M., Ney, M., Charles, S. & Lallich-Boidin, G. (2009) Students' performance and satisfaction with web vs. paper-based practice quizzes and lecture notes. *Computers & Education*, 53, 375–384.

Madenci, E. & Guven, I. (2015) *The Finite Element Method and Applications in Engineering Using ANSYS®*. Springer, New York.

Marín, R., Sanz, P.J. & Del Pobil, A.P. (2003) The UJI online robot: An education and training experience. *Autonomous Robots*, 15, 283–297.

Marín, R., Sanz, P.J., Nebot, P. & Wirz, R. (2005) A multimodal interface to control a robot arm via the Web: A case study on remote programming. *IEEE Transactions on Industrial Electronics*, 52(6), 1506–1520.

Mateos, V., Gallardo, A., Richter, T., Bellido, L., Debicki, P. & Villagrá, V. (2012) Lila booking system: Architecture and conceptual model of a rig booking system for on-line laboratories. *International Journal of Online Engineering* (iJOE), 7(4), 26–35.

Müller, D. & Ferreira, J.M. (2004) MARVEL: A mixed-reality learning environment for vocational training in mechatronics. In: *Proceedings of the Technology Enhanced Learning International Conference (TEL 03)*, Hugony Editore 2004, 65–72.

National Research Council (NRC) (2006) *America's Laboratory Report: Investigations in High School Science,* National Academy Press, Washington, DC, USA.

Nedic, Z., Machotka, J. & Nafalski, A. (2003) Remote laboratories versus virtual and real laboratories. In: *Proceedings of Frontiers in Education Conference of ASEE/IEEE*, 5–8 Nov 2003, Boulder, CO, USA. pp. T3E1–T3E6.

Nersessian, N.J. (1991) Conceptual change in science and in science education. In: Matthews, M.R. (ed) *History, Philosophy, and Science Teaching*. OISE Press, Toronto, Canada. pp. 133–148.

Olmi, C., Cao, B., Chen, X. & Song, G. (2011) A unified framework for remote laboratory experiments. In: *American Society for Engineering Education Conference and Exposition*, 26 June to 29 June 2011, Vancouver, BC, Canada.

Olympiou, G., Zacharia, Z.C., Papaevripidou, M. & Constantinou, C.P. (2008) Effects of physical and virtual experimentation on students' conceptual understanding in heat and temperature. *Journal of Research in Science Teaching*, 45(9), 1021–1035.

Orduña, P., Irurzun, J., Rodriguez-Gil, L., Zubía, J.G., Gazzola, F. & López-de-Ipiña, D. (2011) Adding new features to new and existing remote experiments through their integration in WebLab-Deusto. *International Journal of Online Engineering (iJOE)*, 7(S2), 33–39.

Pidd, M. (1998) *Computer Simulation in Management Science*, 4th ed. Wiley, Chichester.

Rahman, A. (2017) A blended learning approach to teach fluid mechanics in engineering. *European Journal of Engineering Education*, 42(3), 252–259.

Rahman, A. & Al-Amin, M. (2014) Teaching of fluid mechanics in engineering course: A student- centered blended learning approach. In: Alam, F. (ed) *Using Technology Tools to Innovate Assessment, Reporting, and Teaching Practices in Engineering Education*. IGI Global Publisher, Hershey, PA, USA. pp. 12–20.

Richter, T., Tetour, Y. & Boehringer, D. (2011) Library of labs-a European project on the dissemination of remote experiments and virtual laboratories. In: *Multimedia (ISM), Proceedings of 2011 IEEE International Symposium*, IEEE, New York, USA. pp. 543–548.

Rock, M.M., Marx, H., Kane, S.M., Garrick, R., Villasmil Urdaneta, L.A. & Lee, J.H. (2011) Effectively utilizing local and remote Thermo-fluids laboratory experiments to enhance student learning. In: *American Society for Engineering Education Conference and Exposition*, 26 June to 29 June 2011, Vancouver, BC, Canada.

Safaric, R., Debevc, M., Parkin, R.M. & Uran, S. (2001) Telerobotics experiments via Internet. *IEEE Transactions on Industrial Electronics*, 48, 424–431.

Sánchez, J., Dormido, S., Pastor, R. & Morilla, F. (2004) A Java/MATLAB based environment for remote control system laboratories: Illustrated with an inverted pendulum. *IEEE Transaction on Education*, 47(3), 321–329.

Schmid, C. (2001) Virtual control laboratories and remote experimentation in control engineering. *Proceedings of 11th Annual Conference on Innovations in Education for Electrical and Information Engineering*, 213–218.

Shachar, M. & Neumann, Y. (2010) Twenty years of research on the academic performance differences between traditional and distance learning: Summative meta-analysis and trend examination. *MERLOT Journal of Online Learning and Teaching*, 6(2), 318–334.

Stroustrup, B. (1995) *The C++ Programming Language*. Pearson Education India, Noida, India.

Sundararajan, S. & Dautremont, J.J. (2014) Development, assessment and evaluation of remote thermo-fluids laboratory experiments: Results from a pilot study. In: *Mechanical Engineering Conference Presentations, Paper, and Proceedings*, American Society for Engineering Education, Washington, DC, USA. p.53.

Taylor, K. & Dalton, B. (2000) Internet robots: A new robotics niche. *IEEE Robotics and Automation Magazine*, 7, 27–34.

Tiernan, P. (2010) Enhancing the learning experience of undergraduate technology students with LabVIEW software. *Computers & Education*, 55, 1579–1588.

Chapter 19

Technology-enhanced learning for civil engineering education

Use of dynamic and virtual reality-based simulation, online data analysis and optimization tools

R.S.V. Teegavarapu

Department of Civil, Environmental and Geomatics Engineering, Florida Atlantic University

ABSTRACT: The availability and access to internet-based content and simulation and analytical tools provide a tremendous advantage to online instruction geared towards distance learning education in engineering via online teaching methods. Teaching of engineering courses can be enhanced in online delivery mode by providing students with resources that can be easily accessible at any time. This chapter provides discussion about advantages and limitations of technology-enhanced learning by the use of online tools in civil engineering education. Review of the potential utility of tools available on the internet for teaching basic data analysis and engineering optimization courses is provided in this chapter. The limitations of online tools for enhancing student learning and instruction are also discussed. The chapter also critically examines how learning can be supported by technology instead of driven by it and how distance learning in engineering can be supported by online content.

19.1 Introduction

Engineering education today is experiencing a major paradigm shift from teacher-centered learning to a learner-centered system. Engineering education has been revolutionized by recently proposed approaches which include conceiving, designing, implementing and operating (CDIO) approach (Crawley *et al.*, 2014) and technology-supported initiatives (Collins and Halverson, 2009). These approaches supplement already existing methods of teaching, for example situated learning (Lave and Wenger, 1991) and virtual reality-based educational environments (Lee *et al.*, 2007). The advancement of computational power and use of technology in every aspect of classroom education has brought major changes in teaching and learning. A number of tools and approaches now being adopted in teaching engineering courses include the use of smart boards, clicker-based electronic devices, podcasts, dynamic and virtual simulation (including immersive and non-immersive) environments (Liu *et al.*, 2017), tools developed using open source software and supplementary software available in the public domain. Hands-on modeling tools with help of simulation can help understand complex physical processes as reported by Aghakouchak and Habib (2010). Online and offline learning environments are used nowadays in many engineering courses for enhancing the learning process. McGuffin-Cawley (2005) stated that employing technology-enhanced learning by contemporaneously web posting searchable video recordings of engineering core courses offers the possibility of using a lecture recording as reference material in a time-effective manner. Student attendance and participation in class are both positively impacted by the availability of recorded lectures, as the recordings are used as supplemental, rather than alternate primary, material (McGuffin-Cawley, 2005).

Effective and productive teaching lies around empowering students to learn, providing them with avenues to fuel their enthusiasm and desire for knowledge. The teaching environment should provide an intellectually stimulating environment through innovative classroom research project initiatives and incorporate practical examples of science and technology. There is a need to constantly evaluate teaching tools and techniques. The rapid growth of information technology has made education more accessible than ever before. State-of-the-art tools (web-based and -assisted) to provide effective teaching will be common in the future. Online tools, data and interactive environments will provide students with perennial access to information and learner-centered education and eliminate the time-bound classroom atmosphere that may restrict the learning process. Tailoring of teaching techniques for effective instruction is based on the overall course objectives, course content, class enrollment and the level of student participation. With the use of technology, the teaching process promotes and requires learning as well as exploration (research) that balances the teaching-research cycle.

Engineering education needs to be constantly updated with changing industry standards and ethical practices that are always taught along with conceptual and practical advancements in the discipline. Education in engineering should make the transition of scientific concepts to useful applications smooth and transparent, without neglecting the importance of both. New technology-enhanced education techniques and learner-centered education in any course curriculum is critical to the success of future engineering education. Technology-enhanced learning and instruction (TELE) is an emerging paradigm that will help address several issues prevailing in restrictive classroom-based instructor-only driven learning and teaching environments in engineering education. TELE will also encourage inquiry-based learning (IBL) and knowledge discovery process. IBL is a shift away from passive methods of teaching, which involves the transmission of knowledge to students, to more facilitative teaching methods through which students are expected to construct their own knowledge and understanding by engaging in supported processes of inquiry. Online learning is an innovative educational approach where instruction and content are focused on interactivity, design, learner-centered approaches and facilitated learning experiences (Madathil *et al.*, 2017).

19.1.1 Issues for emerging learning environments and engineering education

Any internet-assisted learning and instruction exercise should encourage students to use web-based resources for learning. The improvement in the instruction and learning can be attributed to the availability and utilization of appropriate online interactive tools. Students can benefit from and become resourceful in searching, assessing and evaluating the online tools available to carry out experiments in the learning process. The learning process can be augmented by the use of some of the online tools which are free of cost and are available freely outside the restrictive closed classroom setting. E-learning that uses electronic technologies can also be an extremely beneficial component of merging approaches to learning. Students can feel motivated to reformulate their real-life problems suitable for a solution using a variety of software and solution methods.

Several major issues can be identified and addressed for the development of e-learning systems that use technology and internet-based resources for engineering education. These issues are:

1 Effective development of e-learning systems and delivery systems.
2 Mechanism to promote active student engagement in the learning process.
3 Seek new ways in which learning can be supported by technology but not driven by it.
4 Examine the existing learning environments and consider how improved design may lead to more effective learning.
5 Assess how technology can be used to increase effective use of teaching resources and increase personalization and flexibility in the learning process.

19.1.2 Technology-enhanced and -assisted learning and instruction

Technology-enhanced and -assisted learning and instruction can happen with the use of one of these mechanisms: (1) videos in the classroom environment, (2) web-based videos, (3) offline and online simulation with internet interaction, (4) online data processing and visualization and (5) online/offline optimization. In this chapter, the focus is on the utilization of offline and online simulation environments and a discussion of enhanced learning environments involving virtual reality simulation and online and publicly available software for optimization and statistical analysis topics that are taught in many engineering courses. The emphasis is on the software and content relevant to a few water and hydrologic modeling-related courses in the civil engineering curriculum.

19.2 Dynamic simulation environment

Dynamic simulation environments that replace spreadsheet-based modeling/simulation environments can be useful in several engineering courses. In the next few sections, system dynamics (SD) simulation concepts are introduced for applications as descriptive modeling approaches in water and hydrology-related courses in the civil engineering discipline. The building blocks of SD are so generic that they can be used for modeling any system in any discipline outside of engineering. Behavior of systems that are influenced by a number of processes and factors can be understood by developing models using system dynamics principles (Teegavarapu and Simonovic, 2014). Some of the most recent real-life case studies provide applicability of various contemporary modeling techniques in water resources engineering. A more general and easy performance evaluation of water resource systems can be achieved by developing models that use object-oriented simulation (OOS) environments. Recent applications of OOS to water resource management problems (e.g. Fletcher, 1998; Keyes and Palmer, 1993; Simonovic *et al.*, 1997; Simonovic and Fahmy, 1999; Teegavarapu and Simonovic, 2000, 2014) are benefited by adopting this approach. The principles of system dynamics are well suited for modeling and application to water resources problems. Systems with OOS environments with highly field-specific objects are available in many modeling systems that are aimed at simulating hydrological processes and water management (e.g., HEC-HMS [Hydrologic Engineering Center-Hydrologic Modeling System] and HEC-RAS [Hydrologic Engineering Center-River Analysis System]).

19.2.1 System dynamics

System dynamics (Coyle, 1996; Forrester, 1961; Wolstenholme, 1990) is a concept based on systems thinking where dynamic interaction between the elements of the system is considered to study the behavior of the system as a whole. As the name suggests, the behavior of the system is monitored over time and thus dynamic. The concept of SD was introduced by Forrester (1961). The main idea of system dynamics modeling is to understand the behavior of the system by the use of simple mathematical structures. SD concepts are aimed at understanding the time-dependent behavior of the complex systems. SD concepts can help in (i) describing the system, (ii) understanding the system, (iii) developing quantitative and qualitative models, (iv) understanding how information feedback governs the behavior of the system and (v) developing control policies for better management of the system. One of the main advantages of system dynamics is that the models can be developed for understanding the behavior of the system using simple building blocks. Simulation environments (e.g. STELLA, VENSIM, etc.) that are conceived on the principles of SD and object-oriented simulation environments can be easily used to develop simulation models of varying complexity.

Causal loop diagrams (influence), feedback, shocks

Causal loop diagrams or influence diagrams were first developed for representing the interrelationships between various elements of the system. Flow diagrams are then extracted to develop simulation models. Computer simulations are finally performed using differential equations to integrate stocks and flows and understand the behavior of the system. Much of the behavior of the systems is governed by dominant feedback in the system (negative or positive). In many engineered systems, negative feedback or controlling feedback is a commonly encountered mechanism. Some of the feedback mechanisms are natural or self-induced (for example in case of sustainable ecosystems), and some others are human controlled. In most of the managed water resource systems, managers/operators enforce the feedback control with rules from the decision-makers. Shocks are exogenous system variables that have a sudden or unanticipated impact on the system behavior.

Building blocks of SD

Central to the theme of systems dynamics (Forrester, 1961) are two important building blocks, stocks and flows that can be used to model the elements that govern the water/groundwater resource systems. Two important assumptions are also used while constructing these system dynamics models: (a) processes modeled form a closed loop and (b) the boundary of the system influences the dynamics. The models developed using system dynamics are validated by several tests that include (1) replication, (2) sensitivity and (3) prediction. These tests will confirm the structure of the model with the physical system that is being modeled. Various other tests that validate the dimensional consistency of the equations and robustness of the model in handling extreme conditions should be carried out before the model is adopted for implementation. All the models developed using the concepts of SD should be validated using all these tests.

Governing equations

The governing equations used for modeling different elements in a system are represented by finite difference expressions that are generally solved using standard numerical schemes. For example, in the simulation environment STELLA, three different schemes with varying levels of accuracy are available. The equations are transparent to the user and can help in understanding various mathematical relationships. The governing equations are generic in nature and are associated with the objects. For example, in case of stock (object), a continuity equation for mass balance will be developed considering the inflows and outflows, while a converter carries a functional relationship between different variables that can be represented in a mathematical or a graphical form. The time interval for simulation is also an important aspect that will determine the accuracy of the numerical scheme used for solving the finite difference equations.

MODELING ENVIRONMENT

The environment in which the system dynamics models are developed is referred to as object-oriented simulation environment because a number of objects are being used that possess generic properties and can be used to model elements with specific values. An object-oriented simulation environment, STELLA, is a perfect tool to model any system. The environment is conceived on the principles of system dynamics and uses objects that have specific properties. A set of basic building blocks, "objects," is shown in Figure 19.1. The environment provides three layers (mapping, model building and equation) that can be used to develop a complete model. These are linked to each other, and any modification made in any one of the layers is reflected in all the other layers.

Modeling water resources and environmental systems

The mapping layer provides the user interface, while the modeling layer is used for the construction of the model. Different objects are used in the modeling layer to develop the model. For example, the object (the stock) can be used to model a reservoir while a flow is used to represent inflow, spill and release from a reservoir. For example, to incorporate a flow transport delay between two reservoirs, a conveyor object can be used. Functional relationships and dependencies can be defined using converters and connectors respectively. Examples of the objects that can be used for a reservoir operation problem and the way they are linked are shown in Figure 19.2. It can be noted that the model built using SD principles can help to understand complex systems dominated by a number of inputs and outputs.

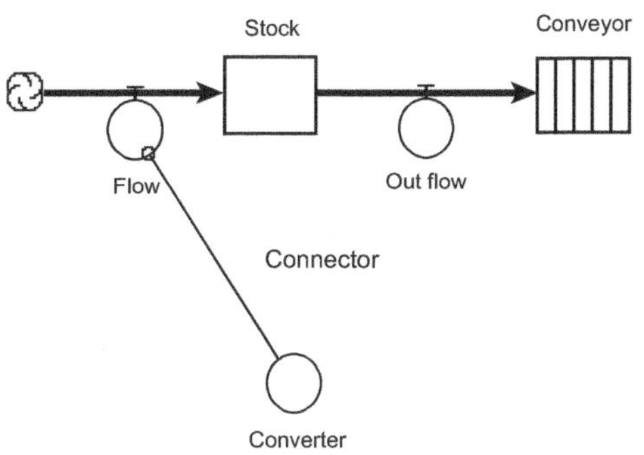

Objects :

Stock : to model storage (e.g. Reservoir)
Flow : to model any time series (e.g. Inflow)
Connector : to carry information about variables
Converter : to model functional relationships
Conveyor : to introduce delay in the system.

Functions :

RANDOM
IF-THEN -ELSE
DELAY.

Graphical Interface :

Graphs, Tables, Slide Sectors.

Figure 19.1 Main objects and functions in the simulation environment

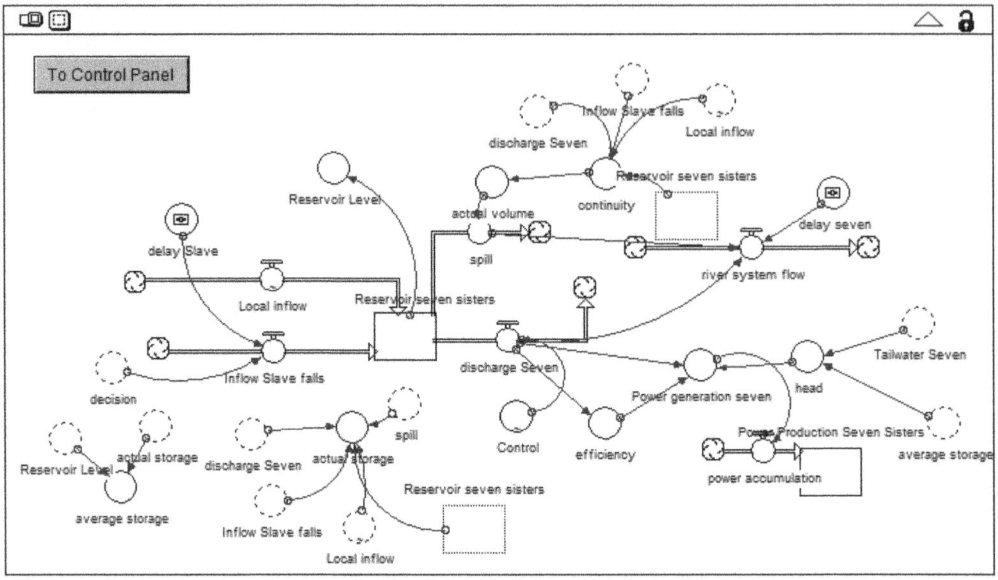

Figure 19.2 Reservoir operation problem represented by stocks and flows using an SD simulation environment

The simulation environment also provides a number of built-in mathematical, logical and statistical functions that can be used in any of the objects. The governing equations based on the model structure are automatically created in the equation layer by the STELLA environment and can be reviewed for accuracy of the structure of the model. Functions such as RANDOM, IF-THEN-ELSE and DELAY can be used to introduce unanticipated changes in the system and to construct rules/options. The environment also provides features such as sensitivity analysis, provision for graphical inputs and a simulation mode in which the inputs can be changed dynamically. The basic building blocks can be easily used to model different components of any groundwater system. For example, a stock can be used as a groundwater system. Flows can be recharged from streams (artificial and natural). The dependencies in the system can be modeled using converters. The boundaries can be defined using sources and sinks.

Applications of SD for water resources engineering problems

System dynamics concepts have been used in the past with the help of object-oriented simulation environments to model a number of water resource systems. A number of models can be envisioned in the field of environmental policy simulation studies in which the different alternatives can be evaluated for efficient management strategies for planned irrigation districts. The advantage of using SD is that it allows the decision-maker to analyze different scenarios for a given management problem and adopt the best strategy that works over a period. It should be noted that SD is a very useful tool for understanding and simulating the behavior of systems and cannot be used for optimizing the operation of the systems.

Groundwater modeling applications

The field of groundwater systems planning provides enormous opportunities for application of SD concepts. A number of simulation models can be developed to study potential problems which include (i) behavior and transport of contaminants, (ii) ground-level subsidence, (iii) pollutant migration, (iv) remediation alternatives, (v) risk assessment of corrective actions, (vi) conjunctive use and (vii) evaluation of recharge, pumping and treatment strategies.

Irrigation planning

Simulation models can be developed that can address several issues in the area of irrigation scheduling and planning. A number of issues can be modeled which include (i) irrigation scheduling options at the field level, (ii) simulation of soil moisture movement and (iii) integrated reservoir-field-level models for management alternatives.

Use in the policy evaluation

System dynamics with a simulation environment can be used to investigate management alternatives associated with the conjunctive use of surface and subsurface water. A simulation model can be developed to evaluate the alternatives. The surface and subsurface water resources components can be easily modeled using an object-oriented simulation environment. Using system dynamics, a number of policy options for groundwater management can be explored, and these options may not always be mutually exclusive. These options can be (i) do nothing at this time, (2) permit groundwater management in areas where groundwater pumping has adversely influenced streamflow, (3) augment water supplies wherever possible and (4) adopt a different approach to conjunctive use. The final policy option would involve a change in the management of water use in urban areas, modification of soil conservation practices and new measures for environmental enhancement. This approach would allow stakeholders and decision-makers to actively participate in the planning process and provide facility to construct "what-if" scenarios. It also separates policy questions from data and makes the results and the model structure functionally transparent and acceptable to a wide group of parties involved in planning.

Sensitivity analysis

Many of the simulation environments (STELLA, VENSIM, etc.) that are conceived on the principles of system dynamics provide additional tools for understanding and gaining insight into the results. Sensitivity analysis will help to evaluate and test the hypothesis made before the building of the model about its behavior. The analysis will also help to carry out prediction, replication and behavior tests on the modeled systems. Sensitivity analysis in a specific form can be used to calibrate models. The idea is similar to any manual calibration method (i.e. trial-and-error approach). The simulation environment facilitates the development of a flexible model structure. This will help the user to modify the parameter/variable values that influence the system.

19.2.2 Virtual reality (VR) learning environments

Visualization and simulation tools are increasingly used to present complex information in an intuitive way to non-specialists and for training exercises and assisting in developing emergency response plans (Sene, 2008). These tools can become valuable enhancements to engineering education. A comprehensive review of VR applications in STEM (science, technology, engineering and mathematics) areas and especially in engineering education are provided by Madathil *et al.* (2017). Readers are also referred to Lee (2011) for an exhaustive exposé on the effectiveness of VR-based learning. In the field of water resources engineering, acquisition of hydrologic and hydraulic data and the modeling and simulation of extreme hydrologic events are key components of disaster and emergency preparedness planning for extreme flooding events. Advances in flood modeling and probability assessment, risk and mapping, and resilience concepts are discussed by many researchers (Ashley *et al.*, 2007; Sene, 2008). Several studies (Dalponte *et al.*, 2007; Horritt and Bates, 2001; Romanowicz and Beven, 2003) in the past have focused on the simulation of flooding events and assessment of inundation areas and risks. Flood visualization with the help of geographical information system (GIS) tools was reported in several studies in the past decade (Bates and De Roo, 2000; Maidment *et al.*, 2005). GIS is increasingly used to assist both with planning for emergencies and in the recovery phase (MacFarlane, 2005; Sene, 2008; Van Oosterom *et al.*, 2005). These studies have focused mainly on representing a static map of the extent of flooding, depiction of water level rise over time and development of animated movies based on the snapshots of the water-level rises.

Virtual environmental planning systems (VEPS) project is one of the leading examples of using virtual reality representation of flooding in a residential area. Geo-visualization (Pajorova *et al.*, 2007) for emergency response applications is an active research area. Gyorfi *et al.* (2006) indicate that simulators for flood response applications should include four functionalities: (1) virtual reality, (2) multimedia, (3) networking and (4) artificial intelligence. Visualization and "fly-by" animation of flooding extent with the incorporation of satellite imagery and 3D structures of different features of the urban and rural landscape are now being used for catastrophic flood impacts assessment (Teegavarapu *et al.*, 2007). Sophisticated modeling environments combined with 4D (four-dimensional – three-dimensional space and time) animation capabilities are extremely useful for such a purpose. A range of virtual, mixed and augmented reality approaches are becoming used to represent the built environment and to make decisions about the design and operation of infrastructure (Whyte and Nikolic, 2018).

Engineering education can also benefit from virtual reality (VR) simulation environments that can help students appreciate concepts that are difficult to understand unless the real-life processes or variations in system behavior are experienced by being part of the system in which these processes occur. VR provides an interactive and immersive environment that motivates the student to develop a sense of the real world through simulations. Several virtual reality applications are now available for educational applications for flood hazard assessment (Chandramouli *et al.*, 2016), solute transport in groundwater (Chandramouli *et al.*, 2010) and others. Use of virtual reality models to help in situated learning in many disciplines is documented in many studies (e.g. Bell and Fogler, 1998; Lee *et al.*, 2007; Manseur, 2005; Sampio *et al.*, 2011). VR, when coupled with improved user interface technology, provides numerous opportunities for enhancing teaching and learning environments (Crumpton and Harden, 1997; Yuen *et al.*, 2011). Ong and Mannan (2004) discussed the use of virtual reality simulation and animations in web-based application in engineering.

Application and evaluation of VR simulation in engineering courses

The author of this chapter was involved in a multi-university research project supported by the US National Science Foundation (NSF) for evaluation of the use of virtual reality simulation for teaching the effects of catastrophic floods and their management in a classroom environment. Complete details of this research project can be obtained from a recent publication by Chandramouli *et al.* (2016). A three-dimensional VR simulation environment developed at Purdue University Northwest, USA, was evaluated at multiple universities for improvement in the learning of flood-related issues in water resources or hydrologic engineering courses. Two important datasets are required for successful visualization of the results from hydrologic simulation models. These are categorized into two data types: (1) spatial and (2) hydrologic data. The spatial data mainly comprise the terrain, aerial imagery, digital elevation model and infrastructure data. The hydrologic data required for simulation relates to specific objectives of the study. In the current context, the data required are mainly the water levels (e.g. stages) obtained from a hydrologic simulation model (or other approximate methods) considering extreme hydro-meteorological inputs. Two hydrologic simulation models developed by the United States Army Corps of Engineers (USACE), namely Hydrologic Engineering Center (HEC) Hydrologic Modeling System (HMS) and River Analysis System (RAS) are combined to evaluate floods in a riverine system. Spatial data from a region prone to flooding and virtual reality environment are integrated along with the numerical results from the hydrologic models to provide the user an idea about the extent of flooding in the area. Students in courses related to water resources and hydrologic engineering at four universities are exposed to this system along with an ability to evaluate different engineering options for flood management.

A comprehensive evaluation of student learning is conducted by taking assistance from an education expert who designed the survey instrument. Students were initially taught the topic of flood plain analysis and management of floods using structural and non-structural measures with no exposure to a 3D VR simulation environment. Students are then allowed to work in the VR environment by understanding the peak flood generation and causative factors that can be modified in the simulation models. The areal extent of flooding in the specific region chosen for the study is available for assessment by the students. Structural measures (i.e. the inclusion of levee and adjustments to channel or river conditions) to alleviate flooding can be evaluated by the students. Students could also explore and evaluate the extent of flooding by experimenting with different options, such as (1) river reaches with or without levees, (2) changes in the heights of the levee at different locations and (3) changes in physical characteristics of the channel with modifications to channel roughness. The VR environment provides students with the ability to virtually port to a specific location to assess the severity of the flooding. A pre- and post-survey is conducted to assess the improvement in the learning and the contribution of 3D VR environment to the students' understanding of flood plain management. The design of survey instrument and nature of questions for evaluation of the efficacy of VR was based on ideas provided by Lee (2011). Figures 19.3 and 19.4 provide the screenshots of the VR environment used for assessing the flooding of a river in Indiana, USA. Results from actual student evaluations of learning from multiple courses in four universities are provided and analyzed in a recent study by Chandramouli *et al.* (2016). Results reported by Chandramouli *et al.* (2016) suggest (1) improvement in student learning, (2) a better appreciation of concepts taught when VR is used and (3) potential utility and benefits of the use of VR technology in teaching other topics in the civil engineering curriculum.

19.2.3 Offline and online learning environments

Two types of learning environments can be identified based on the user interaction with learning systems: (1) offline and (2) online systems mainly classified based on the connectivity to

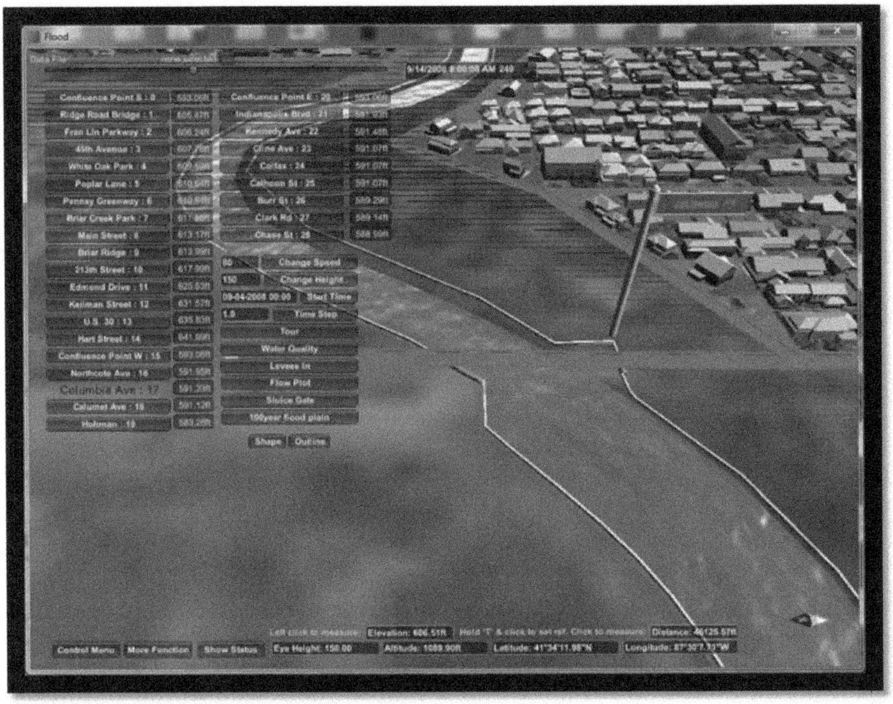

Figure 19.3 Screenshot of the VR simulation environment showing levee
(Source: Purdue University Northwest and FAU, 2016)

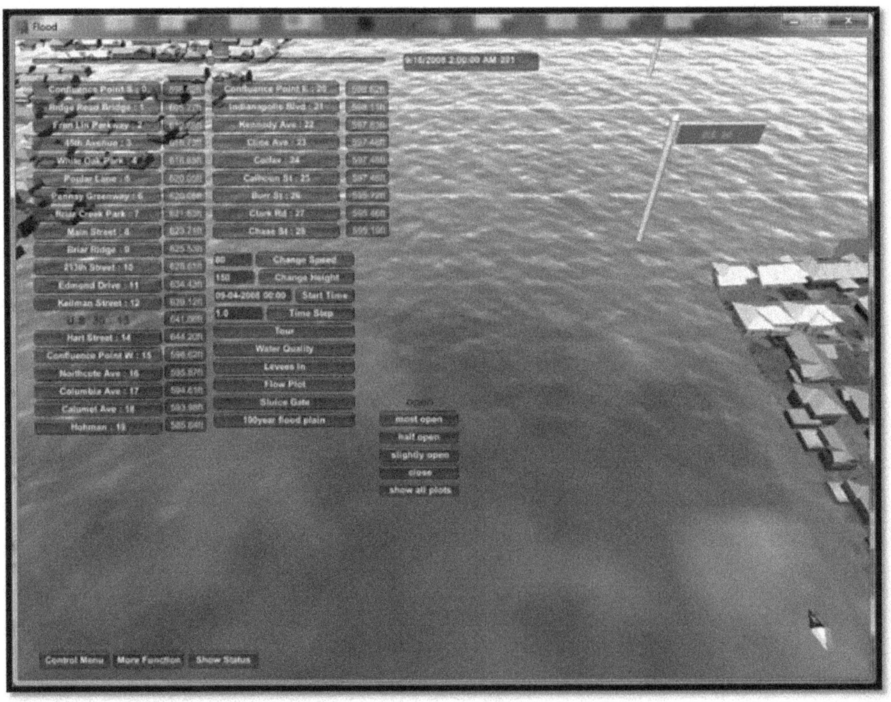

Figure 19.4 Screenshot of the VR simulation environment showing inundation due to flooding
(Source: Purdue University Northwest and FAU, 2016)

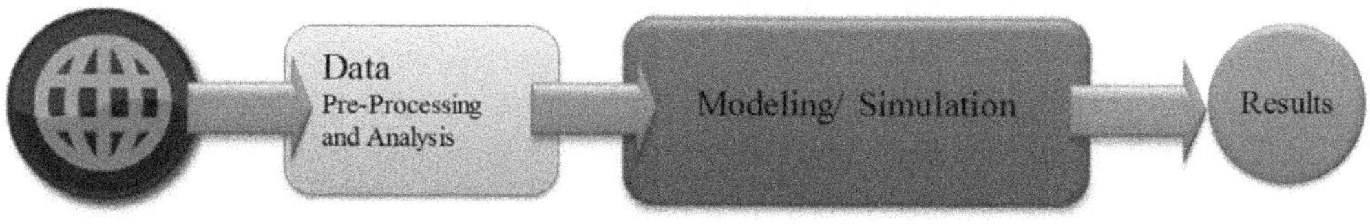

Figure 19.5 Schematic of an offline system learning environment

the internet. In case of an offline system (shown in Fig. 19.5), internet access provides data that can be processed and analyzed before using in a model to obtain results. Tasks involved in such process include (1) data download (internet and other sources), (2) appropriate data pre-processing and analysis, (3) execution of a simulation model and (4) analysis of results using visual statistics and graphical interface that was not linked to model/software.

In case of an online and interactive learning environment, data analysis, modeling, and analysis of results for a system that is being modeled and understood is executed online without any need to carry out any one of these tasks offline. For example, BASINS (Better Assessment Science Integrating Point and Non-point Sources) from the US Environmental Protection Agency (EPA) (BASINS, 2018) allows the users to obtain different datasets (spatial, topographical, hydrological, water quality, etc.) directly into the modeling system when connected to internet. The data download for a particular watershed or basin using BASINS can be used for other purposes also, as the data is available in the public domain.

Online optimization solvers and solutions

Online systems will often involve students interacting with a number of tools online and working interactively to obtain solutions to their problems. An example of such interaction is solving problems online in optimization-based courses in engineering. Optimization approaches are taught as a part of a number of engineering courses. Typically, an optimization solver is used for the solution of different formulations, and the results are explained to the students. However, the need for technology-enhanced instruction is critical in many situations. The main reasons are:

- Optimization software is not easily accessible to students.
- The public domain software available are limited-capability versions of full-blown software.
- Formulation language and syntax is different for different solvers.
- Offline downloadable software tools are available as trial versions for students with limited capabilities.
- Variations in the types of optimization solvers that require users to understand the mechanism of solution generation and explore solutions for different scenarios.

Several online optimization solvers are available to help students in courses that have an optimization component, and different mathematical programming techniques are taught. For example, linear programming (LP) solutions can be visualized using online solvers and interactive environments that display constraint spaces and optimal solutions in in two dimensions. A few web links listed at the end of this chapter provide access to different online optimization solvers. The traveling salesman problem (TSP) is one of the well-known and -researched problems in the discipline of transportation systems analysis. An online software at www.gebweb.net/optimap/ provides a utility for students to optimize trips by calculating the fastest round trips. A concept of network-enabled optimization system (NEOS; http://neos.mcs.anl.gov/) provides a suite of solvers that work offline to provide solutions to formulations provided by users online. The user needs to provide the problem in a specific format as required by the solver, and the solution is obtained by submission of this formulation to one of the NEOS servers.

In an engineering optimization course (e.g. water resource systems) in a graduate-level civil engineering curriculum with emphasis on the use of systems analysis techniques taught by the author, students were typically asked to come up with project ideas/proposals involving:

- Formulation, the definition of system variables, selection of online tool and a standard offline tool to apply optimization techniques to solve real-life problems.
- Problems selected from different disciplines (water distribution, transportation, water quality safety and security, water supply, real-time operation of systems, etc.).
- Experiments with online solvers and reporting their advantages and disadvantages.
- Online solvers and applications/case studies.

19.2.4 Statistics and data analysis

Data analysis and use of statistics are common among engineering courses. In general, one or two specific proprietary software are taught in the classroom to students. Teaching statistical software in class or incorporating it into the curriculum with the help of computer-based laboratory sessions is common. One major limitation of this teaching of the proprietary statistical software is the time spent on teaching the way different modules in the software work. This time can be efficiently used for teaching concepts and approaches of statistical testing in engineering undergraduate and graduate courses. Statistical software that is available in the public domain (i.e. R-package) require students to learn the syntax of the language used for writing scripts or executing scripts at the command-line prompt. A number of resources and testing (i.e. statistical hypothesis testing) environments are available online that can be used effectively for classroom instruction.

The availability of online software and access to students anywhere that the internet is available provide an unrestrictive learning environment to experiment and eliminate the laboratory or classroom-based learning environment. A popular web-based non-commercial usage only and free statistical and forecasting software environment is available online (www.wessa.net). Several publicly available software programs (e.g. ProUCL, HEC-SSP, etc.) from federal agencies in the United States can be used for statistical analysis of data. The software ProUCL and HEC-SSP are standalone systems and can be classified as offline systems. However, HEC-SSP provides a mechanism for downloading data directly from United States Geological Survey (USGS) for analysis when an internet connection is available. Many federal agencies in the US which share data are also provide tools to mine and visually display the data and obtain summary statistics and comparative assessment of data from different periods.

Limitations of internet-assisted learning and instruction

The intra-university access mechanism for software in classes has many problems. The accessibility is sometimes limited due to licensing issues and also lack of adequate versions to replicate the tools on different platforms. The software tools used by students are open to all and platform independent. Accessibility to the internet is much higher than a connection to inter-university systems. However, the online systems that are not supported and endorsed by educational entities or book publishers and authors of books are not always put through a robust testing process. Before embarking on the use of online software or simulation/optimization and data analysis environments, the following tasks should be carried out:

- Evaluate the online software or learning environment for any issues related to privacy and data integrity issues.
- Assess and confirm the validity of results obtained from these environments for multiple benchmark problems.
- Compare and evaluate results from offline and online software and document limitations of the online software.
- Confirm the limitations of the analysis reported by the online software.

Learners from traditional classroom-based environments often complain about the lack of face-to-face learning experiences, rigidity of objective online assessments, security and privacy issues in accessing content and finally incomplete guidance due to limited content in e-learning systems.

Metrics for assessment of student learning

As more and more students are exposed to technology-based teaching methods, assessment of student learning becomes challenging. Students' perception towards online learning should also be evaluated using different factors such as instructor characteristics, social presence, instructional design and trust (Bhagat *et al.*, 2016). The assessment should involve a number of metrics that focus on the quality of learning that has happened. Any criteria established to assess learning should recognize the limitations of the available online and offline learning tools. A clear distinction between student's own creative solutions and those that are already available online needs to be identified. Students also face many challenges in attempting to solve problems assigned to them. Students should be asked to (1) reformulate their project to suit to the requirements of the online tools, (2) question and compare the results from online tools with standard tools available offline, (3) document the results from a number of online tools and differentiate among the limitations and advantages of these tools and (4) assess the applicability of these results for real-life problems. Engineering educators should also devise testing that evaluates a student's skills of independent analysis, practical applicability of solutions proposed and ability to transition from theory to practice. Near-real time, short- and long-term assessments can be used to monitor and document student learning. Near-real time assessment will evaluate focus and attention of students in classroom via short quizzes and discussion about what was learned. Short-term assessment determines the critical and independent thinking of students. As online or VR-based learning exercises in some instances happen in group environments, appropriate testing methods should be devised to assess group learning as a whole. Long-term learning assessment needs to evaluate how students were able to generalize the concepts and their application to different real-life situations, case-study specific design and innovations achieved by access to interactive learning. Surveys devised with approval from an institutional review board (IRB) or similar entity in the university system need to be conducted before and after the students are exposed to new methods of teaching and learning. Student assessment can be improved by:

- Developing electronic assessment or other improved traditional systems.
- Creating mechanisms to assess active student engagement in the learning process.
- Seeking new ways in which learning can be assessed by technology.
- Examining the available strategies for assessing learning environments and considering how improved assessment may lead to more effective learning and instruction.

19.6 Conclusion

A technology-assisted and -enhanced learning paradigm with and without problem-specific tools that are available online is increasingly being used in engineering education. This chapter briefly described the use of system dynamics simulation tools, visualization approaches, offline and online optimization tools for promoting technology-assisted teaching applications in the sub-specialization of water resources in the civil engineering discipline. Experiences of the author in using publicly available software for teaching and improving learning in and outside the classroom are briefly documented. Limitations of these tools are also briefly discussed. An increasing number of studies have shown that learning improves with the adoption of these tools in engineering education. The availability of online learning environments provides a platform for unrestrictive learning and eliminates the need for time-constrained classroom-based education. It is important that a thorough evaluation of the web-based and web-assisted learning environments is conducted before students are engaged in these types of learning activities. Improvement in student learning should also be assessed using appropriate metrics.

Web links

A number of websites that provide online and offline optimization solvers, environmental analysis and statistical analysis are provided for the benefit of readers. The web links were active when they were accessed and are not endorsed by the author, book editors or the publisher of this book for any specific use.

Optimization

- www.gebweb.net/optimap/ (accessed 15 March 2018)
- http://neos.mcs.anl.gov/ (accessed 15 March 2018)
- www.zweigmedia.com/RealWorld/LPGrapher/lpg.html (accessed 15 March 2018)
- https://neos-guide.org/content/linear-programming (accessed 15 March 2018)
- https://neos-guide.org/content/integer-linear-programming (accessed 15 March 2018)
- http://online-optimizer.appspot.com/ (accessed 15 March 2018)
- www.gebweb.net/optimap/ (accessed 15 March 2018)

Statistical analysis

- www.wessa.net/ (accessed 15 March 2018)
- www.itl.nist.gov/div898/handbook/ (accessed 15 March 2018)
- www.hec.usace.army.mil/software/hec-ssp/ (accessed 15 March 2018)
- www.epa.gov/land-research/proucl-software (accessed 15 March 2018)

Environmental and water quality analysis

- www.epa.gov/exposure-assessment-models/basins (accessed 15 March 2018)

References

Aghakouchak, A. & Habib, E. (2010) Application of a conceptual hydrologic model in teaching hydrologic processes. *International Journal of Engineering Education*, 26, 4, 963–973.

Ashley, R., Garvin, S., Pasche, E., Vassilopoulos, A. & Zevenbergen, C. (eds). (2007) *Advances in Urban Flood Management*. Taylor and Francis, The Netherlands.

BASINS (2018) *Better Assessment Science Integrating point and Non-point Sources Modeling Framework*. National Exposure Research Laboratory, North Carolina, USA. Available from: www.epa.gov/exposure-assessment-models/basins [Accessed 15 March 2018].

Bates, P.D. & De Roo, A.P.J. (2000) A simple raster-based model for floodplain inundation. *Journal of Hydrology*, 236, 1–2, 54–77.

Bell, J.T. & Fogler, H.S. (1998) Virtual reality in chemical engineering education. *ASCEE, North Central Section Meeting*, University of Detroit Mercy, Detroit, MI, USA.

Bhagat, K.K., Wu, L.Y. & Chang, C.Y. (2016) Development and validation of the perception of students towards online learning (POSTOL). *Educational Technology and Society*, 19(1), 359–359.

Chandramouli, V.C., Hixon, E., Zhou, C.Q., Moreland, J., Jichao, W., Teegavarapu, R.S.V., Behera, P. & Fox, J. (2016) Evaluating the usefulness of Virtual 3-D lab modules developed for a flooding system in student learning. *ASEE 123rd Annual Conference and Exposition*, New Orleans, LA, USA.

Chandramouli, V., Zhou, C.Q., Jin, L., Narayana, M. & Duvruvai, V.K. (2010) Visualization modules for solute transport in groundwater. *Presented in EWRI Congress, ASCE, Providence*, Rhode Island, 20 May 2010.

Collins, A. & Halverson, R. (2009) *Rethinking Education in the Age of Technology: The Digital Revolution and Schooling in America*. Teachers College, New York, USA.

Coyle, R.G. (1996) *System Dynamics Modeling: A Practical Approach*. Chapman and Hall, London, UK. pp. 413.

Crawley, E.F., Malmqvist, J., Ostund, S., Brodeur, D.R. & Edstrom, K. (2014) *Rethinking Engineering Education: The CDIO Approach*. Springer, Cham, Switzerland.

Crumpton, L.L. & Harden, EL. (1997) Using virtual reality as a tool to enhance classroom instruction. *Computers & Industrial Engineering*, 33(1–2), 217–220.

Dalponte, D., Rinaldi, P., Cazenave, G., Usunoff, E., Vives, L., Varni, M., . . . Clausse, A. (2007) A validated fast algorithm for simulation of flooding events in plains. *Hydrological Processes*, 21(8), 1115–1124.

Fletcher, E.J. (1998) The use of system dynamics as a decision support tool for the management of surface water resources. *1st International Conference on New Information Technologies for Decision Making in Civil Engineering Montreal*, Canada. pp. 909–920.

Forrester, J.W. (1961) *Industrial Dynamics*. Productivity Press, Portland.

Gyorfi, J.S., Buhrke, E.R., Tarlton, M.A., Lopez, J.M. & Valliath, G.T. (2006) *VICC: Virtual Incident Command Center*. Presence 2006. International Society for Presence Research, Philadelphia, USA.

Horritt, M.S. & Bates, P.D. (2001) Predicting floodplain inundation: Raster-based modelling versus the finite element approach. *Hydrological Processes*, 15, 5, 825–842.

Keyes, A.M. & Palmer, R.N. (1993) The role of object-oriented simulation models in the drought preparedness studies. *Proceedings of the 20th Annual National Conference, Water Resources Planning and Management Division of ASCE*, Seattle, WA, USA. pp. 479–482.

Lave, J. & Wenger, E. (1991) *Situated Learning: Legitimate Peripheral Participation*. Cambridge University Press, Cambridge.

Lee, C.H., Liu, A., Del Castillo, S., Bowyer, M., Alverson, D., Muniz, G. & Caudell, T.P. (2007) Towards an immersive virtual environment for medical team training. *Studies in Health Technology and Informatics*, 125, 274–279.

Lee, E.A.L. (2011) *An Investigation into the Effectiveness of Virtual Reality-based Learning*. Doctoral Dissertation, Murdoch University, Perth, Australia.

Liu, D., Chris, D., Huang, R. & Richards, J. (2017) *Virtual, Augmented and Mixed Realities in Education*. Springer, Singapore.

MacFarlane, R. (2005) *A Guide to GIS Applications in Integrated Emergency Management*. Emergency Planning College, Cabinet Office, York, UK.

Maidment, D.R., Robayo, O. & Merwade, V. (2005) *Hydrologic Modeling, GIS, Spatial Analysis and Modeling*. ESRI Press, Redlands, CA, USA.

Manseur, R. (2005) Virtual reality in science and engineering education. *ASEE/IEEE Frontiers in Education Conference*, Indianapolis, IN.

Madathil, K.C., Frady, K., Hartley, R., Bertrand, J., Alfred, M. & Gramopadhye, A. (2017) An empirical study investigating the effectiveness of integrating virtual reality-based case studies into an online asynchronous learning environment. *The ASEE Computers in Education Journal*, 8(3).

McGuffin-Cawley, J.D. (2005) Work in Progress – Technology enhanced learning for engineering core courses. *Proceedings of Frontiers in Education Conference: Pedagogies and Technologies for the Emerging Global Economy*. Available from: www.icee.usm.edu/ICEE/conferences/FIEC2005/papers/1303.pdf [Accessed 15 March, 2018].

Ong, S.K. & Mannan, M.A. (2004) Virtual reality simulations and animations in a web-based interactive manufacturing engineering module. *Computers & Education*, 43(4), 361–382.

Pajorova, E., Hluchy, E. & Astalos, J. (2007) 3D geovisualization service for grid-oriented applications of natural disasters. *Krakow 07 Grid Workshop*, Krakow, Poland.

Romanowicz, R. & Beven, K. (2003) Estimation of flood inundation probabilities as conditioned on event inundation maps. *Water Resources Research*, 39, 3, 1061–1073.

Sampio, A.Z., Cruz, C.O. & Martins, O.P. (2011) Didactic models in civil engineering education: Virtual simulation of construction works, virtual simulation of construction works, 579-598, Jae-Jin Kim (ed), InTech, London, UK.

Sene, K. (2008) *Flood Warning, Forecasting and Emergency Response*. Springer, UK.

Simonovic, S.P. & Fahmy, H. (1999) A new modeling approach for water resources policy analysis. *Water Resources Research*, 35(1), 295–304.

Simonovic, S.P., Fahmy, H. & El-Shorbagy, A. (1997) The use of object-oriented modeling for water resources planning in Egypt. *Water Resources Management*, 11, 243–261.

Teegavarapu, R.S.V., Scarlatos, P. & Kaner, Y. (2007) *A pilot study on catastrophic flood scenario animation for a region in South Florida*. SFWMD Technical Report, Center for Inter-modal Transportation Safety and Security, Boca Raton, FL, USA. pp. 38.

Teegavarapu, R.S.V. & Simonovic, S.P. (2000) System dynamics simulation model for operation of multiple reservoirs. *Proceedings of World Water Congress*, Melbourne, Australia, 11–17 March 2000.

Teegavarapu, R.S.V. & Simonovic, S.P. (2014) Simulation of multiple hydropower reservoir operations using system dynamics approach. *Water Resources Management*, 28(7), 1937–1958.

Van Oosterom, P., Zlatanova, S. & Fendel, E.M. (eds). (2005) *Geo-information for Disaster Management*. Springer, Berlin.

Whyte, J. & Nikolic, D. (2018) *Virtual Reality and the Built Environment*. Routledge,, Oxon, UK.

Wolstenholme, E.F. (1990) *System Enquiry: A System Dynamics Approach*. John Wiley & Sons,

Yuen, S., Yaoyuneyong, G. & Johnson, E. (2011) Augmented reality: An overview and five directions for AR in education. *Journal of Educational Technology Development and Exchange*, 4(1), 119–140.

Subject index